Engineering Materials

This series provides topical information on innovative, structural and functional materials and composites with applications in optical, electrical, mechanical, civil, aeronautical, medical, bio- and nano-engineering. The individual volumes are complete, comprehensive monographs covering the structure, properties, manufacturing process and applications of these materials. This multidisciplinary series is devoted to professionals, students and all those interested in the latest developments in the Materials Science field.

More information about this series at http://www.springer.com/series/4288

Gil Alberto Batista Gonçalves · Paula Marques
Editors

Nanostructured Materials for Treating Aquatic Pollution

 Springer

Editors
Gil Alberto Batista Gonçalves
TEMA—Centre for Mechanical
Technology and Automation
University of Aveiro
Aveiro, Portugal

Paula Marques
TEMA—Centre for Mechanical
Technology and Automation
University of Aveiro
Aveiro, Portugal

ISSN 1612-1317   ISSN 1868-1212   (electronic)
Engineering Materials
ISBN 978-3-030-33747-6   ISBN 978-3-030-33745-2   (eBook)
https://doi.org/10.1007/978-3-030-33745-2

This Springer imprint is published by the registered company Springer Nature Switzerland AG
The registered company address is: Gewerbestrasse 11, 6330 Cham, Switzerland

Acknowledgements

The editors of this book acknowledge CENTRO-01-0145-FEDER-030513 and the Portuguese Foundation for Science and Technology (PTDC/NAN-MAT/30513/2017) for supporting the project: Graphene based materials and water remediation: a sustainable solution for a real problem?

Contents

Removal of Metal Ions Using Graphene Based Adsorbents

Imran Ali, Zeid A. ALOthman and Abdulrahman Alwarthan

Abstract The pollution of water sources with toxic metal ions is a thoughtful subject. Numerous aquatic systems are polluted with dissimilar toxic metal ions from various industrial effluents and anthropogenic activities. Water treatment seems to the chief environmental test. The sorption is the best effective method for metal ions elimination in water. Currently, graphene and its composite materials are attaining significance as new generation nano-sorbents. Graphene is a two dimensional nano-material with single layer of graphite. These have achieved a great reputation in water treatment because of their distinctive physico-chemical features. The present chapter describes the elimination of metal ions using graphene and its composite materials. The emphasis has been made on syntheses, applications, regeneration and recycling and future perspectives. Definitely, this chapter will be valuable tool for researchers, government authorities and academicians.

Keywords Water treatment · Graphene and composite substances · Toxic and rare earth metals · Adsorption

1 Introduction

Water is the most significant commodity for the existence of all the individuals on the earth. Everything in this life need water. Because of that, people must not only take care for water sources but also to keep water sources clean. The international water community continues to explore safe water for securing the future of human

I. Ali (✉)
Department of Chemistry, College of Sciences, Taibah University,
Al-Medina al-Munawara 41477, Saudi Arabia
e-mail: drimran_ali@yahoo.com; drimran.chiral@gmail.com

Department of Chemistry, Jamia Millia Islamia (Central University),
New Delhi 110025, India

Z. A. ALOthman · A. Alwarthan
Department of Chemistry, College of Science, King Saud University, Riyadh
11451, Kingdom of Saudi Arabia

© Springer Nature Switzerland AG 2019
G. A. B. Gonçalves and P. Marques (eds.), *Nanostructured Materials for Treating Aquatic Pollution*, Engineering Materials,
https://doi.org/10.1007/978-3-030-33745-2_1

and aquatic ecosystem health. The quality of water is a significant issuer during last some decades [1]. If water quality will not be controlled, the disease will develop and increase and the life will be miserable with short span. Among various water contaminates, toxic trace metal ions are very serious pollutants owing to their severe toxicity, carcinogenic and non-biodegradable nature [2]. These metal ions incline to collect in tissues and create numerous diseases and health maladies in humans. It is due to their tendency to interact with proteins, enzymes and non-biodegradable nature [3–5]. The contact to the metal ions, even at bit level, is supposed to be a risk for the human [6–9]. Thus, this problem is of major concern for the scientists, academicians and Government authorities. Therefore, their removal is essential for the health point of views.

2   Water Pollution in Different Ecosystems

The quality of our water springs is worsening rapidly because of rapid growing of the world people, civilization and industrialization. Our fixed fraction of fresh water is being polluting continuously by point and non-point sources. These sources can be categorized based on their targeted water resources i.e. ground and surface water (Table 1); classified on their maximum probability of water contamination. The main causes of water contaminations are domestic, industrial and agricultural happenings [10, 11]. Some other ecological activities and geological variations are also accountable for polluting water; especially ground water.

More than one thousand pollutants are present in water and these include organic, inorganic and biological contaminants. The major uses of water are for agriculture, industrial process, and domestic (household) supply. It is quite obvious that the quality of water determines its potential uses. The domestic supply must be free from constituents harmful to health, such as organic, inorganic and biological contaminations, and it should taste and smell good. Similarly, quality of water required for industrial purposes varies widely depending on the process involved. Some processes may require distilled water, whereas, other simply needs normal water. Water for agriculture should be free from boron and possesses 2–10 sodium adsorption ratio (SAR) value. The various international agencies have devised permissible limits for several constituents in water for different purposes like drinking, bathing, recreation, agricultural etc. [12, 13]. The wastewater contaminated with metals is serious aspect in hydrological science. Metals are presented in water systems due to enduring of rocks and soils, from volcanic outbreaks, and from a diversity of man actions with the processing, mining, utilization of metals and other substances that comprise metal contaminants [14, 15]. The most notorious toxic metal contaminants are cadmium, arsenic, copper, chromium, lead, mercury, selenium and nickel. There are dissimilar causes of metal contaminants, that are given in Table 2. The common restricted metal contamination in water comes via industrial effluents of battery, tanning, ceramics, glassware, mining, electroplating, photographic, paints etc. In addition to these industries, acid mine drain typically issues toxic metals from

Table 1 Sources of water pollution

Surface water	Groundwater
• Urban runoff (oil, chemicals, organic matter, etc.) (U, I, M)	• Leaks from waste disposal sites (chemical, radioactive materials etc.) (I, M)
• Agricultural runoff (oil, metals fertilizers, pesticides, etc.) (A)	• Leaks from buried tanks and pipes (gasoline, oil, etc.) (I, A, M)
• Accidental spills of chemicals including oil (U, R, I, A, M)	• Seepage from agricultural activates (nitrates, heavy metals, pesticides, herbicides etc.) (A)
• Radioactive materials (often involving trucks or train accidents) (I, M)	• Saltwater intrusion into coastal aquifers (U, R, I, M)
• Runoff (solvents, chemicals, etc.) from industrial sites (factories, refineries, mines etc.) (I, M)	• Seepage from cesspools and septic systems (R)
• Leak from surface storage tanks and pipelines (gasoline, oil etc.) (I, A, M)	• Seepage from acid rich water from mines (I)
• Sediments from a variety of sources, including agricultural lands and construction sites (U, R, I, A, M)	• Seepage from mine waste pipes (I)
• Air fallout (particles, pesticides, metals etc.) intro river, lake, ocean (U, R, I, A, M)	• Seepage of pesticides, herbicides nutrients, and so on from urban areas (U)
	• Seepage from accidental spills (train or truck accidents, for example) (I, M) • Inadvertent Seepage of solvent and other chemicals including radioactive materials from industrial sites or small business (I, M)

Key: *U* urban; *R* rural; *I* industrial; *A* agricultural; *M* military

ores, as metals are soluble in an acid. After the drain procedure, these arrangements scatter acid solutions on the ground, comprising high points of metals, which may leach into the groundwater. Besides, the concentrations of toxic metals are much greater than the harmless allowable limits [16].

3 Metal Ions Toxic Effects

The metal ions are important for living systems; but these may have toxicity or carcinogenicity at elevated concentrations. Nowadays, the pollution of our atmosphere owing to toxic metals is very stern and puzzling problem. The toxic metals found into the atmosphere in dissimilar physico-chemical forms. The organometallic molecules are more poisonous than the their inorganic forms; with the exemption of arsenic. The heavy metals inhabit a chiefly significant position as these are maximum poisonous. Some toxic metals are of vivacious position for flora and fauna progress. However, these may be toxicant at high concentrations. The most hazardous features of heavy

Table 2 Heavy metal, present in raw materials/processes in major industries

Industry	Cd	Cr	Cu	Fe	Hg	Mn	Pb	Ni	Sn	Zn
Pulp & Paper, board mills		x	x		x		x	x		x
Organic chemicals	x	x		x	x		x		x	x
Inorganic chemicals	x	x		x	x		x		x	x
Fertilizers	x	x	x	x	x	x	x	x		x
Petroleum Refining	x	x	x	x			x	x		x
Basic steel works	x	x	x	x	x		x	x	x	x
Basic non-ferrous metal works	x	x	x		x		x			x
Motor vehicle Aircraft plating	x	x	x		x			x		
Glass, Cement		x								
Textile mill		x								
Leather tanning		x								
Steam generation power plant		x								

metals is their accumulation in living systems even in the neighboring situation. Out of natural occurring elements, fifty three are heavy metals, while seventeen (Ag, As, Co, Cd, Cu, Cr, Hg, Mn, Fe, Mo, Pb, Ni, U, Sb, W, V and Zn) are obtainable to the alive cells and make solvable cations; showing living properties. The men are predictably bared to carcinogens from ecological, medicinal and occupational sources [2]. Generally, the noxiousness of metals is defined by calculating LD_{50} or LC_{50}. Nevertheless, some workers described toxicities in terms of EC_{50} and IC_{50} terminologies. The metal ions defined in the periodic table are not-toxic and, consequently, only poisonous metal ions are deliberated here. Amongst numerous metal ions the maximum shared toxic metals concerned in severe and/or lingering circumstances comprise arsenic, antimony, lead, chromium, selenium, mercury, etc. Nevertheless, high amount of other metal ions like rhodium, aluminum, zinc, copper, vanadium, platinum, etc. may be poisonous [2]. The poisonousness of some notorious metal ions is discussed below.

3.1 Arsenic (As)

Arsenic is a well-known poisons for log time. The poisonousness of water soluble arsenic species is arsenite) > arsenate > monomethylarsonate; > dimethylarsinate. Arsenic in water sows harming all the phases of arsenic appearance [2]. In human body about 50% arsenic is expelled through faeces, urine, skin, nails, hair, lungs etc. [2]. The people consuming arsenic polluted water indicates arsenical skin lesions. Long use of arsenic polluted water can result into hyperkeratosis, conjunctivitis, hyperpigmentation, disorder in the periphral nervous and vascular systems, cardio-vascular diseases, skin cancer and non-pitting swelling, gangrene, leucomelanosis, splenomegaly, and hepatomegaly [2, 17–19]. The effects on the lungs, uterus, genitourinary tract etc. are noticed in the progressive phase of arsenic poisonousness. Moreover, high amount of arsenic can also create escalation of spontaneous and still-births abortions [2, 20]. The additional side effects of arsenic pollution are visual disorders, sensory disorders, energy loss, coma, fatigue, convulsions, shock, paralyses, atrophy, blindness, kidney damage, etc.

3.2 Cadmium (Cd)

Cadmium has severe toxicities including nausea, diarrhea, cramp muscle and harm of bone marrow. Besides, it is supposed accountable to kidney stones formation [2, 21, 22]. The studies indicated cadmium as responsible for human cancer of prostatic, pulmonary and liver, renal, hematopoietic system, stomach and urinary bladder [23–28]. USA International Agency for Research on Cancer and the National Toxicology Program recognized cadmium as group 1 carcinogenic metal ion [24, 25]. Cadmium is considered to be lethal with vomiting and nausea at 15.0 mg/L concentartion. It is acute poisonous metal but not deadly. The cadmium effects are hypertension renal dysfunction, chest pain, anemia, headache, augmented salivation, diarrhea,, throat dryness, pneumonitis, cough, tenesmus, etc. The diverse molecules carcinogenic in nature are Cd, CdO, CdS, $CdCl_2$ and $CdSO_4$.

3.3   Chromium (Cr)

Cr(III)] is crucial for living beings growth and health (at low level) with a nontoxic concentration of 0.20 mg/day while Cr(VI) is a strong carcinogenic and very deadly to human and animals. Thus, the collected Cr in plants parts (food) may characterize possible health threats to human and animals when it is collected in Cr(VI) form at elevated concentration. The chief noxious effects of chromium(VI) related to respiratory organs, kidney, liver, with hemorrhagic possessions, ulceration and dermatitis of skin. Chromium is categorized as group I carcinogenic by International Agency for research on Cancer; for animals and humans. Chromium is found in ground water of industrial area of steel works, plating, corrosion tanning, etc. industries. Besides the anthropogenic actions are also responsible for its contamination; the weathering of rocks [29, 30]. It is carcinogenic with tendency of forming of respiratory cancers. It was found that Cr(VI) exposure led to danger of emerging lung cancers [31, 32].

3.4   Lead (Pb)

Lead is very substantial toxin among heavy metals with neurotoxical and toxicological effects like damage of brain. Lead is a present in inorganic and organic forms. Inorganic lead damage peripheral and central nervous systems, renal, hematopoietic, reproductive, cardiovascular and gastrointestinal systems. Organic lead is poisonous for central nervous system. These sorts of properties are seen at 100.0–200.0 µg/L blood concentration. The tumor development in rats is owing to lead toxicity. Other dangerous effects include loss of cognitive abilities, convulsions, anti-social constipation, anemia, visual disturbances, severe abdominal pain, tenderness nausea, anemia, paralysis, vomiting, etc. The different lead substances carcinogenic in nature are $PbAc_2$, PbO, $Pb_3(PO_4)_2$ and $PbCrO_4$ [2]. Inorganic lead are absorbed via inhalation, ingestion while organic lead salts are engrossed via skin.

3.5   Mercury (Hg)

Mercury is found in inorganic and organic forms. Water contamination due to mercury is because of use of mercury based pesticides [2, 33, 34]. Inorganic mercury substances are poisonous to nervous and nephrological systems. Organic mercury substances are acute poisonous to central nervous system. Also, irritability, learning disabilities, discouragement, behavior changes, jerky gait, tremors, gums and mouth inflammations, swelling of salivary glands, spasms of extremities, teeth loosening and saliva disorder are other menace of mercury toxicity. However, the toxicity issue of mercury is not clear and still in discussion. Mercury also effect DNA activities and is considered as carcinogen.

4 Water Treatment Methods Available

Many water treatment methods are used for removing toxic contaminants in water. These are grouped as chemical, physical, thermal, biological and electrical nature. The most important are filtration, centrifugation, crystallization, screening, gravity and flotation, sedimentation oxidation, separation, precipitation, solvent extraction, coagulation, evaporation, ion exchange, distillation, electrodialysis, reverse osmosis, adsorption electrolysis, etc. Amongst these techniques, sorption is considered as the finest one owing to ease of operation, eco-friendly, inexpensive and capable to remove all types of the pollutants [35, 36].

Adsorption is an exterior process and is described as an augment for species on the edge or surface among two phases. The species (pollutant), which stick to the solid interface is termed as adsorbate and solid material is called as adsorbent. The sorption is controlled by features of the adsorbate and adsorbent, contact time, pH, dose, temperature, adsorbent particle size, other species, etc. The occurrence of suspended materials, greases and oils decreases the efficacy of the phenomenon. Therefore, sometimes filtration is needed. The pollutants get adhered to the solid surface of adsorbent when contaminated water is shacked with adsorbent. After some times, the amounts of the contaminants sorbed gets constant. The relationship pollutant amounts sorbed at fixed time and equilibrium is termed as adsorption isotherm (Fig. 1). Freundlich, Langmuir, Dubnin, Tempkins and other isotherms are well-known and may describe sorption data efficiently [37]. The adsorption process is determined by batch experiments in the laboratory followed by pilot scale and column operations.

Fig. 1 A typical batch adsorption isotherm

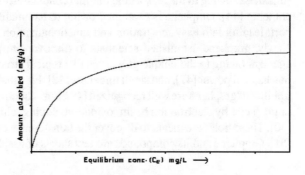

5 Graphene

Graphite is a crystalline material made of carbon in a sequence of fixed parallel sheets. The carbon atoms in every sheet are fixed in even hexagons in a fashion that every carbon atom is linked via three other carbon atoms. The 3.35 Å is a space between two adjacent graphite sheets. The carbon—carbon space in hexagonal collection is 1.42 Å. The intra-planar bonding attractions in every sheet are owing to metallic and covalent bondings, that are greater in comparison to forces of van der Waal. Graphite can react with dissimilar reagents forming intercalation substances; resulting into surface substitution and alteration. The graphite intercalation substances are much essential in water management as showing quite good sorption dimensions for different contaminants. Graphene is one of the most amazing carbon materials in nano size. The widespread graphene use is in numerous areas industrial purposes. A graphene only sheet (classically showing graphene) was achieved over compounds by transferal from only crystal graphite utilizing a glue tape. Much progressive techniques for graphene distinct layer on substrates are described. The even graphene distinct layers are steady on the substrates, while free may be unstable and tending to form a roll. Graphene distinct layer can be made constant due to their contacts with the setting. The surface sorption is utilized to make stable diffusions in water and other organic solvents [38, 39]. Normally, graphene dispersion and powder structures are rarely perfect. Graphene particles are bent to a definite degree with raggedy ends. The basic flaws and outside atoms and elemental groups are committed to graphene frame. Classically graphene is defined as a distinct carbon layer [40]. Nevertheless, recent definition is graphene comprises 10 or less sheets. It is also called as exfoliated graphite nano-sheet or graphite nano-platelet [41]. Graphite nano-platelets fluctuate in the structures dependent on the width. Graphite nano-platelet of lesser than 50.0 nm width does not indicate a cellular construction while graphite nano-platelet more than 50.0 nm thick can or cannot show cellular frame-work.

Graphene distinct layer shows sp^2 carbon hybridization (organized in honeycomb hexagonal lattice) with ~5000 W/mK thermal conductivity (at ambient temp.) in diffusion and hypothetical surface area of 2630 m^2 g^{-1} [42, 43]. The side interactions contain an extreme reduction of 60% in conductivity, which is about 2000 W/mK; at 100 nm graphene sheet thickness. The size of graphene may range 10 to 100 nm (thickness) owing to the reasonable thermal conductivity ($\kappa \approx 1500$ W/mK) at ambient temp [44]. Graphene is examined owing to its tunable and stretchy carbon support; leading into easy integration and functionalization in a range of utilities [45]. Newly, graphene astonished scientists to discover the prospective in water management owing to the unique properties like typical structural properties [46], good mechanical power [47], and small thickness [48]. Elevated tensile power and impenetrability of graphene are well recognized [49]. Sub-nanometer holes may be generated in graphene by electron ion beam, oxidation, doping, cluster/ion bombardment, etc. [50]. These holes are meant to discover the transference of gases and ionic molecules [51]. Graphene and its compounds are few sheets of graphene, monolayer graphene,

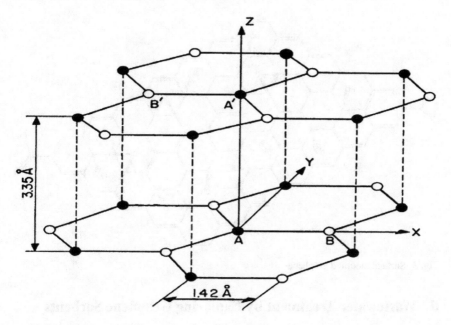

Fig. 2 Crystal structure of graphene

reduced and normal graphene oxides, graphene nano-ribbons, graphene nano-sheets etc.

The dissimilar methods utilized for graphene and nanocomposites preparations are electrochemical, reduction, decomposition exfoliation etc. (thermal, microwave, photochemical) [52, 53]. The laboratory and engineering graphene nano-platelets can be categorized into two categories i.e. multi-layered (10–60 sheets) and few coated (1–10 sheets). The graphene crystal assembly described in Fig. 2. The supra-molecular surface altered graphene structure is represented in Fig. 3. The figure of graphene interpolating compounds is described in Fig. 4. The supra-molecular graphene oxide features are given in Fig. 5. The features of this exceptional material (graphene) is revealed in Table 3. Many methods are described and used to prepare graphene i.e. motorized utility of extremely inclined chemical vapour deposition, pyrolytic graphite, epitaxial growth on SiC, bottom-up preparation from organic moieties, liquid phase utilization carbon nanotubes unzipping, natural reduction of exfoliated graphite oxide [54]. Table 4 shows the profits and disadvantages of some commonly utilized syntheses procedures.

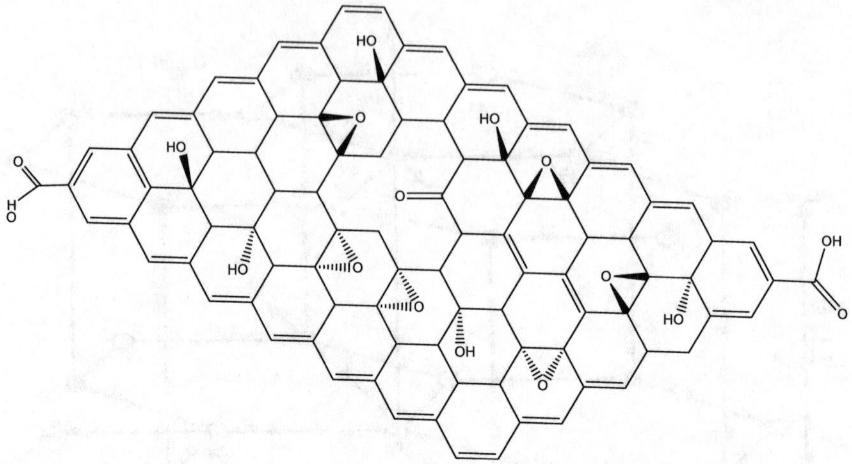

Fig. 3 Surface modified graphene

6 Wastewater Treatment by Exploring Graphene Sorbents

The adsorption is a process where a pollutant sorbed on the sorbent solid surface [55, 56]. It is controlled by a number of factors like pore volume, surface area and functional groups. It is frequently happened with the opposite desorption, which indicates transference of sorbates from sorbent superficial to the solution. The renewal of adsorbent may be tried depending on the sorbate quantity desorbed from sorbent; renewal and desorption processes increase [57]. The sorption can be of physico-chemical type. This being controlled by the types of interactions among the adsorbate and sorbent. For physical sorption, the increase in sorbate quantity at the boundary is because of non-specific van der Waal's interactions. The chemisorptions is developed via chemical reactions between sorbate and the sorbent, that produce covalent or ionic bonds. The latter is specific, weak, reversible with small thermal effect, whereas former irreversible, specific, typically with tens to hundreds of kJ/mol as heat of sorption [58].

The sorption controlling factors are pH, dose, concentration, ionic strength, temperature, the presence of organic moieties, interaction time and shacking stirring speed. The adsorption may be demarcated by normally used Freundlich, Langmuir, Tempkin etc. models [58]. Graphene and its composites are attaining reputation in water treatment through sorption process because of their remarkable features such as high surface area, great quantity of handy groups and outstanding movement of charge species. The graphene and its related materials have been widely explored as adsorbents for water treatment process.

Fig. 4 Schematic diagram
of intercalating compounds
of graphene

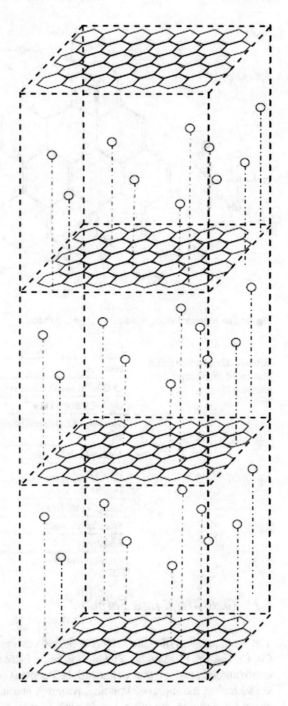

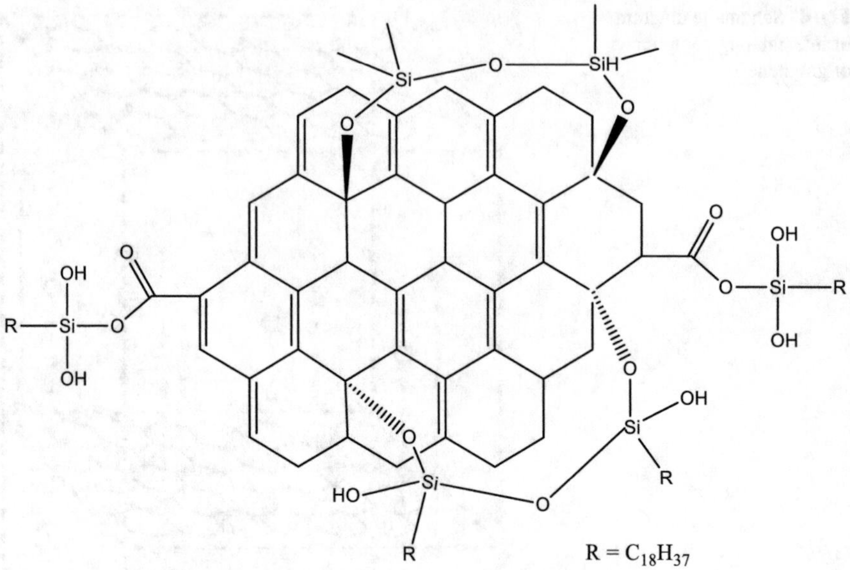

Fig. 5 The supramolecular structure of graphene oxide

Features	Values
C–C bond length thickness	14.2×10^{-2} nm
Optical transparency	97%
Theoretical BET specific surface area	2630 m^2 g^{-1}
Thermal conductivity	5000 W m^{-1} K^{-1}
Density	0.77 mg m^{-3}
Density carrier	10^{12} cm^{-2}
Young's modulus	1100 GPa
Fractural strength	125 GPa
Resistivity	10^{-6} Ω cm
Electron mobility	200,000 cm^2 V^{-1} s^{-1}

Table 3 The characteristics features of single layer graphene [54]

6.1 Toxic Metal Ions Sorption

The contamination of water resources through toxic metal ions like Ag, As, Cd, Cu, Co, Cr, Hg, Fe, Pb, Mn, Ni, Zn etc. It is being increased progressively in last years due to unrestrained releases of industrial effluents. The most contaminated industries are tanneries, mining, metal plating, painting, automobile manufacturing, printing, petroleum refining, smelting, etc. In view of their perseverance in the atmosphere, these metal ions incline to gather in living beings, thus, affecting numerous possible

Table 4 The advantages and disadvantages of some graphene synthetic methods [54]

Route	Advantage	Disadvantage
Mechanical uses of particularly sloping pyrolytic graphite	Less defects, inexpensive, excellent charge mobility, great chemical constancy and mechanical robustness	Problem of large scale manufacture & little yield
CVD	Mass manufacture, high quality, big graphene layers size	Expensive
Epitaxial growing on SiC	Big graphene layers size and good quality	High growth temperature and costly method
Reduction of exfoliated graphite oxide	Large scale production & low cost	Low purity, numerous defects, electrical properties and poor mechanical
Liquid phase exfoliation	Large scale production, low cost	Moderate quality, low yield and many impurities
Carbon nanotubes unzipping	High quality, high yield & potentially low cost	Moderate scalability

unpleasant effects on human and animals at approval levels. Presently, numerous studies have been completed to measure metal removal ability of graphene and its related materials (Table 5).

The graphene ability to pick up Co(II) and Fe(II) from water is described by Chang and co-workers [59]. Graphene was prepared via ionic liquid assisted electrochemical phenomenon as per the procedure described by Liu and co-workers [60]. Co(II) and Fe(II) adsorption was inspected by batch mode. The graphene displayed great sorption capabilities of 370.0 and 299.30 mg g^{-1} for Fe(II) and Co(II), correspondingly. Wu and co-workers [61] tested the adsorption ability of graphene oxide (GO) to remove Cu(II) in water. The finest sets for Cu(II) removal by GO in a batch mode were 1.0 mg mL^{-1} dosage, 5.3 pH and 150.0 min interaction time. Copper(II) equilibrium adsorption information fitted completely to Freundlich isotherm. GO displayed great adsorption ability (117.5 mg g^{-1}) for Copper(II) as per Langmuir constants. The sorption of Copper(II) on GO was primarily accredited to electrostatic attraction, outside ion exchange, and complexation as per equations given below:

$$Cu^{2+} + GO - COOH \rightarrow GO - COO - Cu^{2+} + H^+ \tag{1}$$

$$Cu^{2+} + (GO - COOH)_2 \rightarrow (GO - COO^-)_2 - Cu^{2+} + 2H^+ \tag{2}$$

$$Cu^{2+} + GO - OH \rightarrow GO - O - Cu^{2+} + 2H^+ \tag{3}$$

$$Cu^{2+} + (GO - OH)_2 \rightarrow (GO - O-)_2 - Cu^{2+} + 2H^+ \tag{4}$$

Table 5 The adsorption of heavy metal ions in water with graphene-based nanomaterials (data from Ref. 54 with permission)

Adsorbents	Metal ions	Concs.	pHs	Temp. (K)	Contact times (h)	Adsorption capacity	References
Graphene	Pb(II)	40 mg L^{-1}	4.0	303	15	22.42 mg g^{-1}	[104]
Graphene	Sb(III)	1–10 mg L^{-1}	11.0	303	4	10.919 mg g^{-1}	[71]
Graphene	Pb(II)	40 mg L^{-1}	4.0	303	15	35.46 mg g^{-1}	[70]
Graphene	Pb(II)	40 mg L^{-1}	4.0	303	15	35.21 mg g^{-1}	[104]
Graphene	Fe(II)	20 mg L^{-1}	8.0	–	24	299.3 mg g^{-1}	[59]
Graphene	Co(II)	20 mg L^{-1}	8.0	–	24	370 mg g − 1	[59]
CTAB modified graphene	Cr(VI)	20–100 mg L^{-1}	2.0	293	1	21.57 mg g^{-1}	[67]
Functionalized graphene (GNSPF6)	Cd(II)	–	6.2	–	4	73.42 mg g^{-1}	[68]
Functionalized graphene (GNSPF6)	Pb(II)	–	5.1	–	4	406.4 mg g^{-1}	[68]
Functionalized graphene (GNSC8P)	Cd(II)	–	6.2	–	4	30.05 mg g^{-1}	[68]
Functionalized graphene (GNSC8P)	Pb(II)	–	5.1	–	4	74.18 mg g^{-1}	[68]
GO	Zn(II)	–	5.6	–	–	30.1 ± 2.5 mg g^{-1}	[79]
GO	Pb(II)	5–300 mg L^{-1}	6.8	298 ± 2	24	367 mg g^{-1}	[72]

(continued)

Table 5 (continued)

Adsorbents	Metal ions	Concs.	pHs	Temp. (K)	Contact times (h)	Adsorption capacity	References
GO	Pb(II)	–	5.6	–	–	35.6 ± 1.3 mg g^{-1}	[79]
GO	Cd(II)	–	5.6	–	–	14.9 ± 1.5 mg g^{-1}	[79]
GO	Pb(II)	5–300 mg L^{-1}	7.0 ± 0.5	298 ± 5	24	692.66 mg g^{-1}	[85]
GO	Cu(II)	25–250 mg L^{-1}	5.3	–	2.5	117.5 mg g^{-1}	[61]
GO	Zn(II)	–	5.0	298	2	345 mg g^{-1}	[64, 85]
GO	Cu(II)	–	5.0	298	2	294 mg g^{-1}	[62]
GO	Pb(II)	–	5.0	298	2	1119 mg g^{-1}	[62, 85]
GO	Cd(II)	–	5.0	298	2	530 mg g – 1	[62, 85]
GO	Zn(II)	10–100 mg L^{-1}	7.0 ± 0.1	293	–	246 mg g^{-1}	[63]
Poly(amidoamine) modified GO	Pb(II)	0.0193 mmol L^{-1}	–	Room temp.	24	0.0513 mmol g^{-1}	[69]
Poly(amidoamine) modified GO	Zn(II)	0.0193 mmol L^{-1}	–	Room temp.	24	0.2024 mmol g^{-1}	[69]
Poly(amidoamine) modified GO	Cr(III)	0.0193 mmol L^{-1}	–	Room temp.	24	0.0798 mmol g^{-1}	[69]

(continued)

Table 5 (continued)

Adsorbents	Metal ions	Concs.	pHs	Temp. (K)	Contact times (h)	Adsorption capacity	References
Poly(amidoamine) modified GO	Fe(III)	0.0193 mmol L^{-1}	–	Room temp.	24	0.5312 mmol g^{-1}	[69]
Poly(amidoamine) modified GO	Cu(II)	0.0193 mmol L^{-1}	–	Room temp.	24	0.1368 mmol g − 1	[69]
FGO	Cd(II)	–	6.0 ± 0.1	303	–	106.3 mg g^{-1}	[65]
FGO	Pb(II)	–	6.0	293	24	842 mg g^{-1}	[65]
FGO	Co(II)	–	6.0 ± 0.1	303	–	68.2 mg g^{-1}	[65]
Graphene/δ-MnO_2	Cu(II)	–	6.0	298 ± 2	2	1637.9 μmol g − 1	[91]
Graphene/δ-MnO_2	Ni(II)	10–100 mg L^{-1}	–	298	3	46.55 mg g^{-1}	[90]
GO aerogel	Cu(II)	50–75 mg L^{-1}	6.3	283	0.5	17.73 mg g^{-1}	[64]
EDTA modified GO	Pb(II)	5–300 mg L^{-1}	6.8	298 ± 2	24	525 mg g^{-1}	[72]
Graphene/Fe	Cr(VI)	25–125 mg L^{-1}	4.25	293	4	162 mg g^{-1}	[73]
Graphene/$Mn^{2+}Fe^{3+}$ O^{2-}	As(III)	1–8 mg L^{-1}	7.0 ± 0.1	300 ± 1	2.5	14.42 mg g^{-1}	[104]
Graphene/Fe@Fe2O3@SiSiO	Cr(VI)	1 g L^{-1}	7.0	–	–	1.03 mg g^{-1}	[105]

(continued)

Table 5 (continued)

Adsorbents	Metal ions	Concs.	pHs	Temp. (K)	Contact times (h)	Adsorption capacity	References
SiO$_2$/graphene	Pb(II)	20 mg L − 1	6.0	298	1	113.6 mg g^{-1}	[75]
Graphene/c-MWCNT	Pb(II)	50 mg L^{-1}	–	Room temp.	120	104.9 mg g^{-1}	[77]
Graphene/c-MWCNT	Hg(II)	50 mg L^{-1}	–	Room temp.	120	93.3 mg g^{-1}	[77]
Graphene/c-MWCNT	Cu(II)	50 mg L^{-1}	–	Room temp.	120	33.8 mg g^{-1}	[77]
Graphene/MWCNT	Cu(II)	50 mg L^{-1}	–	Room temp.	120	9.8 mg g^{-1}	[77]
Graphene/c-MWCNT	Ag(II)	50 mg L^{-1}	–	Room temp.	120	64.0 mg g^{-1}	[77]
Graphene/MWCNT	Ag(II)	50 mg L^{-1}	–	Room temp.	120	46.0 mg g^{-1}	[77]
Graphene/MWCNT	Pb(II)	50 mg L^{-1}	–	Room temp.	120	44.5 mg g^{-1}	[77]
Graphene/MWCNT	Hg(II)	50 mg L^{-1}	–	Room temp.	120	75.6 mg g^{-1}	[77]
Graphene/MgAl-layered double hydroxides	Cr(VI)	50–250 mg L^{-1}	2.0	–	24	183.82 mg g^{-1}	[80]
GO/ferric hydroxide	As(V)	0.5–20 mg L^{-1}	4.0–9.0	Room temp.	24	23.78 mg g^{-1}	[76]
GO–iron oxide	Pb(II)	–	6.5 ± 0.1	303	48	588.24 mg g^{-1}	[103]

(continued)

Table 5 (continued)

Adsorbents	Metal ions	Concs.	pHs	Temp. (K)	Contact times (h)	Adsorption capacity	References
Chitosan/GO	Pd(II)	80–500 mg L^{-1}	–	Room temp.	16	216.920 mg g^{-1}	[95]
Chitosan/GO	Au(III)	80–500 mg L^{-1}	–	Room temp.	16	1076.649 mg g^{-1}	[95]
GO/chitosan	Pb(II)	50 mg L^{-1}	–	Room temp.	–	99 mg g – 1	[93]
GO–chitosan composite hydrogel	Pb(II)	0–120 mg L^{-1}	4.9	294 ± 1	4	90 mg g^{-1}	[96]
GO–chitosan composite hydrogel	Cu(II)	0–120 mg L^{-1}	5.1	294 ± 1	10	70 mg g – 1	[96]
Magnetic chitosan/GO	Pb(II)	–	5.0	303 ± 0.2	1	76.94 mg g^{-1}	[98]
Polypyrrole/GO	Cr(VI)	–	–	–	–	9.56 mmol g^{-1}	[74]
GO–TiO$_2$	Cd(II)	–	5.6	–	–	72.8 ± 1.6 mg g^{-1}	[79]
GO–TiO$_2$	Zn(II)	–	5.6	–	–	88.9 ± 3.3 mg g^{-1}	[79]
GO–TiO$_2$	Pb(II)	–	5.6	–	–	65.6 ± 2.7 mg g – 1	[79]
GO–ZrO(OH)$_2$	As(V)	2–80 mg L^{-1}	7.0 ± 0.2	298.5 ± 0.2	0.25	84.89 mg g – 1	[78]
GO–ZrO(OH)$_2$	As(III)	2–80 mg L^{-1}	7.0 ± 0.2	298.5 ± 0.2	0.25	95.15 mg g^{-1}	[78]

(continued)

Table 5 (continued)

Adsorbents	Metal ions	Concs.	pHs	Temp. (K)	Contact times (h)	Adsorption capacity	References
GO–FeOOH	As(V)	–	7.0	298	–	73.42 mg g^{-1}	[82]
Magnetic cyclodextrin–chitosan/GO	Cr(VI)	50 mg L – 1	3.0	303	–	61.31 mg g^{-1}	[84]
Sulfonated magnetic GO composite	Cu(II)	73.71 mg L^{-1}	5.0	283.15	6	50.678 mg g^{-1}	86]
Calcium alginate/GO	Cu(II)	–	–	Room temp.	1.5	60.24 mg g – 1	[81]
Polypyrrole–RGO	Hg(II)	50–250 mg L^{-1}	3.0	293	3	979.54 mg g^{-1}	[106]
Poly(N-vinylcarbazole)–GO	Pb(II)	5–300 mg L^{-1}	7.0 ± 0.5	298 ± 5	24	982.86 mg g^{-1}	[85]
Magnetite–RGO	As(III)	3–7 mg L^{-1}	7.0	293	2	10.20 mg g^{-1}	[102]
RGO–MnO$_2$	Hg(II)	1 mg L^{-1}	–	303 ± 2	–	9.50 mg g^{-1}	[107]
RGO–Fe$_3$O$_4$	As(III)	2–6 mg L^{-1}	7.0	298	1	21.2 mg g^{-1}	[108]
RGO–Fe(0)	As(III)	2–6 mg L^{-1}	7.0	298	1	37.3 mg g^{-1}	[108]
RGO–Fe(0)/Fe$_3$O$_4$	As(III)	2–6 mg L^{-1}	7.0	298	1	44.4 mg g^{-1}	[108]
RGO–iron oxide	Pb(II)	10–15 mg L^{-1}	6.5 ± 0.1	303	48	454.55 mg g^{-1}	[98]
RGO–Fe(0)/Fe$_3$O$_4$	Hg(II)	2–6 mg L^{-1}	7.0	298	1	22.0 mg g^{-1}	[108]
RGO–Fe(0)/Fe$_3$O$_4$	Pb(II)	2–6 mg L^{-1}	7.0	298	1	19.7 mg g^{-1}	[108]
RGO–Fe(0)/Fe$_3$O$_4$	Cr(VI)	2–6 mg L^{-1}	7.0	298	1	31.1 mg g^{-1}	[108]
RGO–Fe(0)	As(III)	2–6 mg L^{-1}	7.0	298	1	37.3 mg g^{-1}	[108]
RGO–Fe(0)/Fe$_3$O$_4$	Cd(II)	2–6 mg L^{-1}	7.0	298	1	1.91 mg g^{-1}	[108]

Sitko et al. [62] described the usage of GO for removal of Cu(II), Cd(II), Zn(II) and Pb(II), that was in Pb(II) > Cd(II) > Zn(II) > Cu(II) in order in one metal arrangements, and Pb(II) > Cu(II) > > Cd(II) > Zn(II) in two metal arrangements. The adsorption kinetic readings showed monolayer adsorption of metal ions considered. Also, the workers defined chemisorption for the process. Wang et al. [63] utilized GO to eradicate Zn(II) ions in water, that depended on dose, pH, and quantity. The data followed Langmuir and pseudo-second-order kinetics isotherms. A worthy sorption capability was 246.0 mg g^{-1} at 293 K with fast exothermic adsorption. Mi and co-workers [64] prepared GO aerogels for the removal of Cu(II) in water. The GO aerogel was extremely soft with numerous oxygen atoms with some groups, and, henceforth, an outstanding sorbent. The adsorption was reliant on pH and the concentrations. The GO aerogel showed quick adsorption rate; with 17.73 mg g^{-1} at 283 K to 29.59 mg g^{-1} at 313 K. The adsorption was through chemosorption as specified by pseudo-second-order calculation.

Zhao et al. [65] described the usage of GO nano-sheets for uptake of Cd(II), and Co(II), that was pH and ionic strength reliant on. It was detected that Langmuir isotherm was followed because of monolayer sorption with 106.3 and 68.2 mg g^{-1} for Cd(II) and Co(II) at pH 6.0; with endothermic and fast sorption, correspondingly. Zhao and co-workers [66] synthesized and utilized some nano-sheets of GO to remove Pb(II) ions in water. The adsorption augmented at pH ranging from 1.0 to 8.0. The outcomes followed Langmuir model with adsorption capability of 1850, 1150, and 842 mg g^{-1} at 333, 313 and 293 K; with endothermic and fast phenomenon, correspondingly. Worthy adsorption was owing to tough outside complexation amid oxygen groups and Pb(II). Yan et al. [67] defined graphene alteration through cetyltrimethylammonium bromide (CTAB) for improved removal of Cr(VI). The removal was concentration and pH reliant on with maximal removal at pH 2.0. The outcomes followed Langmuir and pseudo-second-order kinetics isotherms; with 21.57 mg g^{-1} as adsorption capability. The adsorption procedure was exothermic and fast. Deng and co-workers [68] altered graphene chemically to advance removal capability of metal ions. The workers utilized 1-octyl-3-methylimidazolium hexafluorophosphate and potassium hexafluorophosphate ionic liquids for graphene alteration. The maximal adsorption described was 89.89% succeeding pseudo-second-order Langmuir and Freundlich isotherms. Yuan and co-workers [69] altered graphene with poly-amidoamine for improved removal of Cr(III), Cu(II), Pb(II) Fe(III), and Zn(II).

Huang and co-workers [70] manufactured graphene nanosheets (GNSs) at 473 K temperature via chemical exploitation method in great vacuum situations. The nanosheets were canned at 773 and 973 K to attain altered GNSs (denoted as GNS-700 and GNS-500). Both pristine and thermally canned GNSs proven similar adsorption for Pb(II) in the water. Lead(II) elimination rate declines with the rise in initial Pb(II) quantities, while it increased with a increase in pH from 3.0 to 5.0. High lead(II) adsorption capabilities achieved by Langmuir isotherm were 35.46, 22.42, 35.21, and mg g^{-1} for GNS-700, GNSs, and GNS-500, correspondingly. The chances of using graphene as an adsorbent to eliminate Sb(III) was described by Leng and co-workers [71]. The sorption experimentations were conceded to explain the effects of the noteworthy variables such as Sb(III) amounts, pH, time, and temperature. It was seen that adsorption capability was loosen with the rise in metal ion quantities, whereas it rises with rising temperature. An austere development in metal elimination efficacy was noticed with a growth in pH afar 3.8. High elimination of around 99.5% was attained at pH > 11. The procedure seemed to follow Freundlich model better than Langmuir one. In improved conditions, adsorption capability for Sb(III) was 10.919 mg g^{-1}. The adsorption and kinetic data followed fitted pseudo-second-order equation, thus, showing it likely to suitably deduce total adsorption process and indicated chemisorption as rate determining step. Madadrang and co-workers [72] presented chelating groups to GO and improved its adsorption capacity for Pb(II). The chemicals utilized was N-(trimethoxysilylpropyl) ethylenediamine triacetic acid. The elimination was worthy with 525.0 mg g^{-1} as adsorption capability; succeeding Langmuir and pseudo-second-order kinetics isotherms.

The nano-porous graphene membranes have also been used for metal ions removal in sorption process. Jabeen and co-workers [73] synthesized graphene layers decorated with zero valent iron nano-particles. The industrialized adsorbent was used Cr(VI) removal, that was pH-relayed; with good elimination at pH 2.0–3.0. The workers termed the best adsorption by Langmuir model. Li and co-workers [74] prepared polypyrrolelGO composite by pattern polymerization. The composite was analyzed for removing Cr(VI) and data trailed Langmuir isotherm. Hao et al. [75] prepared SiO$_2$/graphene composite and utilized for Pb(II) elimination; with maximal removal capability as 113.6 mg g^{-1}. Zhang and co-workers [76] synthesized ferric hydroxide/GO composite representing more than 95% adsorption for As(V) in polluted drinking water. Sui and co-workers [77] prepared CNTs-graphene aerogels for elimination of Ag(II), Cu(II), Hg(II), and Pb(II). The workers described worthy adsorption of the metal ions because of the incidence of more oxygens with in CNTs-graphene aerogels. Luo and co-workers [78] prepared zirconium oxide/GO composite and utilized for elimination of As(V) and As(III) in water. The adsorption was very quick with 10–15 min as contact times and up to 95%. The adsorption capabilities were 95.5 and 84.89 mg g^{-1} for As(III) and As(V), correspondingly. Lee and Yang [79] synthesized TiO$_2$-GO substance used for elimination of Pb(II), Cd(II), and Zn(II) in water. The adsorption capabilities were 88.9, 65.6 and 72.8, mg g^{-1} for Zn(II), Pb(II) and Cd(II), correspondingly.

Yuan and co-workers [80] prepared MgAl-graphene double layered hybrid nanomaterial utilizing graphene, aluminium nitrate and magnesium nitrate. The composite was utilized to eliminate Cr(VI) with 183.82 mg g^{-1} as adsorption capability. The sorption was fast and endothermic. The data followed Freundlich and pseudo-second-order kinetics isotherms. The adsorption was organized by chemisorption. Algorithm and co-workers [81] synthesized calcium alginate-GO compound material and utilized for Cu(II) elimination in water. The statistics followed Langmuir and pseudo second order isotherms with maximal adsorption capability 60.2 mg g^{-1}. Peng and co-workers [82] prepared GO-FeOOH mixture material for As(V) elimination in water. The adsorption was quick with maximal capability of 73.42 mg g^{-1}. Zhu and co-workers [83] compared Cr(VI) elimination presentation on pristine graphene and magnetic graphene composites; with later as the best adsorbent. The workers advocated this adsorbent as noteworthy in water management in actual water samples. Hu and co-workers [84] synthesized sulfonated magnetic cross composite and utilized to eliminate Cu(II) in water. The workers improved adsorption procedure by some variables. These too used Box-Behnken experimental design matrix to attain the finest elimination of this metal ion. The ANOVA and regression analyses showed investigational variable dependents adsorption with 106.20 as Fisher's F-value. Lastly, the workers described maximal adsorption as 62.73 mg g^{-1} at the finest investigational conditions i.e. 4.68 pH, 73.71 mg L^{-1} concentration and 323 K temperature. Copper(II) adsorption data designated a decent relationship with the Langmuir model. The data followed pseudo-second order kinetics; with spontaneous and endothermic sorption.

Musico and co-workers [85] synthesized poly(N-vinylcarbazole)-GO hybrid materials and utilized for Pb(II) elimination. The elimination was straight related to the quantity of GO in hybrid because of improved oxygen useful groups. The adsorption capability was 887.98 mg g^{-1} at elevated pH and the statistics followed Langmuir isotherm. The polymeric forms of α, δ, β, γ and λ manganese dioxides are utilized for pollutants elimination [86–89] because of their small price of synthesis and eco-friendly nature. Ren and co-workers [90] synthesized δ-MnO$_2$$^-$graphene nano-sheets and utilized for Ni(II) elimination water. The adsorption was endothermic with Langmuir and pseudo-second-order rate isotherms follow-up. The maximal adsorption capability was 46.55 mg g^{-1}. The desorption procedure was attained by 0.1 M hydrochloric acid with only 9.0% loss. Ren and co-workers [91] utilized GNS/MnO$_2$ for elimination of Cu(II) and Pb(II) and statistics followed Langmuir and pseudo-second-order kinetics isotherms. The maximum adsorption capabilities of GNS/MnO$_2$ for Cu(II) and Pb(II) ions were 1637.9 and 793.65 μmol g^{-1}. The workers described adsorption because of tetradentate surface complexes development of monodentate, multidentate bidentate mononuclear, configurations and bidentate binuclear. Also, outside oxygens with functional were chiefly tangled in the adsorption process as given below:

$$S - OH + M_{2+} \rightarrow (S - O - M) + +H^+ \tag{5}$$

$$S - O - + M_{2+} \rightarrow (S - O - M)+ \tag{6}$$

$$S - OH + M_{2+} + H_2O \rightarrow S - OMOH^+ + 2H^+ \tag{7}$$

where S and M are GNS/MnO$_2$ surface and metal ion.

The chitosan has various functional moieties and is utilized to take up aqueous metal ions [92]. He and co-workers [93] described lead(II) removal using chitosan-graphene oxide composite with 99.0 mg g^{-1} as adsorption ability. Zhang and co-workers [94] produced good elastic and biodegradable type chitosan-GO-gelatin by mono-directional freezed dried method and used for Cu(II) and Pb(II) removal. The composite material showed decent metal elimination ability with numerous times recycling without any severe damage in adsorption. Liu and co-workers [95] also produced GO-chitosan and utilized for lead and gold removal in aqueous solution. The adsorption abilities were 3 to 4 for Pb(II) and 3 to 5 for Au(III). The adsorption was endothermic and spontaneous. The information followed pseudo second order kinetic model, displaying chemical adsorption. The workers defined successive three stints recycles after regenerating the adsorbent without any alteration in the adsorption. Likewise, Chen and co-workers [96] utilized chitosan-GO for Pb(II) and Cu(II) elimination with 70.0 and 90.0 mg g^{-1} as the adsorption abilities. Li and co-workers [97] synthesized chitosan-GO composite exhibit high surface area, magnetic properties and hydroxyl and amino groups. The workers utilized the substance for elimination of chromium(VI) with 67.66 mg g^{-1} as adsorption abilities. The sorption followed pseudo second order kinetics and Freundlich isotherms. The adsorbent might lose merely 5.0% adsorption capability. Fan and co-workers [98] arranged magnetic GO-chitosan for Pb(II) elimination with sorption capability of 76.94 mg g^{-1}. The workers appealed the method as valuable in industrial waters treatment.

Magnetite is iron erosion product having good capability in adsorbing numerous metal ions in aqueous solutions [99, 100]. Some workers prepare composite material by mixing graphene with other chemicals, and the resultant composites were explored for metal ions elimination in water. Liu and co-workers [101] prepared this type composite substance for Co(II) elimination in water. Langmuir and pseudo second order models were applicable; indicating experimental validity. The workers took out sorbent by using high power magnet. Chandra and co-workers [102] synthesized GO-magnetite and used to remove arsenic species i.e. [As(V) and As(III)] in water. The adsorption capability was about 99.9% at 1 µg/L amount. Langmuir and pseudo-second-order kinetic isotherms indicated the experimental validity. Yang and co-workers [103] attached GO on magnetite and utilized for elimination of Pb(II) with 588.24 mg g^{-1} adsorption capability. The applicability Langmuir isotherm confirm the good applicability of experimental information. Nandi and co-workers [104] synthesized graphene-Mn-magnetite cross material and described 99.90% As(III) elimination. The optimum investigational settings were 7.0 pH, 150.0 time and 27.0 °C temperature. Langmuir and pseudo-second-order kinetics isotherms were the best fitted for sorption. As per the workers the synthesized composite might be beneficial for drinking water management.

Zhu and co-workers [105] synthesized magnetic graphene hybrid attached through nanoparticles of core@double shell of iron oxide and inside shells of the amorphous Si-S-O. The workers utilized the described substance for Cr(VI) elimination. The workers described approximately hundred percent elimination within 5.0 min time at dose of 3.0 g/L and pH of 1–3 range. The workers advocated the prepared material useful because of small contact time but as per our knowledge its low pH will limit the application of the materials. Chandra and Kim [106] synthesized RGO-polypyrrole for abolition of Hg(II) in water. The workers described 979.54 mg g^{-1} as adsorption capability. Sreeprasad and co-workers [107] also synthesized Ag-RGO and MnO$_2$-RGO adsorbents and used for elimination of Hg(II) in aqueous solution. The workers described 9.50 and 9.53 as sorption capacities on MnO_2-RGO and Ag-RGO adsorbents, respectively. Bhunia and co-workers [108] synthesized Fe_3O_4-RGO-Fe(0) material for As(III) elimination in water. The sorption phenomenon followed Langmuir and pseudo second order kinetics isotherms. The adsorption capability was 44.40 mg g^{-1}. The workers also analyzed the substance for Pb(II), Cd(II), Hg(II) and Cr(VI) metal ions with sorption capacities in the range of 1.91–19.7 mg g^{-1}. Zhang and co-workers [109] synthesized rGO/NiO material for removal of Cr(VI) in water. The workers described sorption capability of 198 mg g^{-1}. The statistics followed Freundlich Langmuir and pseudo-second-order kinetic isotherms. As per these workers pH of solution influenced the adsorption process.

6.2 Rare Earth Metal Ions Sorption

The rare earth elements are too toxic and existing in waste water at some places. Numerous authors attempted to eliminate rare earth metal ions in water utilizing graphene and related composites as sorbents. Some important findings of the different authors are described herein. Ashour and co-workers [110] utilized GO for the elimination of Nd(III), Y(III), Gd(III) and La(III) metal ions in water. The workers described pretty good adsorption capacity with good fitting of Langmuir model; representing mono layer adsorption. Gu and Fein [111] described the sorption of some metal ions onto GO. The authors advocated sorption process as external complexation; with simulation and linear free energy relations. The sorption capability of these metal ions are summarized in Table 6.

Table 6 A comparison of adsorption of rare earth metal ions on GO surface

Adsorbents	Metal ions	Adsorption capacity (mg/g)	pH	References
Colloidal GO	Eu(III)	89.64	4.5	[123]
	Gd(III)	286.80	5.9	[124]
GO nanosheets	Eu(III)	175.40	6.0	[125]
Magnetic GO	Eu(III)	70.15	4.5	[123]
GO	La(III)	85.67	6.0	[110]
GO	Y(III)	135.070	6.0	[110]
GO	Gd(III)	225.50	6.0	[110]
GO	Nd(III)	188.60	6.0	[110]

7 Regeneration of Used Graphene Materials

The economy at commercial scale application of graphene based materials is determined by their recycling and reuse efficiencies. Consequently, desorption and recycling are precise significant processes for the viable utilization of graphene based materials [112]. Generally, the graphene based materials are separated and regenerated followed by recycling. The separation of graphene materials is a hard task because of their nano size. It is being controlled by nature of materials. The different methods are used to efficiently discrete graphene based materials in water, amongst which the best one are cross-flow filtration, centrifugation, electric field, field-flow fractionation etc. [113–115]. Wang and co-workers [63] reported desorption capacities of Zn(II) as 53.2, 73.4 and 91.6% by using H_2O, 0.1 M nitric acid and 0.1 M hydrochloric acid, correspondingly. For copper(II) it was 74% at pH lower than 1.0 using HCl [61]. Generally, it may be presumed that the desorption proportion of metal ions on GO and its compounds augmented when lessening pH of solution [61, 79]. Furthermore, thiourea-HCl [95] and EDTA [71, 95] were also utilized as actual eluents for extracting the sorbed metal ions in spent graphene sorbents.

The adsorption capability of copper(II) on recycled GO could reserve terminated 95% initial capability after five sequences and more than 90% after ten sequences [61]. The sorption ability of GO membrane for Ni(II), Cu(II) and Cd(II) and after six sequences reduced merely to approximately 21, 10 and 12%, respectively [116]. Sahraei and co-workers [117] defined adsorption of Pb(II) and Cu(II) utilizing tragacanth-GO by 0.1 N nitric acid for 8 h. Also, three successive sorption-desorption sequences were done to examine the reusability of sorbent. The adsorption was initiated to decline 6.5% Cu(II) and 2.75% Pb(II). After three sequences of the sorption procedure, the desorption reduced to 7.0% for Pb(II) and Cu(II). The desorption stages were not comprehensive while the adsorption ability of the adsorbent was nearly completely preserved after three sequences. Chen and co-workers [118] considered copper(II) desorption on cellulose/GO hydrogel utilizing 1.0 M hydrochloric acid 5.3 pH, 298 K temperature and 2 h contact time. Additional renewal with 0.1 M sodium hydroxide was perceived with no noticeable harms to the adsorption capability and desorption efficacy of the sorbent; engaged as some categorizations increased from 1.0 to 5.0. As per the researchers, carboxylic groups were transformed to COO^- in renewal by sodium hydroxide solution, which indicated good attraction to copper ions. The researchers also effectively confirmed the recycling of the sorbent in the event of Fe(III), Pb(II) Zn(II). La and co-workers [119] defined Pb(II) desorption on graphene@teranry oxides material by employing column test. The column was renewed utilizing numerous washing with 2.0 M sodium hydroxide after each lead(II) adsorption sequence. It was detected that lead(II) desorption effectiveness faintly augmented with the eluting solvent amount from 5.0 to 40.0 mL. Lead recovery had become nearly unaffected when the eluting solvent volume was higher than 20.0 mL. The sorbent permitted to extent more than 90.0% extraction of lead in 30 min utilizing 2.0 M sodium hydroxide as removing agent. Zhang and co-workers [109] defined chromium(VI) desorption on a NiO/rGO composite utilizing 0.1 M hydrochloric acid. The workers observed that Cr(VI) elimination capacity decreased gradually at augmenting numbers of cycles, and chromium elimination efficiency continued to be steady at 83% after five sequences.

Ashour and co-workers [110] described of Gd(III), Nd(III), La(III) and Y(III)] by using 0.1 M nitric acid. The workers reported about 99% recovery of these metal ions. Chen and co-workers [120] reported the recovery of catechol on graphene nano-platelets; with assumption desorption as slow. Gan and co-workers [121] regeneration of sulfonated graphene after ex situ soil wash for polycyclic aromatic hydrocarbons. The reduction in elimination was on successive renewal series. Sahraei and co-workers [117] also described desorption of congo red and crystal violet dye on amended gum tragacanth-GO utilizing 0.1 N NaOH and EtOH for about for 8 h. The workers observed that desorption reduced to 4.4 and 7.50% for congo red and crystal violet dyes; after three cycles, whereas the desorption reduced to 5.5 and 9% for congo red and crystal violet dyes after three sequences. Currently, desorption (electrochemical) had been utilized to advance original ecologically pleasant renewal procedures. Pan and co-workers [122] selected graphene aerogel as a perfect conductive sorbent to examine the regeneration (electrochemical) for methylene blue dye and Cu(II) metal ion. These workers observed that the sorbent utilized empowered in situ electrochemistry persuaded scrubbing by electrochemical degradation/desorption of the sorbed organic contaminants and the electro-repulsive interactions and improvement of the sorbed metal ions in the lack of extra substances. The control of the utilized graphene materials is also an important subject. The regenerated sorbents may be used in developing cement, brick, steel and other building materials [115].

8   Future Perspectives

The graphene materials are being utilized for water treatment purposes because of to their extraordinary features. These may be worthy replacements to the other nanomaterials like fullerenes and CNTs. Their utilization in sorption method of water treatment for eliminating numerous organic and inorganic pollutants appears to be hopeful and may resolve pure water disaster in near future. However, in spite optimistic perspectives of graphene materials, there are still trials in achieving the inexpensive and ascendable industrialization. Also, there is a necessity for their surface alteration to increase the utilities in water treatment. Furthermore, renewal of graphene materials is very significant in the economy side. The other concerns restricting their applications are environmental and health risks. Therefore, there is a demand to regulate the leakage of these substances in the water treatment process. Also, these must be suitably predisposed after utilization. Despite of the overhead stated problems and bearing in mind the capability of these substances to eliminate different pollutants; even at low amounts with a little quantity of dose, these must be prepared in cost-effective and ecologically pleasant manners and used in restricted ways to evade any threat to the living beings and ecology.

9 Conclusions

The present time is of graphene and its related composites in water treatment as being used efficiently for the purpose. This is because of the distinctive features such as small particle size, high surface area, high mobility, free active valences and charge carriers etc. These are very effective in removing many metal ions like Ag, As, Al, Cd, Cu, Co, Cr, Pb, Fe, Hg, Mn, Ni, Sn, and Zn in wastewater. These materials can be the best alternatives for the low cost sorbents and may be used for the elimination of all most all sorts of metal ions in water. The outcomes are very effective in terms of elimination capability. Despite of numerous inexpensive wastewater treatment methods, huge world population is not drinking safe water. This chapter will be useful for the researchers, academicians and government authorities for planning and developing new strategies for metal ions removal in water.

Acknowledgements The authors extend their appreciation to the International Scientific Partnership Program ISPP at King Saud University for funding this research work through ISPP# 0037.

References

1. Gupta, V.K., Ali, I.: Environmental Water: Advances in Treatment, Remediation and Recycling. Elsevier, The Netherlands (2012)
2. Ali, I., Aboul-Enein, H.Y.: Instrumental Methods in Metal Ions Speciation: Chromatography, Capillary Electrophoresis and Electrochemistry. Taylor & Francis Ltd., New York, USA (2006)
3. Hu, J., Zhao, D.L., Wang, X.K.: Removal of Pb(II) and Cu(II) from aqueous solution using multiwalled carbon nanotubes/iron oxide magnetic composites. Water Sci. Technol. **63**, 917–923 (2011)
4. Srivastava, S.K., Bhattacharjee, G., Tyagi, R., Pant, N., Pal, N.: Studies on the removal of some toxic metal-ions from aqueous-solutions and industrial-waste. 1. Removal of lead and cadmium by hydrous iron and aluminum-oxide. Environ. Technol. Lett. **9**, 1173–1185 (1988)
5. Ozay, O., Ekici, S., Baran, Y., Aktas, N., Sahiner, N.: Removal of toxic metal ions with magnetic hydrogels. Water Res. **43**, 4403–4411 (2009)
6. Jamil M, Zia MS, Qasim M (2010) Contamination of agro-ecosystem and human health hazards from wastewater used for irrigation. J. Chem. Soc. Pak. **32**: 370–378
7. Khan, S., Cao, Q., Zheng, Y.M., Huang, Y.Z., Zhu, Y.G.: Health risks of heavy metals in contaminated soils and food crops irrigated with wastewater in Beijing. China, Environ Pollut **152**, 686–692 (2008)
8. Singh, A., Sharma, R.K., Agrawal, M., Marshall, F.M.: Health risk assessment of heavy metals via dietary intake of foodstuffs from the wastewater irrigated site of a dry tropical area of India. Food Chem. Toxicol. **48**, 611–619 (2010)
9. Peng, S.H., Wang, W.X., Li, X.D., Yen, Y.F.: Metal partitioning in river sediments measured by sequential extraction and biomimetic approaches. Chemosphere **57**, 839–851 (2004)
10. Snoeyink, V.L., Jenkins, D.: Water Chemistry. Wiley, New York, USA (1980)
11. Hodges, L.: Environmental Pollution, 2nd ed., Holt, Rinehart and Winston, New York, USA (1973)

12. World Health Organization.: Guidelines for Drinking Water Quality. Health Criteria and Other Supporting Information, 2nd ed., WHO, Geneva, Switzerland (1996)
13. US Environmental Protection Agency: Ground Water and Drinking Water, Current Drinking Water Standards, EPA-822-F-97-009. Office of Water, Washington, DC (2002)
14. Okamura, H., Aoyama, I.: Interactive toxic effect and distribution of heavy metals in phytoplankton. Environ. Toxicol. Water Qual. **9**, 7–15 (1994)
15. Martin, T.R., Holdich, D.M.: The acute lethal toxicity of heavy metals to percarid crustaceans (with particular reference to fresh-water asellids and gammarids. Water Res. **20**, 1137–1147 (1986)
16. Moore, J.W.: Inorganic Contaminants of Surface Water Research and Monitoring Priorities. Springer-Verlag, New York, USA (1991)
17. Yamamura, Y., Yamauchi, H.: Arsenic metabolites in hair, blood and urine in workers exposed to arsenic trioxide Industrial Health **18**: 203–210 (1980)
18. World Health Organisation.: Environmental health criteria, 18: Arsenic, world health organisation, Geneva (1981)
19. Pershagen, G.: The Epidemiology of human arsenic exposure. In: Fowler, B.A. (ed.) Elsevier, Amsterdam, The Netherlands, 199 (1983)
20. Csanady, M., Straub, I.: Health damage due to pollution in hungary. In: Proceedings of the Rome Symposium, IAHS Publ No 233 (1995)
21. Johnson, B.B.: Effects of pH, temperature and concentration on the adsorption of cadmium on goethite. Environ. Sci. Technol. **24**, 112–118 (1990)
22. Waseem, M., Mustafa, S., Naeem, A., Koper, G.J.M., Shah, K.H.: Cd^{2+} sorption characteristics or iron coated silica. Desalination **277**, 221–226 (2011)
23. Waalkes, M.P.: Cadmium carcinogenesis in review. J. Inorg. Biochem. **79**, 241–244 (2000)
24. International Agency for Research on Cancer Monographs: Beryllium, Cadmium, Mercury and Exposures in the Glass Industry, vol. 58, p. 119. IARC, Lyon (1993)
25. National Toxicology Program.: Tenth Report on Carcinogens, Department of Health and Human Services, Research Triangle Park, NC, III-42 (2000)
26. Pesch, B., Haerting, J., Ranft, U., Klimpel, A., Oelschlagel, B., Schill, W.: MURC Study Group. Occupational risk factors for renal cell carcinoma: agent-specific results from a case–control study in Germany. Int. J. Epidemiol. **29**, 1014–1024 (2000)
27. Hu, J., Mao, Y., White, K.: Canadian cancer registries epidemiology research group. Renal cell carcinoma and occupational exposure in Canada. Occup. Med. **52**, 157–164 (2002)
28. Waalkes, M.P., Misra, R.R.: Cadmium carcinogenicity and genotoxicity. In: Chang, L. (ed.) Toxicology of Metals, p. 231. CRC Press, Boca Raton, FL (1996)
29. Fantoni, D., Brozzo, G., Canepa, M., Cipolli, F., Marini, L., Ottonello, G., Zuccolini, M.V.: Natural hexavalent chromium in groundwaters interacting with ophiolitic rocks. Environ. Geol. **42**, 871–882 (2002)
30. Ball, J.W., Izbicki, J.A.: Occurrence of hexavalent chromium in groundwater in the Western Mojave desert, California. Appl Geochem. **19**, 1123–1135 (2004)
31. Gibb, H.J., Lees, P.S., Pinsky, P.F., Rooney, B.C.: Lung cancer among workers in chromium chemical production. Am. J. Ind. Med. **38**, 115–126 (2000)
32. Leonard, A., Lauwerys, R.R.: Carcinogenicity and mutagenicity of chromium. Mutat. Res. **76**, 227–239 (1980)
33. Kramer, J.R., Allen, H.E.: Metal speciation: theory, analysis and application, Lewis Chelsea (1988)
34. Krull, I.S.: Trace Metal Analysis and Speciation. Elsevier, Amsterdam, The Netherlands (1991)
35. Mattson, J.S., Mark, H.B.: Activated carbon surface chemistry and adsorption from aqueous solution. Marcel Dekker, New York (1971)
36. Cheremisinoff, P.N., Ellerbush, F.: Carbon Adsorption Hand Book. Ann Arbor Science Publishers, Michigan (1979)
37. Faust, S.D., Aly, O.M.: Chemistry of Water Treatment. Butterworth, Stoneham (1983)

38. Lotya, M., Hernandez, Y., King, P.J., Smith, R.J., Nicolosi, V., Karlsson, L.S.: Liquid phase production of graphene by exfoliation of graphite in surfactant/water solutions. J. Am. Chem. Soc. **131**, 3611–3620 (2009)

39. Pu, N.W., Wang, C.A., Liu, Y.M., Sung, Y., Wang, D.S., Ger, M.D.: Dispersion of graphene in aqueous solutions with different types of surfactants and the production of graphene films by spray or drop coating. J. Taiwan Inst. Chem. Eng. **43**, 140–146 (2012)

40. Boehm, H.: Graphene-how a laboratory curiosity suddenly became extremely interesting. Angew. Chem. Int. Ed. **49**(49), 9332–9335 (2010)

41. Ahmadi-Moghadam, B., Taheri, F.: Effect of processing parameters on the structure and multi-functional performance of epoxy/GNP-nanocomposites. J. Mater. Sci. **49**, 6180–6190 (2014)

42. Abdelkader, A.M., Cooper, A.J., Dryfe, R.A.W., Kinloch, I.A.: How to get between the sheets? A review of recent works on the electrochemical exfoliation of graphene materials from bulk graphite. Nanoscale **7**, 6944–6956 (2015)

43. Chung, D.D.L.: A review of exfoliated graphite. J. Mater Sci. **51**, 554–568 (2015)

44. Segal, M.: Selling grapheme by the ton. Nat. Nanotech. **4**, 612–614 (2009)

45. Zhu, Y., Murali, S., Cai, W., Li, X., Suk, J.W., Potts, J.R., Ruoff, R.S.: Graphene and graphene oxide: Synthesis, properties, and applications. Adv. Mater. **22**, 3906–3924 (2010)

46. Ruan, M., Hu, Y., Guo, Z., Dong, R., Palmer, J., Hankinson, J., Berger, C., de-Heer, W.A.: Epitaxial graphene on silicon carbide: Introduction to structured graphene. MRS Bull. **37**, 1138–1147 (2012)

47. Leenaerts, O., Partoens, B., Peeters, F.M.: Graphene: a perfect nanoballoon. Appl. Phys. Lett. **93**, 193107 (2008)

48. Lu, Q., Huang, R.: Nonlinear mechanics of single-atomic layer graphene sheets. Int. J. Appl. Mech. **1**, 443–467 (2009)

49. Zaib, Q., Fath, H.: Application of carbon nano-materials in desalination processes. Desalin Water Treat **51**, 627–636 (2012)

50. Alsharaeh, E., Ahmed, F., Aldawsari, Y., Khasawneh, M., Abuhimd, H., Alshahrani, M.: Novel synthesis of holey reduced graphene oxide (HRGO) by microwave irradiation method for anode in lithium-ion batteries. Sci. Rep. **6**, 29854 (2016)

51. Suk, M.E., Aluru, N.: Water transport through ultrathin graphene. J. Phys. Chem. Lett. **1**, 1590–1594 (2010)

52. Climent-Pascual, E., Garcia-Velez, M., Álvarez, Á.L., Coya, C., Munuera, C., Diez-Betriu, X.: Large area graphene and graphene oxide patterning and nanographene fabrication by one-step lithography. Carbon **90**, 110–121 (2015)

53. Yang, J., Gunasekaran, S.: Electrochemically reduced graphene oxide sheets for use in high performance supercapacitors. Carbon **51**, 6–44 (2013)

54. Ali, I., Basheer, A.A., Mbianda, X.Y., Burakov, A., Galunin, E., Burakova, I., Mkrtchyan, E., Tkachev, A., Grachev, V.: Graphene based adsorbents for remediation of noxious pollutants from wastewater. Environ. Int. **127**, 160–180 (2019)

55. Ali, I.: New generation adsorbents for water treatment. Chem. Rev. **112**, 5073–5091 (2012)

56. Ali, I., Gupta, V.K.: Advances in water treatment by adsorption technology. Nature London **1**, 2661–2667 (2006)

57. Mishra, S.P.: Adsorption-desorption of heavy metal ions. Curr Sci India **107**, 601–612 (2014)

58. Singh, N., Gupta, S.K.: Adsorption of heavy metals: a review. Int. J. Innovative Res. Sci. **5**, 2267–2281 (2016)

59. Chang, C.F., Truong, Q.D., Chen, J.R.: Graphene sheets synthesized by ionic-liquid-assisted electrolysis for application in water purification. Appl. Surf. Sci. **264**, 329–334 (2013)

60. Liu, N., Luo, F., Wu, H., Liu, Y., Zhang, C., Chen, J.: One-step ionic-liquid-assisted electrochemical synthesis of ionic-liquid-functionalized graphene sheets directly from graphite. Adv. Funct. Mater. **18**, 1518–1525 (2008)

61. Wu, W.Q., Yang, Y., Zhou, H.H., Ye, T.T., Huang, Z.Y., Liu, R., Kuang, Y.: Highly efficient removal of Cu(II) from aqueous solution by using graphene oxide. Water Air and Soil Pollut **224**, 1372 (2013)

62. Sitko, R., Turek, E., Zawisza, B., Malicka, E., Talik, E., Heinman, J., Gagor, A., Feist, B., Wrzalik, R.: Adsorption of divalent metal ions from aqueous solutions using graphene oxide. Dalton Trans. **42**, 5682–5689 (2013)

63. Wang, H., Yuan, X., Wu, Y., Huang, H., Zeng, G., Liu, Y., Wang, X., Lin, N., Qi, Y.: Adsorption characteristics and behaviors of graphene oxide for Zn (II) removal from aqueous solution. Appl. Surf. Sci. **279**, 432–440 (2013)

64. Mi, X., Huang, G., Xie, W., Wang, W., Liu, Y., Gao, J.P.: Preparation of graphene oxide aerogel and its adsorption for Cu^{2+} ions. Carbon **50**, 4856–4864 (2012)

65. Zhao, G., Li, J., Ren, X., Chen, C., Wang, X.: Few-layered graphene oxide nanosheets as superior sorbents for heavy metal ion pollution management. Environ. Sci. Technol. **45**, 10454–10462 (2011)

66. Zhao, G., Ren, X., Gao, X., Tan, X., Li, J., Chen, C., Huang, Y., Wang, X.: Removal of Pb(II) ions from aqueous solutions on few-layered graphene oxide nanosheets. Dalton Trans. **40**, 10945–10952 (2011)

67. Yan, L., Zheng, Y.B., Zhao, F., Li, S., Gao, X., Xu, B., Weiss, P.S., Zhao, Y.: Chemistry and physics of a single atomic layer: strategies and challenges for functionalization of graphene and graphene-based materials. Chem. Soc. Rev. **41**, 97–114 (2012)

68. Deng, X., Lu, L., Li, H., Luo, F.: The adsorption properties of Pb(II) and Cd(II) on functionalized graphene prepared by electrolysis method. J. Hazard. Mater. **183**, 923–930 (2010)

69. Yuan, Y., Zhang, G., Li, Y., Zhang, G., Zhang, F., Fan, X.: Poly(amidoamine) modified graphene oxide as an efficient adsorbent for heavy metal ions. Polym Chem **4**, 2164–2167 (2013)

70. Huang, Z.H., Zheng, X., Lv, W., Wang, M., Yang, Q.H., Kang, F.: Adsorption of lead(II) ions from aqueous solution on low-temperature exfoliated graphene nanosheets. Langmuir **27**, 7558–7562 (2011)

71. Leng, Y., Guo, W., Su, S., Yi, C., Xing, L.: Removal of antimony(III) from aqueous solution by graphene as an adsorbent. Chem. Eng. J. **211**, 406–411 (2012)

72. Madadrang, C.J., Kim, H.Y., Gao, G., Wang, N., Zhu, J., Feng, H., Gorring, M., Kasner, M.L., Hou, S.: Adsorption behavior of EDTA-graphene oxide for Pb (II) removal. ACS Appl. Mater. Interfaces. **4**, 1186–1193 (2012)

73. Jabeen, H., Chandra, V., Jung, S., Lee, J.W., Kim, K.S., Kim, S.B.: Enhanced Cr(vi) removal using iron nanoparticle decorated graphene. Nanoscale **3**, 3583–3585 (2011)

74. Li, S., Lu, X., Xue, Y., Lei, J., Zheng, T., Wang, C.: Fabrication of polypyrrole/graphene oxide composite nanosheets and their applications for Cr(VI) removal in aqueous solution. PLoS ONE **7**, 43328 (2012)

75. Hao, L., Song, H., Zhang, L., Wan, X., Tang, Y., Lv, Y.: SiO_2/graphene composite for highly selective adsorption of Pb(II) ion. J. Colloid Interface Sci. **369**, 381–387 (2012)

76. Zhang, K., Dwivedi, V., Chi, C., Wu, J.: Graphene oxide/ferric hydroxide composites for efficient arsenate removal from drinking water. J. Hazard. Mater. **182**, 162–168 (2010)

77. Sui, Z., Meng, Q., Zhang, X., Ma, R., Cao, B.: Green synthesis of carbon nanotube–graphene hybrid aerogels and their use as versatile agents for water purification. J. Mater. Chem. **22**, 8767–8771 (2012)

78. Luo, X., Wang, C., Wang, L., Deng, F., Luo, S., Tu, X., Au, C.: Nanocomposites of graphene oxide-hydrated zirconium oxide for simultaneous removal of As (III) and As (V) from water. Chem. Eng. J. **220**, 98–106 (2013)

79. Lee, Y.C., Yang, J.W.: Self-assembled flower-like TiO_2 on exfoliated graphite oxide for heavy metal removal. J. Ind. Eng. Chem. **18**, 1178–1185 (2012)

80. Yuan, X., Wang, Y., Wang, J., Zhou, C., Tang, Q., Rao, X.: Calcined graphene/MgAl-layered double hydroxides for enhanced Cr(VI) removal. Chem. Eng. J. **221**, 204–213 (2013)

81. Algothmi, W.M., Bandaru, N.M., Yu, Y., Shapter, J.G., Ellis, A.V.: Alginate-graphene oxide hybrid gel beads: an efficient copper adsorbent material. J. Colloid Interface Sci. **397**, 32–38 (2013)

82. Peng, F., Luo, T., Qiu, L., Yuan, Y.: An easy method to synthesize graphene oxide-FeOOH composites and their potential application in water Purification. Mater. Res. Bull. **48**, 2180–2185 (2013)

83. Zhu, J., Wei, S., Chen, M., Gu, H., Rapole, S.B., Pallavkar, S., Ho, T.C., Hopper, J., Guo, Z.: Magnetic nanocomposites for environmental remediation. Adv. Powder Technol. **24**, 459–467 (2013)

84. Hu, X.J., Liu, Y.G., Wang, H., Chen, A.W., Zeng, G.M., Liu, S.M., Guo, Y.M., Hu, X., Li, T.T., Wang, Y.Q., Zhou, L., Liu, S.H.: Removal of Cu(II) ions from aqueous solution using sulfonated magnetic graphene oxide composite. Sep. Purif. Technol. **108**, 189–195 (2013)

85. Musico, Y.L.F., Santos, C.M., Dalida, M.L.P., Rodrigues, D.F.: Improved removal of lead(ii) from water using a polymer-based graphene oxide nanocomposite. J. Mater. Chem. **1**, 3789–3796 (2013)

86. Tripathy, S.S., Kanungo, S.B.: Adsorption of Co^{2+}, Ni^{2+}, Cu^{2+} and Zn^{2+} from 0.5 M NaCl and major ion sea water on a mixture ofδ-MnO2 and amorphous FeOOH. J. Colloid Interface Sci. **284**, 30–38 (2005)

87. Dong, Y., Yang, H., He, K., Song, S., Zhang, A.: ß-MnO2 nanowires: A novel ozonation catalyst for water treatment. Appl. Catal B: Environ. **85**, 155–161 (2009)

88. Zhao, D.L., Yang, X., Zhang, H., Chen, C.L., Wang, X.K.: Effect of environmental conditions on Pb(II) adsorption on b-MnO2. Chem. Eng. J. **164**, 49–55 (2010)

89. Zhu, M.X., Wang, Z., Xu, S.H., Li, T.: Decolorization of methylene Blue by δ-MnO2-coated montmorillonite complexes: emphasizing redox reactivity of Mn-Oxide coatings. J. Hazard. Mater. **181**, 57–64 (2010)

90. Ren, Y., Yan, N., Wen, Q., Fan, Z., Wei, T., Zhang, M., Ma, J.: Graphene/δ-MnO2 composite as adsorbent for the removal of nickel ions from wastewater. Chem. Eng. J. **175**, 1–7 (2011)

91. Ren, Y., Yan, N., Feng, J., Ma, J., Wen, Q., Li, N., Dong, Q.: Adsorption mechanism of copper and lead ions onto graphene nanosheet/d-MnO2. Mater. Chem. Phys. **136**, 538–544 (2012)

92. Gerente, C., Lee, V.K.C., Le-Cloirec, P., McKay, G.: Application of chitosan for the removal of metals from wastewaters by adsorption—mechanisms and models review. Crit. Rev. Env. Sci. Technol. **37**, 41–127 (2007)

93. He, Y.Q., Zhang, N.N., Wang, X.D.: Adsorption of graphene oxide/chitosan porous materials for metal ions. Chin. Chem. Lett. **22**, 859–862 (2011)

94. Zhang, N., Qiu, H., Si, Y., Wang, W., Gao, J.: Fabrication of highly porous biodegradable monoliths strengthened by graphene oxide and their adsorption of metal ions. Carbon **49**, 827–837 (2011)

95. Liu, L., Li, C., Bao, C., Jia, Q., Xiao, P., Liu, X., Zhang, Q.: Preparation and characterization of chitosan/graphene oxide composites for the adsorption of Au (III) and Pd (II). Talanta **93**, 350–357 (2012)

96. Chen, Y., Chen, L., Bai, H., Li, L.: Graphene oxide–chitosan composite hydrogels as broad-spectrum adsorbents for water purification. J. Mater. Chem. A **1**, 1992–2001 (2013)

97. Li, L., Fan, L., Sun, M., Qiu, H., Li, X., Duan, H., Luo, C.: Adsorbent for chromium removal based on graphene oxide functionalized. Colloids Surf B: Biointerfaces **107**, 76–83 (2013)

98. Fan, L., Luo, C., Sun, M., Li, X., Qiu, H.: Highly selective adsorption of lead ions by water-dispersible magnetic chitosan/graphene oxide composites. Colloids Surf B: Biointerfaces **103**, 523–529 (2013)

99. Shin, S., Jang, J.: Thiol containing polymer encapsulated magnetic nanoparticles as reusable and efficiently separable adsorbent for heavy metal ions. Chem. Commun. **41**, 4230–4232 (2007)

100. Hua, M., Zhang, S., Pan, B., Zhang, W., Lv, L., Zhang, Q.: Heavy metal removal from water/wastewater by nanosized metal oxides: a review. J. Hazard. Mater. **211**, 317–331 (2012)

101. Liu, M., Chen, C., Hu, J., Wu, X., Wang, X.: Synthesis of magnetite/graphene oxide composite and application for cobalt(II) removal. J. Phys. Chem. C **115**, 25234–25240 (2011)

102. Chandra, V., Park, J., Chun, Y., Lee, J.W., Hwang, I.C., Kim, K.S.: Water-dispersible magnetite-reduced graphene oxide composites for arsenic removal. ACS Nano **4**, 3979–3986 (2010)

103. Yang, X., Chen, C., Li, J., Zhao, G., Ren, X., Wang, X.: Graphene oxide-iron oxide and reduced graphene oxide-iron oxide hybrid materials for the removal of organic and inorganic pollutants. RSC Adv. **2**, 8821–8826 (2012)

104. Nandi, D., Gupta, K., Ghosh, A.K., De, A., Banerjee, S., Ghosh, U.C.: Manganese-incorporated iron(III) oxide-graphene magnetic nanocomposite: synthesis, characterization, and application for the arsenic(III)-sorption from aqueous solution. Nanotech Sustain. Dev. 149–162 (2012)

105. Zhu, J., Wei, S., Gu, H., Rapole, S.B., Wang, Q., Luo, Z., Haldolaarachchige, N., Young, D.P., Guo, Z.: One-pot synthesis of magnetic graphene nanocomposites decorated with core@double-shell nanoparticles for fast chromium removal. Environ. Sci. Technol. **46**, 977–985 (2012)

106. Chandra, V., Kim, K.S.: Highly selective adsorption of Hg^{2+} by a polypyrrole-reduced graphene oxide composite. Chem. Commun. **47**, 3942–3944 (2011)

107. Sreeprasad, T.S., Maliyekkal, S.M., Lisha, K.P., Pradeep, T.: Reduced graphene oxide-metal/metal oxide composites: facile synthesis and application in water purification. J. Hazard. Mater. **186**, 921–931 (2011)

108. Bhunia, P., Kim, G., Baik, C., Lee, H.: A strategically designed porous iron-iron oxide matrix on graphene for heavy metal adsorption. Chem. Commun. **48**, 9888–9890 (2012)

109. Zhang, K., Li, H., Xu, X., Yu, H.: Synthesis of reduced graphene oxide/NiO nanocomposite for the removal of Cr(VI) from aqueous water by adsorption. Microporous Mesoporous Mater. **255**, 7–14 (2018)

110. Ashour, R.W., Abdelhamid, H.N., Abdel-Magied, A.F., Abdel-Khalek, A.A., Ali, M.M., Uheida, A., Muhammed, M., Zhou, X., Dutta, J.: Rare earth ions adsorption onto graphene oxide nanosheets. Solvent Extr. Ion Exch. **35**, 91–103 (2017)

111. Gu, D., Fein, J.B.: Adsorption of metals onto graphene oxide: surface complexation modeling and linear free energy relationships. Colloids Surf A: Phys. Chem Eng Asp. **481**, 319–327 (2015)

112. Peng, W., Li, H., Li, Y., Song, S.: A review on heavy metal ions adsorption from water by graphene oxide and its composites. J. Mol. Liq. **230**, 496–504 (2017)

113. Moeser, G.D., Roach, K.A., Green, W.H., Hatton, T.A.: High-gradient magnetic separation of coated magnetic nanoparticles. AIChE J. **50**, 2835–2848 (2004)

114. Kim, S., Marion, M., Jeong, B.H., Hoek, E.M.V.: Crossflow membrane filtration of interacting nanoparticle suspensions. J. Membr. Sci. **284**, 361–372 (2006)

115. Ali, I., Alharbi, O.M.L., Tkachev, A., Galunin, E., Burakov, A., Grachev, V.: Water treatment by new generation graphene materials: Hope for bright future. Environ. Sci. Pollut. Res. **25**, 7315–7329 (2018)

116. Tan, P., Sun, J., Hu, Y., Fang, Z., Bi, Q., Chen, Y., Cheng, J.: Adsorption of Cu^{2+}, Cd^{2+} and Ni^{2+} from aqueous single metal solutions on graphene oxide membrane. J. Hazard. Mater. **297**, 251–260 (2015)

117. Sahraei, R., Pour, Z.S., Ghaemy, M.: Novel magnetic bio-sorbent hydrogel beads based on modified gum tragacanthin/graphene oxide: removal of heavy metals and dyes from water. J. Cleaner Prod. **142**, 2973–2984 (2017)

118. Chen, X., Zhou, S., Zhang, L., You, T., Xu, F.: Adsorption of heavy metals by graphene oxide/cellulose hydrogel prepared from NaOH/urea aqueous solution. Mater **9**, 582–596 (2016)

119. La, D.D., Nguyen, T.H.P., Nguyen, T.A., Bhosale, S.V.: Effective removal of Pb(II) using a graphene ternary oxides composite as an adsorbent in aqueous media. New J. Chem. **41**, 14627–14634 (2017)

120. Chen, L., Li, X., Tanner, E.E.L., Compton, R.G.: Catechol adsorption on graphene nanoplatelets: Isotherm, flat to vertical phase transition and desorption kinetics. Chem. Sci. **8**, 4771–4778 (2017)

121. Gan, X., Teng, Y., Ren, W., Ma, J., Christie, P., Luo, Y.: Optimization of ex-situ washing removal of polycyclic aromatic hydrocarbons from a contaminated soil using nano-sulfonated graphene. Pedosphere **27**, 527–536 (2017)

122. Pan, M., Shan, Ch., Zhang, X., Zhang, Y., Zhu, Ch., Gao, G., Pan, B.: Environmentally friendly in situ regeneration of graphene aerogel as a model conductive adsorbent. Environ. Sci. Technol. **52**, 739–746 (2018)
123. Li, D., Zhang, B., Xuan, F.: The sorption of Eu(III) from aqueous solutions by magnetic graphene oxides: a combined experimental and modeling studies. J. Mol. Liq. **211**, 203–209 (2015)
124. Yusan, S., Gok, C., Erenturk, S., Aytas, S.: Adsorptive removal of thorium (IV) using calcined and flux calcined diatomite from Turkey: evaluation of equilibrium, kinetic and thermodynamic data. Appl. Clay Sci. **67**, 106–116 (2012)
125. Greaves, M.J., Elderfield, H., Klinkhammer, G.P.: Determination of the rare earth elements in natural waters by isotope-dilution mass spectrometry. Anal. Chim. Acta **218**, 265–280 (1989)

Nanostructured Carbon-Based Materials for Adsorption of Organic Contaminants from Water

Roosevelt D. S. Bezerra, Paulo R. S. Teixeira, Edson C. da Silva-Filho, Anderson O. Lobo and Bartolomeu C. Viana

Abstract Graphene is currently one of the most promising carbon-based materials being studied in the world due to its novel electronic, thermal and optical properties. Nowadays, graphene can be considered the basis for the whole family of carbon nanomaterials, except for diamond structure. Because of that, several methods of graphene production have been studied; however, such methods need to be improved and the scaling is still a bottleneck for the productive sector. This article presents the main way to produce graphene and some techniques to modify its surface. In addition, here we present a review about graphene and its derivatives in the adsorption from the aqueous medium of the most diverse types of organic contaminants in the environment, such as pharmaceuticals, dyes, surfactants, pesticides, etc. We have shown that graphene and its derivatives are efficient adsorbents for the removal of emerging organic pollutants from the environment.

R. D. S. Bezerra
Federal Institute of Piauí, Teresina-Central Campus,
IFPI, Teresina, PI 64000-040, Brazil

P. R. S. Teixeira
Federal Institute of Piauí, Teresina-Zona Sul Campus,
IFPI, Teresina, PI 64018-000, Brazil

E. C. da Silva-Filho · A. O. Lobo · B. C. Viana
Interdisciplinary Laboratory for Advanced Materials (LIMAv), Federal University of Piaui,
Teresina,
PI 64049-550, Brazil

B. C. Viana (✉)
Materials Physics Laboratory (FisMat), Department of Physics, Federal University of Piaui,
Teresina,
PI 64049-550, Brazil
e-mail: bartolomeu@ufpi.edu.br

© Springer Nature Switzerland AG 2019
G. A. B. Gonçalves and P. Marques (eds.), *Nanostructured Materials for Treating Aquatic Pollution*, Engineering Materials,
https://doi.org/10.1007/978-3-030-33745-2_2

1   Introduction

The rapid growth of industrial and agricultural activities has led to a growing increase in the release of pollutants into the environment, such as persistent organic pollutants, for example, the industrial dyes [27, 122]. This has led to several environmental concerns, because the accumulations of toxic waste are potentially harmful to human health and the environment [17, 84, 119].

Therefore, efficient solutions are needed to lessen the effects caused by these pollutants in the environment. Thus, several methods have been used to remove these pollutants from the wastewater, where it is possible to highlight the adsorption, degradation, coagulation, flocculation, ion exchange, membrane filtration, electrochemical and catalytic oxidation, catalytic ozonization, all these technologies have limitations and restrictions [29, 75]. Among the remediation options, adsorption is one of the most used techniques due to its simplicity, ease of operation, low cost and efficient removal in low concentrations [121].

Among the materials used as adsorbents in the removal of organic compounds in wastewater, the carbon nanomaterials, such as graphene [49, 79, 99], which have recently been received special attention because of their specific surface area associated to relatively high adsorption capacity [21, 79, 107].

Therefore, graphene has been widely studied as an adsorbent for water decontamination because of its physico-chemical properties, which include, besides the high specific surface area, its mechanical, thermal and chemical stabilities [16, 36, 68, 120]. Therefore, these characteristics make graphene a potential material to be used in the treatment of effluents [35, 93].

Although recent studies have shown that graphene is being used as an adsorbent for emerging organic pollutants from the aquatic environment, there are few studies with this approach and there is a need for further studies and applicability on an industrial scale [33].

2   Graphene

The term "graphene" was first used in 1947, but the official definition was given by the International Union of Pure and Applied Chemistry (IUPAC) in 1994. In 1947, Walace in 1947 first studied to understand the structure of graphite, which consists of superimposed and weakly bonded graphene layers to each other, contrasting with the interatomic bonds in each plane, which are the strongest known bonds in a crystal [112].

In 2004, researchers were able to isolate small fragments exfoliated from graphite and after characterization verified that it was a single-layer structure of carbon atoms [77]. This material has attracted much attention from the scientific community because of its specific physical and chemical properties, large outer surface area, and

Table 1 Some of the main properties of graphene Huang et al [40, 52, 57]

Rupture force	~42 Nm^{-1}
Tensile strength (Young modulus)	1.0 TPa
Specific surface area	2630 m^2 g^{-1}
Intrinsic mobility	200 000 cm^2 V^{-1} s^{-1}
Thermal conductivity	~5000 W^{-1} K^{-1}

ease of modification [110]. Some of the main properties of graphene are shown in Table 1.

These properties provide graphene with properties that translate into greater mechanical strength than steel, higher electron mobility than silicon, higher thermal conductivity than copper, higher surface area than graphite and lighter material than many others [98].

2.1 Structure

Graphene is a two-dimensional (2D) network, composed of a hexagonal structure, which looks like a honeycomb [34]. Graphene has a sp^2 hybridization, where the carbon atoms form covalent bonds with each other (δ bonds), where the distances of the C–C bonds are approximately 1.42 Å, with a thickness of one carbon atom (approximately 1 Angstrom = 10^{-8} cm) (Fig. 1) [88].

In the graphene structure, the sp^2 hybrid orbital is a result of the bonding in the plane of the pure orbitals s, p$_x$ and p$_y$, while the pure orbital p$_z$ is free and perpendicular to the plane. This produces a sharing of sp^2 orbital hybridization of a carbon with three neighboring carbon atoms forming the hexagonal and 2D structure of graphene, as shown in Fig. 2 [105].

2.2 Synthesis of Graphene

Graphene production can be performed through two different strategies, which are designated as Top-Down and Bottom-Up [30].

2.2.1 Top Down Approach

The Top Dow strategy (top to bottom) is characterized by the attack of graphite powder, which will eventually separate its layer to generate graphene sheets [11]. Among Top Dow methods, we can highlight mechanical and chemical exfoliation and chemical reduction.

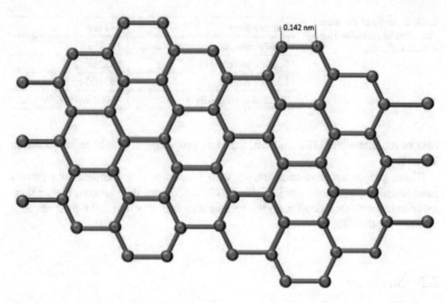

Fig. 1 Structure of a Graphene Blade. Reproduced with permission from Ref. [80]. Copyright: 2017 Elsevier

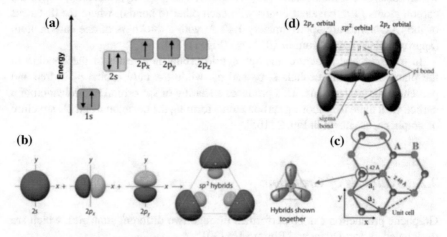

Fig. 2 **a** Energy levels of outer electrons in carbon atoms. **b** The formation of sp² hybrids. **c** The crystal lattice of graphene, where A and B are carbon atoms belonging to different sub-lattices, a_1 and a_2 are unit-cell vectors. **d** Sigma bond and pi bond formed by sp² hybridization. Reproduced with permission from Ref. [105]. Copyright: 2019 Elsevier

Mechanical Exfoliation

Graphite forms when the layers of mono-atomic graphene are stacked together by weak forces of van der Waals, the method of exfoliating the inverse of the stacking

[87, 125]. Hummers and Offman were the first to synthesize graphene in 1958, they obtained graphene through graphite exfoliation, although they did not obtain only carbon graphite as it is now known as graphene [41].

Andre Geim and Konstantin Novoselov reported in 2004, the synthesis of graphene nanosheets through the mechanical exfoliation. It was observed for the first time that it was possible to isolate a sheet of carbon atoms at room conditions [73]. Mechanical exfoliation is an ancient method and the most popular, since it was the process that led to the discovery of graphene [74].

This method involves exfoliating the graphite sheets until a single sheet is obtained. Exfoliation has the purpose of removing the graphite layers, this can be done using a variety of agents, such as adhesive tape [73], ultrasound [25] and electric field [58], and this technique does not require special equipment, but has a low performance disadvantage [100].

Chemical Exfoliation

Chemical exfoliation is the best route to be used to obtain graphene. This method consists in using the graphite to produce the graphene through the continuous reduction of the forces of Van der Waals that occurs through the insertion of special chemicals in the atomic planes of the graphite [113].

It is important to stress that chemical exfoliation is performed in two steps. Initially, occurs the reduction of the interlayer van der waals forces to produce compounds interlaced with graphenes [113]. Then, graphene is exfoliated with one or a few layers by rapid heating or sonication [6, 64].

Chemical Reduction of Graphite Oxide

In 1859, graphite oxide was first reported by Brodie, this one also presented the possibility of exfoliation, through the use of functional groups introduced in the oxidative graphite [14]. Graphite oxidation can be produced by the oxidation of graphite using oxidants such as concentrated sulfuric acid, nitric acid and potassium permanganate based on the method of Brodie [14], Staudenmaier method [102], Hummers method [41].

The oxidized graphite disperses easily in polar solvents, and the sonication process separates the leaves to obtain a smaller number of leaves than in graphite [7]. The number of leaves varies according to the experimental conditions of oxidation and sonication [24].

2.2.2 Bottom-Up Approach

The Bottom-Up strategy explore carbon dioxide to generate graphene [11]. Among the methods used in this strategy are: pyrolysis, epitaxial growth and CVD.

Pyrolysis

The graphene can be obtained through the pyrolysis of sodium ethoxide via sonication. This process improves the performance of graphene sheet detachment. Thus, the generated graphene sheets are measured up to 10 μm [22].

Epitaxial Growth

The term "epitaxy" derives from the Greek, the prefix epi means "on" and taxis means "order" or "agreement". When deposition of a single crystalline film on a single crystalline substrate produces epitaxial film and the process is known as epitaxial growth [9]. In this process, graphene is produced in the substrate (semiconductor) and the growth process is followed by lithography to make electronic devices based on graphene [9, 114]. The graphene produced by this process is called epitaxial graphene that can be grown on various substrates, such as silicon carbide [31], iron (Fe) [71] and copper (Cu) [32].

A promising substrate used in the growth of epitaxial graphene is silicon carbide. The preparation of graphene can be accomplished by applying heat and cooling of a silicon crystal (SiC). Usually, on the Si face of the crystal, there will be single layer or double layer graphene, however, on face C, graphene of few layers is cultivated [18].

Chemical Vapor Deposition—CVD

Chemical vapor deposition (CVD) is defined as the placement of a solid on a heated surface from a chemical reaction in the vapor phase where the deposition species are atoms or molecules. The chemical energy required to initiate a specific chemical reaction is provided by heat, light or electric charge [56].

High quality graphene deposition from the CVD process is usually done on various transition metal substrates such as Ni [45], Cu [90] and Pd [22], where graphene growth by CVD has been produced mainly on copper and nickel substrates [108].

3 Modifications

One of the great limitations in the use of graphene is its low reactivity, which hinders its interaction with different chemical species. Thus, there is a constant search for techniques that can be used to increase the reactivity of graphene with the introduction of different functional groups, these groups can be added by functionalization, doping and defects [72].

Among these techniques, chemical functionalization has attracted attention due to the possibility of increasing the solubility, processability and interactions of graphene with organic polymers [48]. An example is the introduction of oxygen-containing groups (−COOH, −CO and −OH) on the surface of the graphene. These groups occupy up to 60% of the surface area of graphene oxide (GO) [92].

Graphene oxide (GO), an oxidized version of graphene, can be synthesized by the oxidation of natural graphite [38, 41, 96]. This reaction occurs through the oxidation and exfoliation of graphite, usually in aqueous solution. This produces carbon-based hydrophilic plates decorated with various oxygenated functional groups (hydroxyl, epoxide and carboxyl), as shown in Fig. 3 [66].

The continuous, atomically thin, two-dimensional (2D) arrangement of carbon atoms that is functionalized with epoxy and hydroxy groups in the basal carbon plane and carboxy groups around the edges is the most accepted structure of GO leaves [66]. The GO may be reduced, partially, to graphene-like sheets by withdrawing the oxygen-containing groups and causing the recovery of a conjugated structure. Reduced GO (rGO) sheets are generally considered as a chemically derived graphene type (Fig. 3) [76]. The reduction of the GO does not completely remove oxygenated functional groups, only promotes a significant reduction of them. One way to accomplish GO reduction is to use high temperature treatment or use chemical agents such as hydrazine to form rGO [66, 76].

When performing the oxidative removal of the functional groups of oxygen from the GO there is the formation of another derivative of the graphene, named of graphene acid. The experimental procedure is based on the multiple oxidation of graphite with potassium permanganate in acidic medium. This procedure leads to the formation of a graphene derivative with an approximate composition of carboxyl

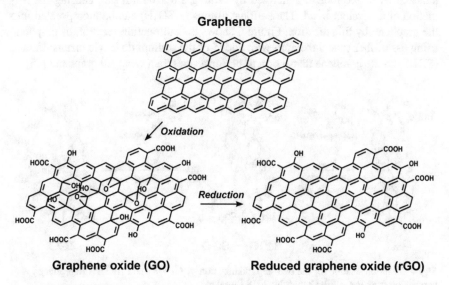

Fig. 3 Idealised structures of graphene, graphene oxide and reduced graphene oxide. Reproduced with permission from Ref. [66]. Copyright: 2019 Elsevier

groups, $[C_1(COOH)_1]_n$, of reaches 30 wt%, and simultaneously causing the oxidative elimination of other functional oxygen groups [42].

GO has a negative surface when in aqueous medium due to the formation of COO^-. Thus, when there is functionalization on the GO surface with a functional group which in aqueous medium produces a positive charge occurs the formation of the zwitterionic graphene oxide (ZGO). Zwitterionic compound containing both anionic and cationic groups would convert charge with the pH values. One way of producing ZGO is through the introduction of silane groups. For example, the preparation of the ZGO nanosheets with two silanes ((3-glycidyloxypropyl)trimethoxysilane (GPTMS) and (3-aminopropyl)triethoxysilane (APTES)), as shown in Fig. 4. Covalent bonds are responsible for grafting the silane molecules onto the GO surface and when the pH of the aqueous solution is changed from 12 to 2, there is also the variation of the potential (ζ) of ZGO of -29.0 mV to $+28.5$ mV [26].

Another important functionalization of graphene is the incorporation of thiol groups on its surface. An example of this reaction is shown in Fig. 5. From the Figure, it can be observed that, initially, the graphene is oxidized and forms graphene oxide (GO), which is then heated to above 400 °C in order to exfoliate the GO structure in layers in thermally reduced graphite oxide (TRGO) sheets. TRGO contains different oxygen groups directly bound to the carbon skeleton of a two-dimensional graphene-derived backbone (Fig. 5a). Soon after, The TRGO is placed to react with propylene sulfide forming the TRGO-SH, as shown in Fig. 5b [65].

Pristine graphene can be functionalized by organic functionalities such as free radicals and dienophiles. For example, sp^2 carbon atoms of graphene are easily attacked by a free radical generated by heating a diazonium salt, causing the formation of a covalent bond. The sulfonic groups ($-SO_3H$) can be incorporated into the graphene by this reaction. Figure 6 shows the sulfonation reaction of graphene using 4-sulfobenzenediazonium with added of the sodium dodecylbenzensulfonate (SDBS) to the graphene dispersion to perform surfactant wrapped graphene [85].

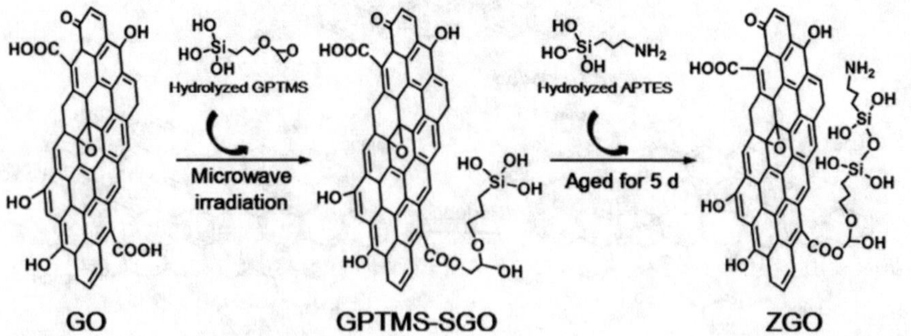

Fig. 4 Reaction scheme for the two steps silanization by GPTMS and APTES. Reproduced with permission from Ref. [26]. Copyright: 2018 Elsevier

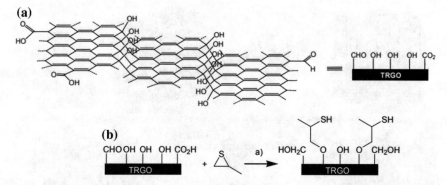

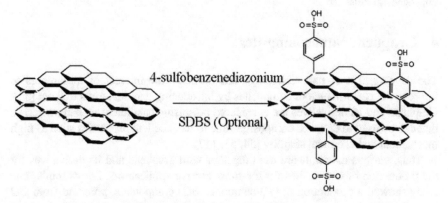

Fig. 5 **a** Schematic presentation of a model structure for thermally reduced graphite oxide (TRGO) with its functional carboxyl, aldehyde and especially hydroxyl groups according to recent studies. Reproduced with permission from Ref. [65]. Copyright: 2014 Elsevier. **b** Schematic presentation of the ring opening reaction of propylene sulfide (2-methylthiirane) with deprotonated hydroxyl groups of TRGO to yield the material TRGO-SH. Reproduced with permission from Ref. [65]. Copyright: 2014 Elsevier

Fig. 6 The reaction of sulfonation of the graphene using 4-sulfobenzenediazonium. Reproduced with permission from Ref. [85]. Copyright: 2018 Elsevier

In addition, on the graphene surface can also be incorporated phosphate groups. One way to accomplish this incorporation is the phosphating of GO from the covalent attachment of triethylphosphite on the GO surface via the Arbuzov reaction (Fig. 7). This reaction occurs at about 150 °C and is characterized by the reaction of triethylphosphite with electrophilic site of graphene oxide like epoxide or near to carbonyl group. After the non-localized attack of the GO surface by the triethylphosphite is formed the diethyl phosphonate. Then the ethyl group reacts with the bromide obtained from the LiBr to form the ethyl bromide. This process is carried out until the formation of the phosphene oxide graphene. Finally, hydrolyzing the ethyl groups present in the phosphonated GO is carried out to give hydroxyl groups and to form the phosphated graphene oxide (PGO) [2].

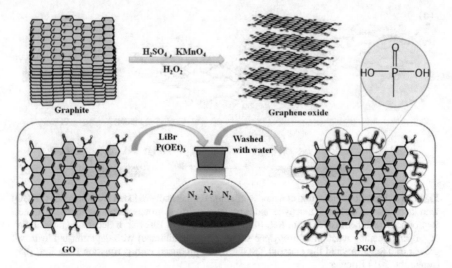

Fig. 7 Schematic diagram for the synthesis of PGO. Reproduced with permission from Ref. [2]. Copyright: 2018 Elsevier

4 Graphene Nanocomposites

The development of nanotechnology has attracted the interest of researchers to improve the performance of composites by introducing nanomaterials to them [54]. Graphene and its derivatives are considered promising candidates as nanomaterials that can be used to enhance composites, due to its excellent properties such as high mechanical and chemical stability [54, 55, 117].

Thus, various materials are used together with graphene and its derivatives for the production of composites for the most diverse applications. For example, one study showed a production of GO/bentonite (BG) compounds, prepared from GO interspersed with an intermediate layer of bentonite, as shown in Fig. 8. In these composites, it can be observed that after the addition of the GO there was an increase of the BET surface area, because the presence of GO between the BG layers formed an exfoliated structure. In addition, BG composites exhibited a greater adsorption capacity of toluidine blue dye (TB) from aqueous solutions than normal bentonite due to the synergistic effect between bentonite and GO. Finally, it was concluded that the adsorption of TB in BG composites occurs mainly through ion exchange, electrostatic interaction and intermolecular interactions (Fig. 8) [116].

The GO may also be introduced onto the surface of silica gel to decorate/functionalize GO through specific functional groups. To prepare this composite ($SiO_2@GO–PO_3H_2$) four steps are required which are schematically presented in Fig. 9. (i) Silica gel (SG) is grafted using 3-aminopropyltriethoxysilane (APTS) producing the compound SG-APTS. (ii) The GO is added to the surface of the SG-APTS forming $SiO_2@GO$. (iii) An aminosilane group is added to the surface of the $SiO_2@GO$ generating the $SiO_2@GO-NH_2$. (iv) The vinylphosphonic acid is added

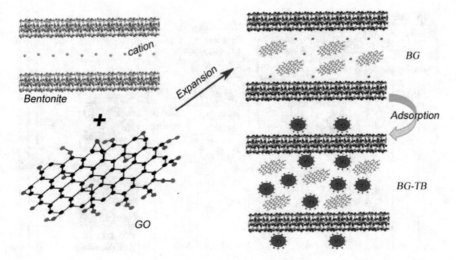

Fig. 8 Illustration of the synthesis of BG composites and TB adsorption using BG composites. Reproduced with permission from Ref. [116]. Copyright: 2019 Elsevier

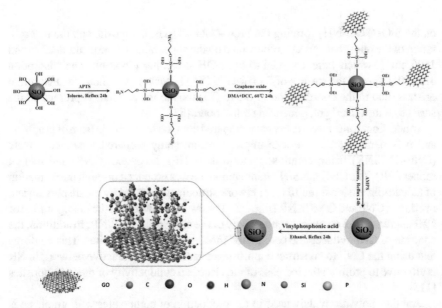

Fig. 9 The scheme for preparation of $SiO_2@GO-PO_3H_2$. Reproduced with permission from Ref. [53]. Copyright: 2019 Elsevier

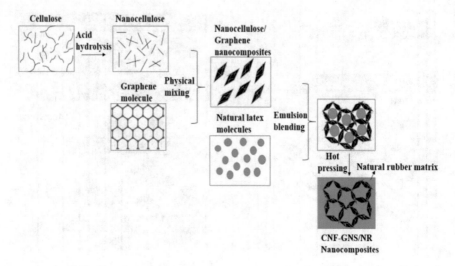

Fig. 10 Schematic of the CNF-GNS/NR fabrication process. Reproduced with permission from Ref. [115]. Copyright: 2019 Elsevier

on the $SiO_2@GO-NH_2$ forming the $SiO_2@GO-PO_3H_2$. composite and the incorporation of the phosphate group is confirmed by the appearance of the peaks at 2325 and 1695 cm^{-1}, which were assigned to P(O)-OH stretching vibration. The adsorption of ln (III) ions on the surface of $SiO_2@GO-PO_3H_2$ occurs by a chemical adsorption process, due to the coordination of the ln (III) ions to the element P, present on the surface of the adsorbent, generating a lnP bond [53].

In addition, some researchers have prepared composite materials from of graphene and polymeric materials. For example, a recent study prepared a nanocomposite (CNF-GNS/NR) using cellulose nanofibrils (CNF), graphene (GNS) and natural rubber (NR). The CNF-GNS/NR nanocomposite was produced through the formation of a three-dimensional multilayer network structure generated by the dispersion and overlap of CNF and GNS in NR (Fig. 10). The results showed that the incorporation of CNF increased the electrical conductivity of the composite GNS/NR, in addition, the addition of CNF reduced the percolation threshold of the composites. This confirms that using the CNF to construct a multi-layer multi-layer conductive network in NR is effective to promote the increase of the electrical conductivity of these composites [115].

Another polymer widely used in the production of composites with graphene is chitosan (CS). As an example, it may be mentioned a research recent which produced a composite using CS and graphene. For the production of this composite SiO_2 was used to improve the dispersibility of the grafene in chitosan suspensions. The CS/SiO_2-loaded graphene composite beads were cross-linked with ethylene glycol diglycidyl ether (EGDE) and prepared by the phase inversion method (Fig. 11). The results showed that the spheres Cs and Cs/graphene-SiO_2 presented spherical

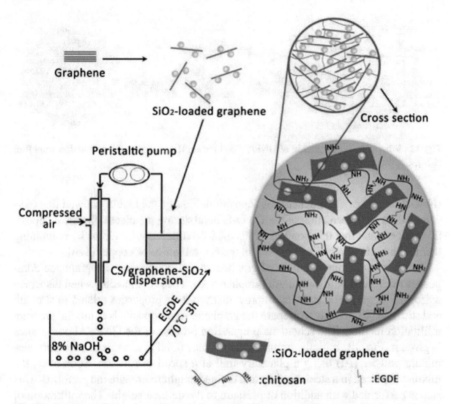

Fig. 11 Schematic illustration of the preparation of CS/graphene-SiO$_2$ composite beads. Reproduced with permission from Ref. [19]. Copyright: 2019 Elsevier

shapes and the incorporation of graphene improved the mechanical strength of the composites compared to the CS spheres [19].

In addition to studies related to graphene/polymer composites, research related to metal matrix graphene composites (CMMs) has also gained prominence. An example is the graphene/Cu composites, these composites attracted attention because of their electrical, thermal and mechanical properties, which made them promising materials for a wide range of applications [23].

The two main difficulties in the production of reinforced copper based graphene composites are the non-uniform dispersion of the graphene flakes inside the copper matrix and the low bond strength of the graphene/Cu matrix interface. An effective preparing approach of graphene-strengthening copper-based composites is an easy electrochemical deposition and atmosphere sintering. Figure 12 shows the schematic of preparation of graphene-nanoplatelets/copper (GNP/Cu) composite by electrochemical deposition and atmosphere sintering. The results showed that when the graphene content is 1.8 wt%, it is dispersed homogeneously in the matrix and forms a coherent interface in the GNP/Cu composites. In addition, it is observed that Cu-O-C bond formation is promoted by nickel (Ni) in the formation of graphene nanoplatelets

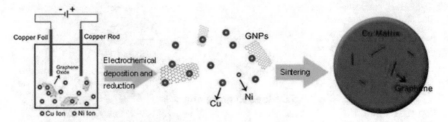

Fig. 12 Schematic of preparation of GNP/Cu composite. Reproduced with permission from Ref. [126]. Copyright: 2018 Elsevier

(Ni-GNPs) during electrochemical deposition. Finally, the results showed that oxygen mediates the formation of the Cu-O-C bond during the electrochemical deposition process and that this bond modifies the interstitial oxygen atoms by combining them with ionic or covalent bonding in the crystal lattice of copper [126].

Another interesting example of graphene/metal composites are graphene composites reinforced with aluminum (graphene/Al). These composites when dispersed uniformly in graphene show an improvement in their properties related to strength and stiffness [43]. A way to prepare the graphene/Al composite is to use the pressure infiltration method. The schematic preparation process of the GNPs/Al composites is given in Fig. 13. Initially, graphene nanoplates (GNPs) are mixed with pure aluminum powder (Al) using a planetary mill at a speed of rotation. Soon after, the mixture is placed in a steel mold up to a defined height to prepare the preforms. This step is performed with addition of pressure to the defined height. The infiltration of molten Al is carried out at low temperature and with higher pressure (15 MPa) to inhibit the reaction and overcome the poor wettability between the graphene and the Al matrix. No Al_4C_3 phase was detected in the composites, while GNPs have been well bonded with the Al matrix. It is also suggested that the use of graphene with less defect may inhibit the reaction between the pure Al matrix and graphene. In addition, it has been found that the addition of graphene significantly improves the mechanical properties of the composites [118].

5 Adsorption of Organic Compounds

In recent years, graphene-based nanomaterials have sparked great interest into potential applications in a wide range of industrial or agricultural fields. Among these applications, it is possible to mention the adsorption of polluting organic compounds from aqueous medium. These materials can interact with organic pollutants and alter the transport, fate and bioavailability of these contaminants in the environment [91].

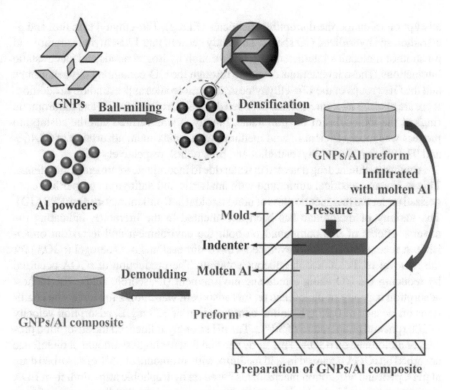

Fig. 13 The schematic preparation process of the GNPs/Al composites. Reproduced with permission from Ref. [118]. Copyright: 2018 Elsevier

5.1 Pharmaceutical Drugs

Pharmaceutical drugs are organic pollutants that have attracted increasing attention from recent research. These pollutants are resistant and have become a hazard to the aquatic ecosystem. An example of this type of contaminant is the antibiotic ciprofloxacin (CIP), a type of quinolone. Recent research has produced a hydrogen and graphene adsorbent (GH) from a hydrothermal reduction for the removal of CIP from the aqueous medium. In the adsorption process of CIP on GH, The interactions responsible for the adsorption process were based on the π–π interaction of electron donor-receptor (EDA), which are proven interactions as responsible for the adsorption of organic pollutants with benzene rings on carbon materials [103].

Another example of organic pollutants of pharmaceutical origin are the substances that can affect the endocrine system, called endocrine disrupting chemicals (EDCs). These contaminants can generate serious environmental and health problems. In recent years, EDCs have been found in effluents from different sources, which makes their removal an important issue. Thus, the process of removal of this contaminant type has been well reported in the literature. For example, researchers performed the adsorption of endocrine disrupting chemicals (EDCs), 17α-ethynyl estradiol and β-estradiol, sp^2 hybridized GO sheets. The study showed that 17α-ethynylestradiol and β-estradiol molecules interacted with GO through hydrogen bonds and electrostatic interactions. These interactions originate through the GO-containing oxygen groups and the OH groups of the 17α-ethynyloestradiol molecules as β-estradiol. In addition, there are also the π–π interactions between the π electrons of the GO and the aromatic rings of the molecules of the pollutants. Finally, it was verified that the adsorption process was more efficient in acid medium, with a maximum adsorption of 98.46% and 97.19% for 17α-ethinyl estradiol and β-estradiol, respectively [12].

Another pollutant drug detected in water due to incomplete treatment is diclofenac. This is a pharmaceutical compound with analgesic and antipyretic properties, categorised under the therapeutic class of non-steroidal anti-inflammatory drug (NSAID). The toxicity of diclofenac has been documented in the literature, indicating the adverse effects of this compound on both the environment and terrestrial organisms. A recent research has developed a three-dimensional rGO aerogel (rGOA) for the removal of diclofenac in aqueous solution. The production of rGOA occurred by reducing the GO using a reducing environment (L-ascorbic acid). The highest adsorption capacity of diclofenac on this adsorbent was 596.71 mg under the conditions of: dosage of 0.25 g/L, initial concentration of 325 mg/L, adsorption velocity of 200 rpm and temperature of 30 °C. The pH strongly influenced the adsorption process of diclofenac on rGOA. This study showed that the highest amount of diclofenac adsorbed on rGOA occurred in acid medium, with an amount of 96% of adsorbed drug at pH < pKa and adsorption mechanisms based on hydrophobic attraction, π–π EDA and precipitation. Under basic conditions, it was observed that the amount of drug adsorbed decreased to 81% when the pH reached the value of 9. In this medium, hydrogen bonds and hydrophobic interactions are responsible for the removal of diclofenac using rGOA [37]. Table 2 shows other pharmaceutical pollutants that are removed through graphene and its derivatives.

5.2 Organic Dyes

In addition of drugs, dyes are also found in the wastewater. The dyes present high biological toxicity, decrease the transparency of the water and consumption of high quantity of oxygen. This causes a reduction in the capacity of water bodies to perform self-purification, in addition to affecting the growth of water organisms and microorganisms. Like this, the removal of the dyes from the aqueous medium is necessary to reduce water pollution. This can be done using the adsorption process which is one of the most effective ways to turn polluted water into clean water. The use of graphene and its derivatives as adsorbents for dyes has been reported in recent research. For example, researchers prepared graphene-tannic acid hydrogel using GO and tannic acid (TA) through one-step hydrothermal method, as shown in Fig. 14 [106].

The hydrogels produced from graphene were used in adsorption of the organic dye methylene blue (MB). The adsorption process was improved by the presence of

Table 2 Types of pharmaceutical organic pollutants removed using graphene and its derivatives

Adsorbent	Adsorbate	Maximum adsorbed amount (mg g^{-1})	pH	Type of interaction	References
MnO$_2$/graphene nanocomposite	Tetracycline	168	5	Electrostatic interaction, H-bonding, coordination effect and π–π interactions	[101]
Graphene nanoplatelets (GNPs)	Aspirin Caffeine	12.98 19.72	8	Non-electrostatic interactions	[4]
Phosphorous-doped microporous carbonous material (PPhA)	Ibuprofen Carbamazepine	22.04 17.91	6–7	H-bonding, π–π and n–π electron donor–acceptor interactions	[97]
Nitrogen-doped reduced graphene oxide/Fe$_3$O$_4$ nanocoposite	Norfloxacin Ketoprofen	158.1 468	10	Electrostatic attraction, π–π interaction and hydrophobic interaction	[78]
Graphene oxide nanosheet (GOS)	Doxycycline	117	6–7	Cation–π bonding and π–π interaction	[94]
Double-oxidized graphene oxide (DGO)	Acetaminophen	704	8	Hydrophobic interactions	[69]
Graphene oxide (GhO)	Atenolol Propranolol	116 67	2	Electrostatic forces, pi–pi interactions and van der Waals forces	[50]
Graphene oxide nanosheet (GOS)	Sulfamethoxazole	122	6	Electrostatic interaction	[95]
Reduced graphene oxides (rGO)	Sulfapyridine Sulfathiazole	117 142	5	Hydrophobic interaction, π–π EDA interaction and electrostatic interaction	[62]
Magnetic graphene nanoplatelets (M-GNPs)	Amoxicillin	106.38	5	π–π stacking and electrostatic interaction	[44]
Graphene oxide (GO)-based composite	Triclosan	14.5	4	hydrogen bonding, electrostatic, and π–π interactions	[28]

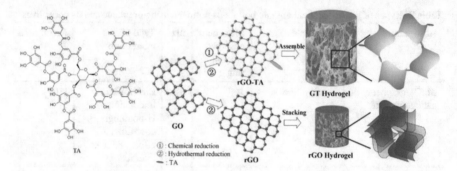

Fig. 14 Structure of TA molecule and formation of the hydrogels. Reproduced with permission from Ref. Tang et al. [106]. Copyright: 2018 Elsevier

tanic acid on the surface of the graphene sheets and microstructures of the hydrogels. In addition, the results showed that the adsorption of the dye on the geis occurs more effectively at alkaline pH. The main mechanism of adsorption of the dye on the surface of the hydrogels is the electrostatic attraction, which increases with increasing pH, reaching a maximum value in pH 10 (the maximum adsorption capacity = 563 mg g^{-1}), due to the presence of tanic acid on the surface of the hydrogels [106]. Thus, graphene and its derivatives can be used as adsorbents for removal of dyes from the aqueous medium. Table 3 shows other examples of dyes removed by graphens and their derivatives.

5.3   Surfactants

Surfactants play a key role in many industrial processes and are chemical substances with The presence of surfactants in aquatic environment is also a major public health concern. the most varied applications. These compounds being biodegraded generate by-products that have been found in aquatic sediments, groundwater, surface water and drinking water. Thus, the adsorption process is also used in the removal of these contaminants from the aqueous medium. Among the adsorbents that can be used in these adsorption processes are compounds derived from graphene, such as GO and rGO. Recently, the GO and rGO were used in the adsorption of the non-ionic surfactant Triton X-100 (TX-100) from water. The adsorption process of TX-100 on rGO was based on hydrophobic interactions and π-stacking. In addition, GO and rGO revealed superior removal capabilities when compared to other adsorbents [82]

Table 3 Examples of dyes removed from aqueous medium using graphene and its derivatives

Adsorbent	Adsorbate	Maximum adsorbed amount	pH	Type of interaction	References
Graphene oxide	Rhodamine B	88%	Not included	Electrostatic interactions	[67]
Graphene nanosheets	Methyl orange	22%	neutral pH	Van der Waals interactions	[5]
Partly reduced graphene oxide aerogels induced by proanthocyanidins	Neutral red Amino black Congo red	382.40 mg g^{-1} 313.37 mg g^{-1} 308.86 mg g^{-1}	Not included	Interaction force hydrophilic	[59]
Graphene oxide intercalated montmorillonite nanocomposite	Crystal violet	575 mg g^{-1}	6.2	Interactions π–π and hydrogen bonding	[83]
Poly(methyl methacrylate)/graphene oxide-Fe$_3$O$_4$ magnetic nanocomposite	Malachite green	92%	12	Electrostatic interaction	[86]
Graphene oxide	Proflavine	~90%	9	Electrostatic interaction	[10]
Graphene oxide-doped nano-hydroxyapatite	Trypan Blue	~100%	pH < 11	Electrostatic and π–π interactions	[81]
Modified magnetic graphene oxide by metformin	Methyl violet Acid red 88	239 mg g^{-1} 173 mg g^{-1}	7 4	Electrostatic and π–π interactions	[1]
Graphene oxide modified with polystyrene	Reactive Blue 19 Direct Red 81 Acid Blue 92	13.13 mg g^{-1} 19.84 mg g^{-1} 15.44 mg g^{-1}	7	Not included	[8]
Graphene oxide	Basic Yellow 28 Basic Red 46	64.50 mg g^{-1} 81.40 mg g^{-1}	11	Electrostatic interactions, π–π interactions and hydrogen bonds	[46]
Graphene oxide/chitosan aerogel	Metanil yellow	430.99 mg g^{-1}	~6.8	Electrostatic attraction, π–π stacking and hydrogen bonding	[51]

(continued)

Table 3 (continued)

Adsorbent	Adsorbate	Maximum adsorbed amount	pH	Type of interaction	References
Graphene oxide (GO)	Acid Orange 8 Direct Red 23	27.80 mg g^{-1} 23.30 mg g^{-1}	3	Electrostatic interaction, H-bonding and π–π stacking interactions	[47]
Graphene oxide (GO)	Basic Red 12	56.30 mg g^{-1}	9	Electrostatic interactions	[68]
Polypyrrole/ Chitosan/Graphene oxide (PPy/CS/GO) nanocomposite	Ponceau 4R	6.27 mg g^{-1}	2	Electrostatic interactions, hydrophobic effect and π–π bonds	[15]
dithiocarbamate-functionalized graphene oxide	Basic Blue 41	~80%	4.5	Electrostatic interactions	[63]

5.4 Pesticides

In addition, another serious environmental problem is the contamination of water with pesticides. Pesticides are classified according to their biological function and target organism (herbicides, fungicides, rodenticides, insecticides, etc.). The presence of pesticides in the environment can cause serious damage to human health, such as headaches and nausea to chronic impacts such as cancer, reproductive damage and endocrine disruption. The pesticides are very soluble and cause adverse effects even at levels of μg/L and are therefore considered primary pollutants for aquatic ecosystems. An efficient process for the removal of pesticides from the aqueous medium is to use graphene based nanocomposites as adsorbent through the adsorption process. For example, a study used Fe_3O_4/rGO nanocomposite for the adsorption of the pesticide ametryn in an aqueous medium. The adsorption process of this pesticide on the Fe_3O_4/rGO nanocomposite showed was efficient of the 93.61% due to electrostatic, hydrophobic and π–π interactions of composite towards the heterocyclic conjugation of pesticide molecule [13].

Another organic pollutant in the waters is bisphenol A (BPA). BPA is an example of biphenol compounds (BCs) being one of the most produced in the world. BCs exhibit various adverse effects, such as endocrine disruption, cytotoxicity, genotoxicity, reproductive toxicity, dioxin-like effects, and neurotoxicity. These effects caused several materials to be used for the removal of BPA from the aqueous medium, among them we can mention the adsorption process using nanomaterials derived from graphene. For example, a study showed which the graphite (GP), GO and rGO can be used as adsorbents for BPA. The rGO has the highest adsorption capacity

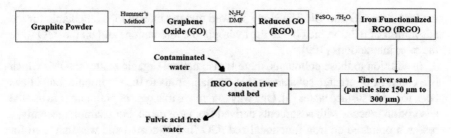

Fig. 15 Flow diagram of fulvic acid removal by iron-functionalized reduced graphene oxide (fRGO). Reproduced with permission from Ref. [89]. Copyright: 2017 Royal Society of Chemistry

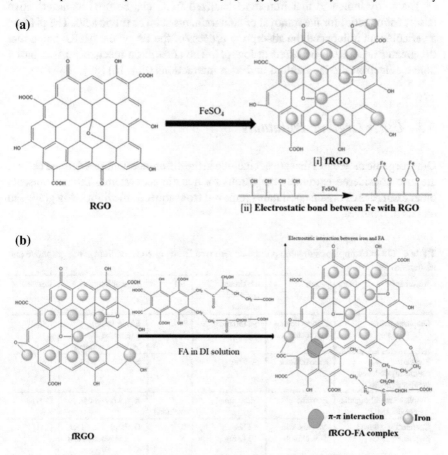

Fig. 16 **a** RGO to fRGO conversion by adding $FeSO_4$ **b** Proposed FA-fRGO interactions complex. Reproduced with permission from Ref. [89]. Copyright: 2017 Royal Society of Chemistry

for BPA than GO and GP. The study showed that the interactions reponsible for the adsorption of BPA on the GP are the hydrophobic interactions and on the rGO were the $\pi-\pi$ interactions [109].

In addition to these pollutants, there is also natural organic matter (NOM) which is a complex of organic substances that are dangerous to the environment and have been found in surface water [3]. One way to remove this type of pollutant is to use the adsorption process with adsorbents derived from graphene. For example, recently, a research prepared an iron-functionalized rGO (frGO)-coated and was that used for the adsorption of natural organic matter, such as fulvic acid (FA), from synthetic water. The production process of the iron-functionalized frGO is shown in Fig. 15.

The study indicated that iron-functionalized frGO can be used as an effective adsorbent material for the removal of contaminants such as fulvic acid. The pH was a variable that influenced the adsorption process of the FA on the fRGO, being that the greater adsorption occurred in low pH. The adsorption mechanism was based both by electrostatic interactions and $\pi-\pi$ interactions (Fig. 16) [89].

5.5 Other Organic Pollutants

Other organic molecules that are not include in the different categories defined before, are also considered extremely dangerous for aquatic ecosystems. Table 4 presents other examples of organic pollutants removed from aqueous medium using graphene and its derivatives.

Table 4 Other examples of organic pollutants removed from aqueous medium using graphene and its derivatives

Adsorbent	Adsorbate	Maximum adsorbed amount	pH	Type of interaction	References
Graphene oxide	Dodecylamine Hydrochloride	~1150 mg g^{-1}	Not included	Electrostatic interaction and hydrogen bonding	[20]
Magnetic chemically-reduced graphene (MCRG)	Phenanthrene	~29 mg g^{-1}	5	$\pi-\pi$ interactions	[39]
Graphene-metal-organic framework (MOF)	Benzene	24.50 mmol g^{-1}	Not included	$\pi-\pi$ interactions	[104]
Chemically reduced graphene (CRG)	Naphthalene 1-Naphthol	~142 ~247 mg g^{-1}	5 5	Hydrophobic effect, $\pi-\pi$ interaction and Sieving effect $\pi-\pi$ interaction, n-π EDA interaction, H-bonding and Lewis acid-base interaction	[111]

(continued)

Table 4 (continued)

Adsorbent	Adsorbate	Maximum adsorbed amount	pH	Type of interaction	References
Graphene oxide/montmorillonite nanocomposite	p-Nitrophenol	15.54 mg g^{-1}	6	Not included	[123]
Defective graphene oxide (GO-COOH)	Nicotine	~125 mg g^{-1}	8	π–π interaction and cation-π bonding	[60]
Sodium bisulfite reduced graphene oxide aerogels	Tetrabromo bisphenol A	128.37 mg g^{-1}	7	π–π interaction	[124]
Graphene oxide–silver nanocomposite	Ethyl violet	39.59 mg g^{-1}	9	Electrostatic attractions	[70]
Fe$_4$O$_3$–graphene oxide (GO)–β-cyclodextrin (β-CD) nanocomposite	Thiamethoxam Imidacloprid Acetamiprid Nitenpyram Dinotefuran Clothianidin Thiacloprid	2.88 mg g^{-1} 2.96 mg g^{-1} 2.88 mg g^{-1} 2.56 mg g^{-1} 1.77 mg g^{-1} 2.88 mg g^{-1} 3.11 mg g^{-1}	Not included	Hydrogen bonding, hydrophobic interactions, electrostatic interactions, and π–π stacking/interactions	[61]
Graphene oxide (GO)	Metformin	~47 mg g^{-1}	6	π–π interactions and hydrogen bonds	[127]

5.6 Conclusions

Thus, it can be observed that there are several ways of producing graphene and an extremely versatile nanomaterial. This versatility causes graphene to undergo various chemical modifications on its surface to improve its physical, chemical and mechanical properties. In addition, graphene can form several composites with a myriad of materials (clays, polymers, metals, etc.). Finally, it can be concluded that graphene and its derivatives have been used as efficient adsorbents in the removal of emerging organic contaminants (pharmaceuticals, dyes, surfactants, pestiscides, etc.) from the aqueous medium, and nanomaterials have become promising for the decontamination of this pollutants of the environment.

References

1. Abdi, G., Alizadeh, A., Amirian, J., Rezaei, S., Sharma, G.: Polyamine-modified magnetic graphene oxide surface: feasible adsorbent for removal of dyes. J. Mol. Liq. **289**, 111118 (2019)
2. Achary, L.S.K., Kumar, A., Rout, L., Kunapuli, S.V.S., Dhaka, R.S., Dash, P.: Phosphate functionalized graphene oxide with enhanced catalytic activity for Biginelli type reaction under condition. Chem. Eng. J. **331**, 300–310 (2018)
3. Algamdi, M.S., Alsohaimi, I.H., Lawler, J., Ali, H.M., Aldawsari, A.M., Hassan, H.M.A.: Fabrication of graphene oxide incorporated polyethersulfone hydrid ultrafiltration membranes for humic acid removal. Sep. Purif. Technol. **223**, 17–23 (2019)

4. Al-Khateeb, L.A., Almotiry, S., Salam, M.A.: Adsorption of pharmaceutical pollutants onto graphene nanoplatelets. Chem. Eng. J. **248**, 191–199 (2014)
5. Ali, M.E.A.: Preparation of graphene nanosheets by electrochemical exfoliation of a grafite-nanoclay composite electrode: application for the adsorption of organic dyes. Colloids Surf., A **570**, 107–116 (2019)
6. Allen, M.J., Tung, V.C., Kaner, R.B.: Honeycomb carbon: a review of graphene. Chem. Rev. **110**, 132–145 (2009)
7. Ayan-Varela, M., Paredes, J.I., Rodil, S.V., Rozada, R., Martinez Alonso, A., Tascon, J.M.D.: A quantitative analysis of the dispersion behavior of reduced graphene oxide in solvents. Carbon **75**, 390–400 (2014)
8. Azizi, A., Moniri, E., Hassani, A.H., Panahi, H.A., Miralinaghi, M.: Polymerization of graphene oxide with polystyrene: non-linear isotherms and kinetics studies anionic dyes. Microchem. J. **145**, 559–565 (2019)
9. Berger, C., Song, Z., Li, T., Li, X., Ogbazghi, A.Y., Feng, R., First, P.N.: Ultrathin epitaxial graphite: 2D electron gas properties and a route toward graphene-based nanoelectronics. J. Phys. Chem. B **108**, 19912–19916 (2004)
10. Bhattacharyya, A., Mondal, D., Roy, I., Sarkar, G., Saha, N.R., Rana, D., Ghosh, T.K., Mandal, D., Chakraborty, M., Chattopadhyay, C.: Studies of the kinetics and mechanism of the removal process of proflavine dye through adsorption by graphene oxide. J. Mol. Liq. **230**, 696–704 (2017)
11. Bhuyan, M.S.A., Uddin, M.N., Islam, M.M., Bipasha, F.A., Hossain, S.S.: Synthesis of graphene. Int. Nano Lett. **6**, 65–83 (2016)
12. Borthakur, P., Boruah, P.K., Das, M.R., Kulik, N., Minofar, B.: adsorption of 17α-ethynyl estradiol and β-estradiol on graphene oxide surface: An experimental and computational study. J. Mol. Liq. **269**, 160–168 (2018)
13. Boruah, P.K., Sharma, B., Hussain, N., Das, M.R.: Magnetically recoverable Fe_3O_4/graphene nanocoposite towards efficient removal of triazine pesticides from aqueous solution: investigation of the adsorption phenomenon and specific ion effect. Chemosphere **168**, 1058–1067 (2017)
14. Brodie, B.C.: On the atomic weight of graphite. Philos. Trans. R. Soc. Lond. **149**, 249–259 (1859)
15. Burakov, A., Neskoromnaya, E., Babkin, A.: Removal of the Alizarin Red S anionic dye using graphene nanocomposites: a study on kinetic under dynamic conditions. Mater. Today: Proc. **11**, 392–397 (2019)
16. Burakov, A.E., Galunin, E.V., Burakova, I.V., Kucherova, A.E., Agarwal, S., Tkachev, A.G., Gupta, V.K.: Adsorption of heavy metals on conventional and nanostructured materials for wastewater treatment purposes: a review. Ecotoxicol. Environ. Saf. **148**, 702–712 (2018)
17. Chappell, G.A., Rager, J.E.: Epigenetics in chemical-induced genotoxic carcinogenesis. Curr. Opin. Toxicol. **6**, 10–17 (2017)
18. Chaste, J., Saadani, A., Jaffre, A., Madouri, A., Alvarez, J., Pierucci, D., Aziza, Z.B., Ouerghi, A.: Nanostructures in suspended mono- and bilayer epitaxial graphene. Carbon **125**, 162–167 (2017)
19. Chen, J., Ma, Y., Wang, L., Han, W., Chai, Y., Wang, T., Li, J., Ou, L.: Preparation of chitosan/SiO_2-loaded graphene composite beads for efficient removal of bilirubin. Carbon **143**, 352–361 (2019)
20. Chen, P., Li, H., Song, S., Weng, X., He, D., Zhao, Y.: Adsorption of dodecylamine hydrochloride on graphene oxide in water. Result Phys. **7**, 2281–2288 (2017)
21. Chen, Y., Chen, L., Bai, H., Li, L.: Graphene oxide–chitosan composite hydrogels as broad-spectrum adsorbents for water purification. J. Mater. Chem. A **1**, 1992–2001 (2013)
22. Choucair, M., Thordarson, P., Stride, J.A.: Gram-scale production of graphene based on solvothermal synthesis and sonication. J. Mater. Chem. A **4**, 30–33 (2009)

23. Chu, K., Wang, F., Wang, X., Huang, D.: Anisotropic mechanical properties of graphene/copper composites with aligned graphene. Mater. Sci. Eng., A **713**, 269–277 (2018)
24. Chua, C., Pumera, M.: Chemical reduction of graphene oxide: a synthetic chemistry viewpoint. Chem. Soc. Rev. **43**, 291–312 (2014)
25. Ci, L.J., Song, L., Jariwala, D., Elias, A.L., Gao, W., Terrones, M., Ajayan, P.M.: Graphene shape control by multistage cutting and transfer. Adv. Mater. **21**, 4487–4491 (2009)
26. Cui, J., Li, J., Qiu, H., Yang, G., Zheng, S., Yang, J.: Zwitterionic graphene oxide modified with silane molecules for multiple applications. Chem. Phys. Lett. **706**, 543–547 (2018)
27. Deblonde, T., Cossu-Leguille, C., Hartemann, P.: Emerging pollutants in wastewater: a review of the literature. Int. J. Hyg. Environ. Health **214**(6), 442–448 (2011)
28. Delhiraja, K., Vellingiri, K., Boukhvalov, D.W., Philip, L.: Development of highly water stable graphene oxide based composites for the removal of pharmaceuticals and personal care products. Ind. Eng. Chem. Res. **58**, 2899–2913 (2019)
29. Du, Q., Sun, J., Li, Y., Yang, X., Wang, X., Wang, Z., Xia, L.: Highly enhanced adsorption of congo red onto graphene oxide/chitosan fibers by wet-chemical etching off silica nanoparticles. Chem. Eng. J. **245**, 99–106 (2014)
30. Edwards, R.S., Coleman, K.S.: Graphene synthesis: relationship to applications. Nanoscale **5**, 38–51 (2013)
31. Emtsev, K., Speck, F., Seyller, T., Ley, L., Riley, J.D.: Interaction, growth, and ordering of epitaxial graphene on SiC 0001 surfaces: a comparative photoelectron spectroscopy study. Physcal Rev. B **77**, 155303 (2008)
32. Gao, L., Guest, J.R., Guisinger, N.P.: Epitaxial graphene on Cu (111). Nano Lett. **10**, 3512–3516 (2010)
33. Gautam, R.K., & Chattopadhyaya, M.C.: Graphene-based nanocomposites as nanosorbents. Nanomater. Wastewater Remediat. 49–78 (2016)
34. Geim, A.K., Novoselov, K.S.: The rise of graphene. Nat. Mater. **6**, 183–191 (2007)
35. González, J.A., Villanueva, M.E., Piehl, L.L., Copello, G.J.: Development of a chitin/graphene oxide hybrid composite for the removal of pollutant dyes: adsorption and desorption study. Chem. Eng. J. **280**, 41–48 (2015)
36. Goodwin Jr., D.G., Adeleye, A.S., Sung, L., Ho, K.T., Burgess, R.M., Petersen, E.J.: Detection and quantification of graphene-family nanomaterials in the environment. Environ. Sci. Technol. **52**, 4491–4513 (2018)
37. Hiew, B.Y.Z., Lee, L.Y., Lai, K.C., Gan, S., Thangalazhy-Gopakumar, S., Pan, G., Yang, T.C.: Adsorptive decontamination of diclofenac by three-dimensional graphene-based adsorbent: Response surface methodology, adsorption equilibrium, kinetic and thermodynamic studies. Environ. Res. **168**, 241–253 (2019)
38. Hong, Y.L., Ryu, S., Jeong, H.S., Kim, Y.: Surface functionalization effect of graphene oxide on its liquid crystalline and assembly behaviors. Appl. Surf. Sci. **480**, 514–522 (2019)
39. Huang, D., Xu, B., Wu, J., Brookes, P.C., Xu, J.: Adsorption and desorption of phenanthrene by magnetic graphene nanomaterials from water: roles of pH, heavy metal ions and natural organic metter. Chem. Eng. J. **368**, 390–399 (2019)
40. Huang, X., Yin, Z., Wu, S., Qi, X., He, Q., Zhang, Q., Zhang, H.: Graphene-based materials: synthesis, characterization, properties, and applications. Small **7**(14), 1876–1902 (2011)
41. Hummers, W., Offeman, R.: Preparation of graphitic oxide. J. Am. Chem. Soc. **80**, 1339–1339 (1958)
42. Jankovský, O., Nováček, M., Luxa, J., Sedmidubský, D., Fila, V., Pumera, M., Sofer, Z.: A new member of the graphene family: graphene acid. Chem. Eur. J. **22**, 17416–17424 (2016)
43. Jiang, Y., Xu, R., Tan, Z., Ji, G., Fan, G., Li, Z., Xiong, D., Guo, Q., Li, Z., Zhang, D.: Interface-induced strain hardening of graphene nanosheet/aluminum composites. Carbon **146**, 17–27 (2019)
44. Kerkez-Kuyumcu, O., Bayazit, S.S., Salam, M.A.: Antibiotic amoxicillin removal from aqueous solution using magnetically modified graphene nanoplatelets. J. Ind. Eng. Chem. **36**, 198–205 (2016)

45. Kim, K.S., Zhao, Y., Jang, H., Lee, S.Y., Kim, J.M., Kim, K.S., Ahn, J.-H., Kim, P., Choi, J.-Y., Hong, B.H.: Large-scale pattern growth of graphene films for stretchable transparent electrodes. Nature **457**, 706–710 (2009)
46. Konicki, W., Aleksandrzak, M., Mijowska, E.: Equilibrium, kinetic and thermodynamic studies on adsorption of cationic dyes from aqueous solutions using graphene oxide. Chem. Eng. Res. Des. **123**, 35–49 (2017)
47. Konicki, W., Aleksandrzak, M., Moszynski, D., Mijowska, E.: Adsorption of anionic azo-dyes from aqueous solutions onto graphene oxide: equilibrium, kinetic and thermodynamic studies. J. Colloid Interdace Sci. **496**, 188–200 (2017)
48. Kuilla, T., Bhadra, S., Yao, D., Kim, N.H., Bose, S., Lee, J.H.: Recent advances in grapheme based polymer composites. Prog. Polym. Sci. **35**, 1350–1375 (2010)
49. Kyzas, G.Z., Deliyanni, E.A., Bikiaris, D.N.: Mitropoulos, A.C. Graphene composites as dye adsorbents Review. Chem. Eng. Res. Des. **129**, 75–88 (2018)
50. Kyzas, G.Z., Koltsakidou, A., Nanaki, S.G., Bikiaris, D.N., Lambroupoulou, D.A.: Removal of beta-blockers from aqueous media by adsorption onto graphene oxide. Sci. Total Environ. **537**, 411–420 (2015)
51. Lai, K.C., Hiew, B.Y.Z., Lee, L.Y., Gan, S., Thangalazhy-Gopakumar, S., Chiu, W.S., Khiew, P.S.: Ice-templated graphene oxide/chitosan aerogel as an effective adsorbent for sequestration of metanil yellow dye. Biores. Technol. **274**, 134–144 (2019)
52. Lee, C., Wei, X., Kysar, J.W., Hone, J.: Measurement of the elastic properties and intrinsic strength of monolayer graphene. Science **321**(5887), 385–388 (2008)
53. Li, M., Meng, X., Huang, K., Feng, J., Jiang, S.: A Novel composite adsorbent for the separation and recovery of indium from aqueous solutions. Hydrometallurgy **186**, 73–82 (2019)
54. Li, G., Yuan, J.B., Zhang, Y.H., Zhang, N., Liew, K.M.: Microstructure and mechanical performance of graphene reinforced cementitious composites. Compos. A **114**, 188–195 (2018)
55. Li, Z., Fu, X., Guo, Q., Zhao, L., Fan, G., Li, Z., Xionf, D., Su, Y., Zhang, D.: Graphene quality dominated interface deformation behavior of graphene-metal composite: the defective is better. Int. J. Plast **111**, 253–265 (2018)
56. Li, X., Magnuson, C.W., Venugopal, A., Na, J., Suk, J.W., Han, B., Zhu, Y., Fu, L., Vogel, E.M., Voelkl, E., Colombo, L., Ruoff, R.S.: Graphene films with large domain size by a two-step chemical vapor deposition process. Nano Lett. **10**, 4328–4334 (2010)
57. Li, X., Zhu, Y., Cai, W., Borysiak, M., Han, B., Chen, D., Piner, R.D., Colomb, L., Ruoff, R.S.: Transfer of large-area graphene films for high-performance transparent conductive electrodes. Nano Lett. **9**(12), 4359–4363 (2009)
58. Liang, X., Chang, A.S.P., Zhang, Y., Harteneck, B.D., Choo, H., Olynick, D.L., Cabrini, S.: Electrostatic force assisted exfoliation of prepatterned few-layer graphenes into device sites. Nano Lett. **9**, 467–472 (2008)
59. Liu, C., Liu, H., Zhang, K., Dou, M., Pan, B., He, X., Lu, C.: Partly reduced graphene oxide aerogels induced by proanthocyanidins for efficient dye removal. Biores. Technol. **282**, 148–155 (2019)
60. Liu, S.-H., Tang, W.-T., Yang, Y.-H.: Adsorption of nicotine in aqueous solution by a defective graohene oxide. Sci. Total Environ. **643**, 507–515 (2018)
61. Liu, G., Li, L., Xu, D., Huang, X., Xu, X., Zheng, S., Zhang, Y., Lin, H.: Metal-organic framework preparation using magnetic graphene oxide-β-cyclodextrin for neonicotinoid pesticide adsorption and removal. Carbohyd. Polym. **175**, 584–591 (2017)
62. Liu, F.-F., Zhao, J., Wang, S., Xing, B.: Adsorption of sulfonamides on reduced graphene oxides as effected by pH and dissolved organic matter. Environ. Pollut. **210**, 85–93 (2016)
63. Mahmoodi, N.M., Ghezelbash, M., Shabanian, M., Aryanasab, F., Saeb, M.R.: Efficient removal of cationic dyes from colored wastewaters by dithiocarbamate-functionalized graphene oxide nanosheets: from synthesis to detailed kinetics studies. J. Taiwan Inst. Chem. Eng. **81**, 239–246 (2017)

64. Marcano, D.C., Kosynkin, D.V., Berlin, J.M., Sinitskii, A., Sun, Z., Slesarev, A., Alemany, L.B., Lu, W., Tour, J.M.: Improved synthesis of graphene oxide. ACS Nano **4**, 4806–4814 (2010)

65. Marquardt, D., Beckert, F., Pennetreau, F., Tolle, F., Mulhaupt, R., Riant, O., Hermans, S., Barthel, J., Janiak, C.: Hybrid materials of platinum nanoparticles and thiol-functionalized graphene derivatives. Carbon **66**, 285–294 (2014)

66. McCoy, T.M., Turpin, G., Teo, B.M., Tabor, R.F.: Graphene oxide: a surfactant or particle? Curr. Opin. Colloid Interface Sci. **39**, 98–109 (2019)

67. Molla, A., Li, Y., Mandal, B., Kang, S.G., Hur, S.H., Chung, J.S.: Selective adsorption of organic dyes on graphene oxide: theorical and experimental analysis. Appl. Surf. Sci. **464**, 170–177 (2019)

68. Moradi, O., Gupta, V.K., Agarwal, S., Tyagi, I., Asif, M., Makhlouf, A.S.H., Sadegh, H., Shahryari-ghoshekandi, R.: Characteristics and electrical conductivity of graphene and graphene oxide for adsorption cationic dyes from liquids: Kinetic and thermodynamic study. J. Ind. Eng. Chem. **28**, 294–301 (2015)

69. Moussavi, G., Hossaini, Z., Pourakbar, M.: High-rate adsorption of acentaminophen from the contaminated water onto double-oxidized graphene oxide. Chem. Eng. J. **287**, 665–673 (2016)

70. Naeem, H., Ajmal, M., Qureshi, R.B., Muntha, S.T., Farooq, M., Siddiq, M.: Facile synthesis of graphene oxide silver nanocomposite for decontamination of water from multiple pollutants by adsorption, catalysis and antibacterial activity. J. Environ. Manage. **230**, 199–211 (2019)

71. N'Diaye, A.T., Coraux, J., Plasa, T.N., Busse, C., Michely, T.: Structure of epitaxial graphene on Ir(111). New J. Phys. **10**(4), 043033 (2008)

72. Nigar, S., Wang, H., Imtiaz, M., Yu, J., Zhou, Z.: Adsorption mechanism of ferrocene molecule on pristine and functionalized graphene. Appl. Surf. Sci. **481**, 1466–1473 (2019)

73. Novoselov, K.S., Geim, A.K., Morozov, S.V., Jiang, D., Zhang, Y., Dubonos, S.V., Grigorieva, I.V., Firsov, A.A.: Electric field effect in atomically thin carbon films. Science **306**, 666–669 (2014)

74. Novoselov, K.S.: Nobel lecture: graphene: materials in Flatland. Rev. Mod. Phys. **83**, 837–849 (2011)

75. Oliveira, E.H.C., Mendonça, É.T., Barauna, O.S., Ferreira, J.M., Da Motta Sobrinho, M.A.: Study of variables for optimization of the dye indosol adsorption process using red mud and clay as adsorbents. Adsorption **22**, 59–69 (2016)

76. Pei, S., Cheng, H.: The reduction of graphene oxide. Carbon **50**, 3210–3228 (2010)

77. Pei, Q.X., Zhang, Y.W., Shenoy, V.B.: A molecular dynamics study of the mechanical properties of hydrogen functionalized graphene. Carbon **48**, 898–904 (2010)

78. Peng, G., Zhang, M., Deng, S., Shan, D., He, Q., Yu, G.: Adsorption and catalytic oxidation of pharmaceuticals by nitrogen-doped reduced graphene oxide/Fe_3O_4 nanocoposite. Chem. Eng. J. **341**, 361–370 (2018)

79. Perreault, F., Fonseca De Faria, A., Elimelech, M.: Environmental applications of graphene-based nanomaterials. Chem. Soc. Rev. **44**, 5861–5896 (2015)

80. Phiri, J., Gane, P., Maloney, T.C.: General overview of graphene: Production, properties and application in polymer composites. Mater. Sci. Eng. **215**, 9–28 (2017)

81. Prabhu, S.M., Khan, A., Farzana, M.H., Hwang, G.C., Lee, W., Lee, G.: synthesis and characterization of graphene oxide-doped nano-hydroxyapatite and its adsorption performance of toxic diazo dyes from aqueous solution. J. Mol. Liq. **269**, 746–754 (2018)

82. Prediger, P., Cheminski, T., Neves, T.F., Nunes, W.B., Sabino, L., Picone, C.S.F., Oliveira, R.L., Correia, C.R.D.: Graphene oxide nanomaterials for the removal of non-ionic surfactant from water. J. Environ. Chem. Eng. **6**(1), 1536–1545 (2018)

83. Puri, C., Sumana, G.: Highly effective adsorption of crystal violet dye from contaminated water using graphene oxide intercalated montmotillonite nanocomposite. Appl. Clay Sci. **166**, 102–112 (2018)

84. PuvaneswarI, N., Muthukrishnan, J., Gunasekaran, P.: Toxicity assessment and microbial degradation of azo dyes. Indian J. Exp. Biol. **44**, 618–626 (2006)

85. Radnia, H., Rashidi, A., Nazar, A.R.S., Eskandari, M.M., Jalilian, M.: A novel nanofluid based on sulfonated graphene for enhanced oil recovery. J. Mol. Liq. **271**, 795–806 (2018)

86. Rajabi, M., Mahanpoor, K., Moradi, O.: Preparation of PMMA/GO and PMMA/GO-Fe_3O_4 nanocomposites for malachite green dye adsorption: kinetic and thermodynamic studies. Compos. B Eng. **167**, 544–555 (2019)

87. Rao, C.N.R., Maitra, U., Matte, H.S.S.R. Synthesis.: Characterization, and selected properties of graphene. In: Rao, C.N.R., Sood, A. K. (Eds.) Graphene: Synthesis, Properties, and Phenomena, 1st ed., pp. 1–47, (2012)

88. Rao, C.N.R., Sood, A.K.: Graphene: Synthesis, Properties, and Phenomena, p. 436. Wiley-VCH, Verlag GmbH, Weinheim (2012)

89. Ray, S.K., Majumder, C., Saha, P.: Functionalized reduced graphene oxide (FRGO) for removal of fulvic acid contaminant. RSC Advances **7**, 21768–21779 (2017)

90. Reina, A., Jia, X.T., Ho, J., Nezich, D., Son, H., Bulovic, V., Mildred Dresselhaus, S., Kong, J.: Large area, few-layer graphene films on arbitrary substrates by chemical vapor deposition. Nano Lett. **9**, 30–35 (2009)

91. Ren, W., Chang, H., Mao, T., Teng, Y.: Planarity effect of polychlorinated biphenyls adsorption by graphene nanomaterials: the influence of graphene characteristics, solution pH and temperature. Chem. Eng. J. **263**, 160–168 (2019)

92. Ren, H., Kulkarni, D.D., Kodiyath, R., Xu, W., Choi, I., Tsukruk, V.V.: Competitive adsorption of dopamine and rhodamine 6G on the surface of graphene oxide. Appl. Mater. Interfaces **6**(4), 2459–2470 (2014)

93. Robati, D., Rajabi, M., Moradi, O., Najafi, F., Tyagi, I., AgarwaL, S., Gupta, V.K.: Kinetics and thermodynamics of malachite green dye adsorption from aqueous solutions on graphene oxide and reduced graphene oxide. J. Mol. Liq. **214**, 259–263 (2016)

94. Rostamian, R., Behnejad, H.: A comprehensive adsorption study and modeling of antibiotics as a pharmaceutical waste by graphene oxide nanosheets. Ecotoxicol. Environ. Saf. **147**, 117–123 (2018)

95. Rostamian, R., Behnejad, H.: A comparative adsorption study of sulfamethoxazole onto graphene and graphe oxide nanosheets through equilibrium, kinetic and thermodynamic modeling. Process Saf. Environ. Prot. **102**, 20–29 (2016)

96. Sahu, M., Raichur, A.M.: Toughening of high performance tetrafunctional epoxy with poly(allyl amine) grafted graphene oxide. Compos. B Eng. **168**, 15–24 (2019)

97. Sekulic, M.T., Boskovic, N., Slavkovic, A., Garunovic, J., Kolakovic, S., Pap, S.: Surface functionalized adsorbent for emerging pharmaceutical removal: adsorption performance and mechanisms. Process Saf. Environ. Prot. **215**, 50–63 (2019)

98. Segundo, J.E.D.V., Vilar, E.O.: Grafeno: Uma revisão sobre as propriedades, mecanismos de produção e potenciais aplicações em sistemas energéticos. Rev. Eletrônica Mater. E Process. **11**, 54–57 (2016)

99. Sham, A.Y.W., Notley, S.M.: Adsorption of organic dyes from aqueous solutions using surfactant exfoliated graphene. J. Environ. Chem. Eng. **6**, 495–504 (2018)

100. Singh, V., Joung, D., Zhai, L., Das, S., Khondaker, S.I., Seal, S.: Graphene based materials: past, present and future. Prog. Mater Sci. **56**, 1178–1271 (2011)

101. Song, Z., Ma, Y.-L., Li, C.-E.: The residual tetracycline in pharmaceutical wastewater was effectively removed by using MnO_2/graphene nanocomposite. Sci. Total. Environ. **168**, 580–590 (2019)

102. Staudenmaier, L.: Verfahren zur Darstellung der Graphitsaure. Eur. J. Inorg. Chem. **31**, 1481–1487 (1898)

103. Sun, Y., Yang, Y., Yang, M., Yu, F., Ma, J.: Response surface methodological evaluation and optimization for adsorption removal of ciprofloxacin onto graphene hydrogel. J. Mol. Liq. **284**, 124–130 (2019)

104. Szczesniak, B., Choma, J., Jaroniec, M.: Ultrahigh benzene adsorption capacity of graphene-MOF composite fabricated via MOF crystallization in 3D mesoporous graphene. Microporous Mesoporous Mater. **279**, 387–394 (2019)
105. Tahriri, M., Del Monico, M., Moghanian, A., Yaraki, M.T., Torres, R., Yadegari, A., Tayebi, L.: Graphene and its derivatives: opportunities and challenges in destistry. Mater. Sci. Eng., C **102**, 171–185 (2019)
106. Tang, C., Yu, P., Tang, L., Wang, Q., Bao, R., Liu, Z., Yang, M., Yang, W.: Tannic acid functionalized graphene hydrogel for organic dye adsorption. Ecotoxicol. Environ. Saf. **165**, 299–306 (2018)
107. Terracciano, A., Zhang, J., Christodoulatos, C., Wu, F., Meng, X.: Adsorption of Ca^{2+} on single layer graphene oxide. J. Environ. Sci. **57**, 8–14 (2017)
108. Viculis, L.M., Mack, J.J., Kaner, R.B.: A chemical route to carbon nanoscrolls. Science **299**, 1361 (2003)
109. Wang, P., Zhang, D., Tang, H., Li, H., Pan, B.: New insights on the understanding of the high adsorption of bisphenol compounds on reduced graphene oxide at high pH values via charge assisted hydrogen bond. J. Hazard. Mater. **371**, 513–520 (2019)
110. Wang, J., Chen, B., Xing, B.: Wrinkles and folds of activated graphene nanosheets as fast and efficient adsorptive sites for hydrophobic organic contaminants. Environ. Sci. Technol. **50**, 3798–3808 (2016)
111. Wang, J., Chen, B.: Adsorption and coadsorption of organic pollutanrs and a heavy metal by graphene oxide and rediced graphene materials. Chem. Eng. J. **281**, 379–388 (2015)
112. Wallace, P.R.: The Band Theory of Graphite. Physical Review Journal **71**, 622–634 (1947)
113. Wu, Y.H., Yu, T., Shen, Z.X.: Two-dimensional carbon nanostructures: Fundamental properties, synthesis, characterization, and potential applications. J. Appl. Phys. **108**(7), 071301 (2010)
114. Wu, X., Li, X., Song, Z., Berger, C., de Heer, W.A.: Weak antilocalization in epitaxial graphene: evidence for chiral electrons. Phys. Rev. Lett. **98**, 136801 (2007)
115. Xiong, X.-Q., Bao, Y.-L., Liu, H., Zhu, Q., Lu, R., Miyakoshi, T.: Study on mechanical and electrical properties of cellulose nanofibrils/graphene-modified natural rubber. Mater. Chem. Phys. **223**, 535–541 (2019)
116. Xu, W., Chen, Y., Zhang, W., Li, B.: Fabrication of graphene oxide/bentonite composites with excellent adsorption performances for toluidine blue removal from aqueous solution. Adv. Powder Technol. **30**, 493–501 (2019)
117. Yadav, A., Upadhyaya, A., Gupta, S.K., Verma, A.S., Negi, C.M.S.: Poly-(3-hexylthiophene)/graphene composite based organic photodetectors: the influence of graphene insertion. Thin Solid Films **675**, 128–135 (2019)
118. Yang, W., Zhao, Q., Xin, L., Qiao, J., Zou, J., Shao, P., Yu, Z., Zhang, Q., Wu, G.: Microstructure and mechanical properties of graphene nanoplates reinforced pure Al matrix composites prepared by infiltration method. J. Alloy. Compd. **732**, 748–758 (2018)
119. Yamjala, K., Nainar, M.S., Ramisetti, N.R.: Methods for the analysis of azo dyes employed in food industry–a review. Food Chem. **192**, 813–824 (2016)
120. Yi, H., Huang, D., Zeng, G., Lai, C., Qin, L., Cheng, M., Ye, S., Song, B., Ren, X., Guo, X.: Selective prepared carbon nanomaterials for advanced photocatalytic application in environmental pollutant treatment and hydrogen production. Appl. Catal. B: Environ. B **239**, 408–424 (2018)
121. Yu, L., Wang, L., Xu, W., Chen, L., Fu, M., Wu, J., Ye, D.: Adsorption of VOCs on reduced graphene oxide. J. Environ. Sci. **67**, 171–178 (2018)
122. Zare, E.N., Motahari, A., Sillanpää, M.: Nanoadsorbents based on conducting polymer nanocomposites with main focus on polyaniline and its derivatives for removal of heavy metal ions/dyes: a review. Environ. Res. **162**, 173–195 (2018)
123. Zhang, C., Luan, J., Yu, X., Chen, W.: characterization and adsorption performance of graphene oxide-montmorillonite nanocomposite for the simultaneous removal of Pb^{2+} and p-nitrophenol. J. Hazard. Mater. **378**, 120739 (2019)

124. Zhang, W., Chen, J., Hu, Y., Fang, Z., Cheng, J., Chen, Y.: Adsorption characteristics of tetrabromobisphenol A onto sodium bisulfite reduced graphene oxide aerogels. Colloids Surf. A **538**, 781–788 (2018)
125. Zhang, Y., Small, J.P., Pontius, W.V., Kim, P.: Fabrication and electric-field-dependent transport measurements of mesoscopic graphite devices. Appl. Phys. Lett. **86**(7), 073104 (2005)
126. Zhao, X., Tang, J., Yu, F., Ye, N.: Preparation of graphene nanoplatelets reinforcing copper matrix composites by electrochemical deposition. J. Alloy. Compd. **766**, 266–273 (2018)
127. Zhu, S., Liu, Y.-G., Liu, S.-B., Zeng, G.-M., Jiang, L.-H., Tan, X.-F., Zhou, L., Zeng, W., Li, T.-T., Yang, C.-P.: Adsorption of emerging contaminant metformin using graphene oxide. Chemosphere **179**, 20–28 (2017)

Nanostructured Materials for the Photocatalytic Degradation of Organic Pollutants in Water

Luzia M. C. Honorio, Pollyana A. Trigueiro, Bartolomeu C. Viana, Alessandra Braga Ribeiro and Josy A. Osajima

Abstract The heterogeneous photocatalysis combined with the use of nanostructured materials has proven to be a viable and highly efficient technology in the degradation of numerous organic compounds. This chapter present the fundamental aspects of nanometric semiconductors related to photodegradation and their treatment conditions for effluents contaminated. The role of the predominant species involved in each process has been studied using radical scavengers. In photocatalytic studies, clay minerals have recently been used as supports for the anchoring of semiconductors, developing photocatalysts with appropriate crystalline structure, which are more ecologically compatible and also have a high surface area. Additionally, the operational devices and influence on the degradation effectiveness such as dyes, pesticides and drugs were discussed for environmental treatment.

Keywords Semicondutor · Oxide · Photodegradation · Clay minerals

1 Introduction

Uncontrolled growth of contamination and discharge of micropollutants in the natural environment is a current major global concern [1, 2]. The majority of these contaminants are derived from human activities, whether from industrial, agricultural and/or urban processes (e.g., pharmaceuticals, dyes, pesticides, plastics). They are a threat to all living beings in terms of toxicity, carcinogenic and mutagenic effects [2–4]. For this reason, the investigation of new methods and materials for effective wastewater treatment is extremely important.

L. M. C. Honorio · P. A. Trigueiro · B. C. Viana · J. A. Osajima (✉)
Universidade Federal do Piauí, UFPI, Laboratório Interdisciplinar de Materiais
Avançados – LIMAV, Teresina, PI 64049-550, Brazil
e-mail: josyosajima@ufpi.edu.br

A. B. Ribeiro
Escola Superior de Biotecnologia, CBQF – Centro de Biotecnologia e Química
Fina – Laboratório Associado, Universidade Católica Portuguesa,
Rua Diogo Botelho 1327, 4169-005 Porto, Portugal

© Springer Nature Switzerland AG 2019   65
G. A. B. Gonçalves and P. Marques (eds.), *Nanostructured Materials*
for Treating Aquatic Pollution, Engineering Materials,
https://doi.org/10.1007/978-3-030-33745-2_3

Efforts have been made to develop ecologically sound and economically viable technologies that guarantee good quality of water resources from the degradation or immobilization of organic pollutants in natural waters [5–9]. Among the various approaches to water treatment are biological (aerobic, anaerobic and enzymatic), physical (decantation, filtration and adsorption) and chemical processes (advanced oxidative processes, ozonation, photocatalysis, etc.) [4, 7, 10, 11]. Advanced oxidative processes (AOPs) have been shown to be an efficient technology for the removal and degradation of organic pollutants due to the production of hydroxyl radicals through different formation routes, generating strong oxidizing power, so effectively degrading toxic and/or recalcitrant compounds, which provides great versatility in the treatment of aquatic pollutants [9, 11, 12]. By definition, AOPs are based on the generation of highly reactive and oxidizing compounds, such as hydroxyl radicals ($^{\cdot}OH$), ozone (O_3), and hydrogen peroxide (H_2O_2), which react non-selectively, and rapidly, with the great majority of organic compounds, mineralizing them and producing carbon dioxide (CO_2), water (H_2O) and inorganic ions (SO_4^{2-}, NO_3^-, Cl^-) [3, 11–13]. Among the types of AOPs, photocatalysis was highlighted due to its heterogeneous nature that allows for the treatment of waters and industrial effluents over a wide range of pH [14–17]. Based on this context, the next section will describe the fundamental aspects of nanometric semiconductors related to photodegradation and their treatment conditions for effluents contaminated with defined species, such as synthetic dyes, drugs, and pesticides.

2   Nanosemiconductors Applied to Heterogeneous Photocatalysis

The phenomenon of heterogeneous photocatalysis has been studied since the 1970s, originating from the work of Fujishima and Honda that described the process of water splitting, using TiO_2 as a photocatalyst, in an electrochemical cell, that promoted its oxidation generating H_2 and O_2 [18–21]. Since then many researches have demonstrated new ideas and innovations in the field of photocatalysis, for example the oxidation of water and pollutant organic compounds [22–24] promoted by semiconductors, especially on a nanometric scale, due to their high surface area and size-induced effects. Figure 1 shows the accelerated growth in several areas, and countries, of the use of photocatalytic technology. The main advantages of this technology include, in particular, the ability to use solar energy to control the temperature and ambient pressure conditions, so favoring economically viable photoresistance and low waste production, i.e. "green treatment", allowing the reuse of the photocatalysts. Finally, this chapter discussed the degradation of a diverse variety of pollutant species [14–17, 19, 24, 25].

The excitation of a semiconductor (e.g., TiO_2, ZnO, CdS, WO_3, ZnS) involves, in principle, the absorption of photons of visible or ultraviolet light [26–28]. In the semiconductor there is a region of energy discontinuity between the electronic levels

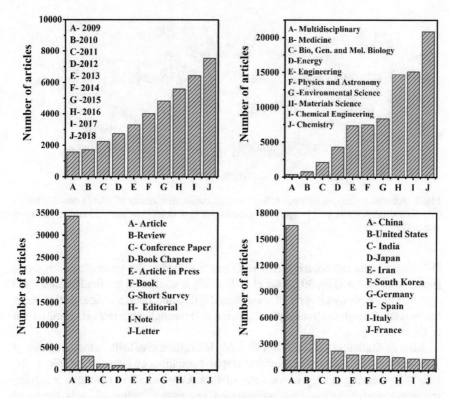

Fig. 1 Percentage related to the growing number of publications using heterogeneous photocatalysis as a technology for environmental remediation, in the last decade, by the Scopus database. Evolution of the number of articles over the years (**a**), areas of concentration involving the use of photocatalysis (**b**), the types of documents that are targeted (**c**), and the countries that concentrate articles on this technology (**d**)

called the band gap, this region being composed of two energy bands, the first called the valence band (VB—lower energy region), and the second, the conduction band (CB—energy), the band gap being the energy difference between the bands VB and CB [9, 19]. When the energy of the photon (hυ) is equal to, or greater than, the band gap energy, the electrons (e) are promoted from VB to CB, leaving a positive hole (h^+) in VB, thus giving an electron/hole pair (e^-_{CB}/h^+_{VB}), the exciton [16, 24], the process of which is illustrated in Fig. 2, which showed the CdS photocatalytic scheme and the photodegradation of RR141 under the irradiation of natural sunlight [29].

These excitons can recombine directly, or indirectly, via surface defects, by radioactive or non-radiative processes, without causing a chemical reaction (recombination of charges), or by migrating to the surface of the semiconductor, so inducing oxidation-reduction reactions that subsequently degrade the organic contaminants [9, 14, 30–32]. Normally h^+_{VB} readily reacts with H_2O from the surface and produces ˙OH (oxidation reaction), whereas, e^-_{CB} most often reacts with O_2 to produce

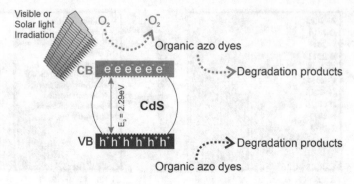

Fig. 2 Schematic diagram of a possible photodegradation mechanism of RR141 azo dye over a CdS photocatalyst under visible light, and natural solar light irradiation. Adapted from Senasu and Nanan [29]

O_2^- (superoxide anion), triggering reactions that lead to the generation of radicals (reduction reactions) [9, 30]. Rauf et al. [15, 33] in studies on the fundamentals and application of photocatalysis, showed that degradation does not necessarily occur exclusively through the hydroxyl radical, but also through other radical species, such as O_2, and $HO_2\cdot$.

After excitation, photocatalytic efficiency is an alternative to the removal of energy at the surface of the semiconductor and a rapid recombination between the pairs e^-/h^+ [26, 34], releasing energy in the form of heat. Oxy-reduction reactions involving semiconductor excitation, load maintenance, and recombination of loads, are listed below [15, 26, 31].

$$\text{Photocatalyst} + h\upsilon \rightarrow e^-{}_{CB} + h^+{}_{VB} \tag{1}$$

$$H_2O_{(ads)} + h^+ \rightarrow {}^\cdot OH_{(ads)} + H^+{}_{(ads)} \tag{2}$$

$$h^+ + OH^- \rightarrow {}^\cdot OH_{(ads)} \tag{3}$$

$$O_2 + e^- \rightarrow O_2{}^{\cdot-}{}_{(ads)} \tag{4}$$

$$O_2{}^{\cdot-}{}_{(ads)} + H^+ \rightarrow HO_2{}^\cdot{}_{(ads)} \tag{5}$$

$$HO_2{}^\cdot{}_{(ads)} \rightarrow O_2 + H_2O_{2(ads)} \tag{6}$$

$$H_2O_2 + {}^\cdot O_2{}^- \rightarrow OH^- + {}^\cdot OH + O_2 \tag{7}$$

$$H_2O_{2(ads)} + h\nu \rightarrow 2\,OH_{(ads)} \tag{8}$$

$$\text{Pollutant } (h^+, \, OH, HO_2\, \text{ or } O_2{}^{\cdot-}) \rightarrow \text{pollutant degradation} \tag{9}$$

The efficiency of a photocatalyst is understood by indirect and direct mechanisms [14, 15, 19, 33, 35]. The indirect mechanism can be explained, based on the indirect oxidation photogenerated by the hole, on the surface of the semiconductor. In the case of direct mechanism, the action of the photogenerated holes occurs through the transfer of direct charge on the surface of the semiconductor, a fact explained by the adsorption (chemisorption) on the surface of the semiconductor [14, 15, 33]. Furthermore, it is known that the knowledge of these steps, in agreement with the kinetics and stoichiometry established for a given reaction, can generate a better understanding of the processes, facilitating the identification of the generated products, as well as facilitating the reactive phases and species important for the photocatalytic efficiency improvement [36].

3 Understanding the Role of Scavengers in Photocatalysis Using Nano-semiconductors

Not only is the efficiency of the degradation process important, but also the influence of each species on the degradation, and its interaction from the use of inhibitors and/or radical scavengers avoiding recombination between the charge. Because of this, the understanding and adjustment of the process efficiency are very important [26, 37], monitoring the mechanisms involved in photodegradation more reliable [37–40]. Recently, there have been several papers in the literature involving the use of radical scavengers to better understand the reactions involved. These works are distinguished by the wide variety of pollutants studied, and the structural modifications of different materials [38, 41, 42]. Some of these works are presented in chronological form as follows:

Samsudin et al. [43] synthesized $BiVO_4$ via the solid-liquid state reaction and evaluated its photocatalytic potential in the oxidation of MB dye (methylene blue). The study emphasized the role of scavengers of reactive species, such as hydrogen peroxide (H_2O_2), silver nitrate ($AgNO_3$), and methanol (CH_3OH), to elucidate the photocatalytic degradation. According to the authors, the presence of $AgNO_3$ generated an intensification in the photocatalytic activity, resulting in 100% degradation of MB in 120 min under visible light irradiation. This high activity is attributed to the action of the scavenger, its performance being strongly dependent on the surface area of the crystalline planes {010} and {110} of the synthesized $BiVO_4$.

Sanad et al. [38], synthesized ZnO, ZnS and ZnO-ZnS nanocomposites by the sol-gel method and evaluated the degradation of methylene blue, and eosin, dyes using electron scavenging agents ($AgNO_3$), holes (KI), hydroxyl radicals (isopropanol), and superoxides (benzoquinone), under UV light. They found that the incorporation of ZnO, on the surface of the ZnS, induces the stabilization of the oxide particles in order to reduce the recombination rate of the electron-hole pair, so causing the complete destruction of the organic pollutants. However, the hydroxyl radicals, and holes played predominant roles in the degradation of the dyes.

Liu et al. [42], by a simple solvothermic method, synthesized AgTiS from TiO_2/sepiolite for the photocatalytic system in the degradation of methyl orange. Specifically, the 5% AgTiS composite exhibited the best photocatalytic activity, with 100% of the dye degraded after 50 min UV light and 260 min visible light. For the entrapment of the active species, t-butanol (TBA, 1 mM), p-Benzoquinone (p-BQ, 1 mM) and ammonium oxalate (AO, 1 mM), were the scavengers of hydroxyl radicals ($^\cdot OH$), superoxide ($^\cdot O_2^-$), and holes (h^+), respectively. The degradation efficiency of methyl orange was slightly reduced by the addition of TBA, however the degradation was suppressed significantly in the presence of β-BQ and AO, indicating that $^\cdot O_2$ and h^+ play important roles in the degradation process. The results of this research highlighted the AgTiS composite as a promising catalyst for environmental protection applications.

Mohammadi et al. [44], when evaluating the degradation of MB dye using rGO-TiO_2 synthesized in a hydrothermal approach, pointed out how the electrons are the main active species in the photodegradation of MB by rGO-TiO_2. This behavior is taken into account, since, among the numerous reactive species scavengers evaluated, the addition of $AgNO_3$ considerably decreases the degradation efficiency, indicating that the electrons are the main active species during the degradation process, a result confirmed by the use of a electron scavenger. In addition, the oxygen rich species $^\cdot O_2^-$ also plays an important role in the photocatalytic process and contribute in the degradation of the molecule adsorbed on the surface of the TiO_2 nanocomposite.

Numerous semiconductors are applied in heterogeneous photocatalytic processes, with titanium dioxide (TiO_2–P25 from Degussa) being the most commonly employed. As already mentioned above, due to its excellent physico-chemical properties, such as chemical and photochemical sensitivity, high thermal stability, chemical stability over a wide range of pH, possibility of immobilization on solids, possibility of activation by sunlight, insolubility in water, non-toxic, and low cost, TiO_2 serves as reference material for the evaluation of other photocatalysts [26, 45–47]. However, due to the band gap of 3.2 eV (= 385 nm), its application is limited to the UV range, which represents only 4–5% of solar irradiation at sea level [23, 48]. In order to reverse this situation, different strategies, such as doping [48], chemical modification with transition metals [49], and nitridation [50], have been applied in order to modify the TiO_2 for use in the visible region [45, 46, 51, 52]. Among the classes of the alternative compounds are, for example, the titanates [52], the ferrites [53], the niobates [54], and stannates [55–57]. Another alternative is the combination of oxide-clays [58–61], used in the degradation of a variety of organic pollutants, which will be given a more specific approach in this work.

4 Nanosemiconductors Supported on Clays

Clays have a different, and interesting, set of characteristics and properties that make them promising in the catalytic processes [62]. In photocatalytic studies, the clay minerals have been used, recently, as support for the anchoring of semiconductors, developing more ecological photocatalysts with an appropriate crystalline structure and with a high surface area. Therefore, the combination of oxides with these clay substrates can improve the photo efficiency, compared to pure semiconductors, resulting in the presence of new energy levels favoring the occurrence of a greater number of photogenerated charges, so obtaining effective catalysts under sunlight, and decreasing the electron/hole recombination and the formation of toxic intermediates [59, 63].

The metallic oxide nanoparticles, such as TiO_2, are mainly studied for the photodecomposition of organic pollutants. However, it is difficult to recover them in solution which prevents adequate reuse, and there is the possibility of agglomeration in solution so reducing their active surface area. Thus, the immobilization of semiconductors in substrates can provide viable solutions to overcome these problems [64]. The intercalation of metal oxide nanoparticles in the interlamellar spacing of clay minerals could affect the photocatalytic activity of the semiconductor due to the difficulty of access to the surface of the intercalated semiconductor [65]. Thus, the incorporation of semiconductors in the structures of clay minerals is still a challenging alternative as a strategy to increase the photocatalytic activity of nanomaterials and their subsequent reuse [66]. However, these porous materials possess favorable properties for their substrate use, such as high adsorption capacity, non-corrosiveness, non-toxic, accessibility, availability, and good thermal and chemical stability [67], which make them an appropriate alternative in the field of environmental remediation. In recent years, different porous silicates such as montmorillonite [68], sepiolite [69], kaolinite [70], diatomita [71], and zeolites [72], have been used as a support to obtain photocatalytic composites.

Montmorillonite is a lamellar clay with a di-octahedral structure of type 2:1, expansivity, high surface area, and cation exchange capacity, being a material widely applied as a support for the dispersion of semiconductors in heterogeneous catalysis. Samples prepared by the precipitation of small crystals of TiO_2 on the montmorillonite surface were used in the photodegradation of the herbicide dimethachlor [65], and phenol [73], in aqueous solution, and exhibited remarkable catalytic efficiency by the mass of TiO_2 dispersed in the nanocomposites.

Among the lamellar clay minerals, vermiculite has a high load between the lamellae, and excellent physicochemical properties, such as high cation exchange capacity, interaction capacity with organic molecules, and variable interlamellar distance. Silicates, derived from the vermiculite form of materials are essentially exfoliated and chemically functionalized by an acid leaching process. TiO_2 nanoparticles were dispersed on the surface of the support to obtain highly efficient functional photocatalysts for the degradation of organic dyes in solution [74].

Functional compounds based on lamellar clay minerals still favor the separation of charges, after immobilization of the semiconductors, so favoring the non-recombination of the excitons. $Zr-TiO_2$/cloisite [63], and $W-TiO_2$/cloisite [75], were synthesized by the sol-gel method and their catalytic performance, under sunlight and visible light, was evaluated by the photodegradation of antipyrine and atrazine. The antipyrine photodegradation results, after six hours of testing, indicated the formation of small organic acids as intermediates that were finally decomposed into CO_2 so achieving a moderate mineralization of the drug. In the experiments for atrazine removal, the results showed that the nanocomposite has higher photocatalytic efficiency under sunlight than under visible light. TiO_2-cloisite heterostructures [59], and $Ce-TiO_2$/cloisite [58], made using the dispersion of semiconductor nanoparticles on the surface of exfoliated lamellar clay, were used as a photocatalyst in the degradation of rhodamine B and phenol under the irradiation of sunlight. The formed nanocomposites were shown to be promising for photodegradation of the dye and the phenolic compound. The results indicated that after 24 h of reaction, the total mineralization of the phenol, and the intermediate aromatic compounds formed, had been observed, thus reducing their initial toxicity. Figure 3 shows the nanostructured materials formed from the presence of the TiO_2 nanoparticles between the delaminated layers of the cloisite clay.

Different techniques of immobilization of semiconductors are used, such as the incorporation of species among the lamellar clay minerals, and subsequent calcination, that gives rise to thermally stable materials by the pillarization process, these materials being promising for the catalytic removal of organic pollutants dissolved in chains, due to the presence of acid sites and their permanent porosity [68, 76]. The thermal stability, adsorptive properties and catalytic activity of the clay minerals can be enhanced by their intercalation with precursor solutions of one or more metal cations. Generally, the most easily incorporated cation (e.g. Al^{3+}) may act alone as a pillarizing agent, or may be followed by the addition of a second cation forming a mixed pillar [77].

Abdennouri et al. [61] investigated the photocatalytic potential of a titanium oxide pillar clay for the decomposition of 2,4-dichlorophenoxyacetic acid (2,4-D) and 2,4-dichlorophenoxypropionic acid (2,4-DP) pesticides in solution, which showed a high photocatalytic activity of the nanomaterial for the selective degradation of the pesticides. González et al. [68] investigated the photocatalytic activity for degradation of trimethoprim using titanium oxide doped with Cr^{3+} or Fe^{3+} doped clay. The Cr^{3+} doped solid presented better results in the degradation of the antibiotic in solution, reaching up to 76% of degradation in 180 min. Therefore, the pillar clays have a two-dimensional heterostructure that can prevent the agglomeration of oxide nanoparticles when incorporated in the interlamellar spacing, and exhibit high surface area and porosity, increasing the photocatalytic activity of the semiconductors [78].

Clays with fibrous morphology, such as sepiolite, have a structure consisting of blocks formed by an octahedral sheet of magnesium sandwiched between two tetrahedral silicon sheets, presenting cavities or tunnels along the c-axis, this structure being formed by channels of the zeolite type [69, 79]. Their structure, microporosity, high surface area, presence of silanol groups, and negative charge on the surface,

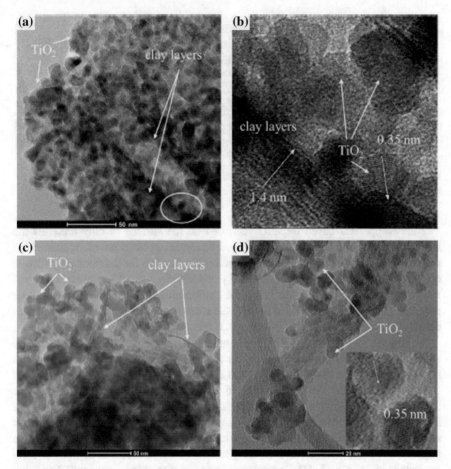

Fig. 3 Transmission electron micrographs (TEM) of the Zr-TiO$_2$/cloisite catalyst [63]

make the sepiolite materials attractive as supports for the anchoring of metallic oxides and applications in the photodegradation of organic pollutants. TiO$_2$/sepiolite photocatalysts show that sepiolite has a positive synergistic effect on TiO$_2$ photocatalytic activity for the degradation of phenolic compounds, which can be attributed to the high surface area obtained from the nanocomposite [69, 79].

Palygorskite is a fibrous silicate of magnesium, containing a significant presence of Al and Fe in its structure, with large crystals and a predominantly di-octahedral character. Nanostructured photocatalysts based on palygorskite were successfully prepared from the incorporation of different oxides in the microfibrous structure of this clay [80–82]. The studies showed the production of an inorganic-inorganic nanocomposite photocatalyst, in the visible spectrum, with high efficiency for the degradation of organic pollutants in aqueous solution.

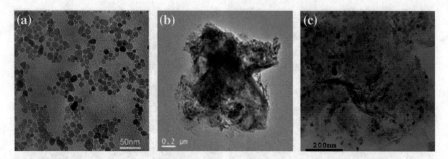

Fig. 4 TEM of the ZnO nanoparticles (**a**), ZnO/cloisite (**b**) and ZnO/Tunisian smectite (**c**) nanoarchitectures [85]

Nanocomposites based on ZnO anchored on the surface of clay minerals are widely reported in the literature for photocatalysis applications of organic compounds. Nanocomposites of ZnO/clays are highly reactive for the photodegradation of dyes in solution, showing good stability and activity, even after several cycles of reuse, with the ZnO: clay ratio being the determining factor of the photocatalysis potential [83, 84]. Figures 4 and 5 show the micrographs of the ZnO nanoparticles distribution on lamellar and fibrous clays, respectively.

Nanoparticles of TiO_2 and ZnO are generally the most used in the photocatalytic studies of nanomaterials based on clay minerals, but other nanocomposites based on different transition metals, for application in environmental remediation, can also be found in the literature [86–88]. Zhang and Yang [89] investigated the photocatalytic activity of cobalt oxide nanoparticles incorporated in halloysite nanotubes (HNTs). Figure 6 shows the scheme of the deposition of the nanoparticles on the clay surface and the proposed mechanism for the photodegradation of methylene blue dye. The authors investigated different oxide:clay ratios. The results indicated that the ratio by weight of 1:6 Co_3O_4 HNTs produced better results, reaching up to 97% of dye degradation. Hence, the combination of the catalytic properties of the cobalt oxide, with the adsorptive properties of the inorganic matrix, formed a nanomaterial with excellent properties for use in the environmental field for the removal of organic pollutants from aquatic effluents.

The various techniques applied for wastewater treatment have their advantages and disadvantages, such as cost, removal capacity, regeneration capacity, and reuse [90]. With advances in nanotechnology, the scientific community uses a lot of photocatalytic degradation to treat aquatic environments and this method has the main advantage, in relation to adsorption and other traditional methods, of the mineralization of the organic pollutant in CO_2 and water as complete photo-oxidation products [91–93]. But the formation of reaction intermediates may occur during the photocatalytic oxidation process, which, in many cases, may be even more harmful organic compounds than the initial pollutants themselves [94]. The TiO_2/zeolite photocatalyst promoted the rapid degradation, and made possible the elimination, of secondary pollution generated by the emission of harmful intermediates during the photocatalytic oxidation reactions of diethyl sulfate [94]. Zeolites are used as a strategy for the

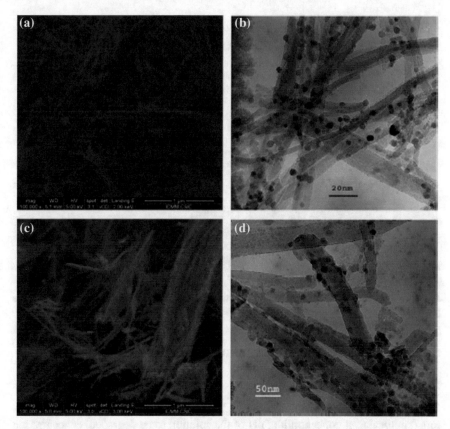

Fig. 5 SEM (**a** and **c**) and TEM (**b** and **d**) micrographs of the sepiolite based nanoarchitectures showing presence of ZnO nanoparticles covering the clay fibers [85]

development of greener catalysts, where the semiconductor, incorporated within the crystals of the zeolite, functions as an active site and the micropores of the support can be selective to diffuse the organic micropollutants in solution [95].

Therefore, the increased photocatalytic efficiency of the semiconductors incorporated in lamellar or fibrous clay minerals, and the zeolites, showed an excellent alternative material that can minimize the disadvantages such as rapid recombination of photoinduced electron-hole pairs, and the limited visible light absorption capacity semiconductors traditionally studied. Furthermore, they significantly improve the photo-efficiency of the functional materials, due to the increase of active sites, surface area, and a decrease in the formation of toxic intermediates during the photocatalysis process of organic pollutants, thus generating new selective and ecologically correct catalysts.

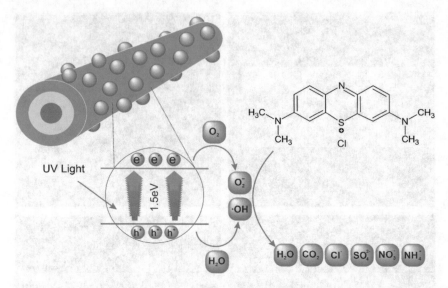

Fig. 6 Proposed mechanism for the photocatalytic activity of Co_3O_4 nanocomposite/HNTs in the photodegradation of MB under UV irradiation. Adapted from Zhang and Yang [89]

5 Operational Devices and Influence on the Degradation Effectiveness

The photocatalytic oxidation rates are highly dependent on several operational factors (photocatalyst concentration, pH of the solution to be degraded, concentration of pollutants, calcination temperature, intensity of light and irradiation time, dissolved oxygen, electrolyte effect and dopant) that affect the yield and efficiency of the process directed to the degradation of the organic compounds. In addition, another important factor is the structural characteristics of the materials and the potentials of the valence and conduction bands, which interfere directly in their performance [14, 15, 96–106]. In this sense, such parameters can be optimized in order to guarantee greater catalytic conversion and lower economic costs.

5.1 Effect of the Initial Concentration of the Organic Pollutant

The initial concentration of the organic contaminant is an important factor from the kinetic point of view, since, it interferes directly in the rate of photocatalytic oxidation. Initially, the rate of degradation is proportional to the increase in substrate concentration. As the oxidation proceeds, the rate of decomposition reaches a saturation that is related to the active surface of the semiconductor. In contrast, the photocatalytic rate is decreased with increasing irradiation time, while maintaining a fixed catalyst amount. The explanation for this decrease depends greatly on the semiconductor properties and the projected operating conditions [14, 15, 100, 104]. For Malato et al. [100], a series of tests, at different initial concentrations, were conducted to confirm whether or not the experimental results could be adjusted to the appropriate optimum pollutant concentration model.

5.2 Effect of the Amount of Catalyst

Normally the increase in the dosage of the photocatalyst gives the medium an increase of the active (OH, etc.), which in turn, react quickly, favoring the maximum of degradation and efficiency of the process. Although the direct ratio of efficiency and dosage guarantees maximum yields, after achieving the equilibrium of degradation at the optimized limit, the excess of particles of the photocatalyst decreases the rate of degradation as there is the dispersion and reduction of light penetration such that the solution blocks the radiation, so preventing the reaction from proceeding, and therefore, promotes a decreased effect of the photocatalytic performance [15, 101–103]. This behavior was also observed by Haque et al. [104] during the degradation of the antibiotic norfloxacin (0.25 mM), being investigated in different concentrations of Degussa P25, ranging from 0.5 to 3 g L^{-1}, in pH 6.8, for a duration 80 min.

5.3 Effect of pH

pH is a parameter that significantly affects the surface charge properties of the photocatalyst, so it is a very important variable in the photocatalytic processes. The analysis of the effects of pH on the efficiency in the photodegradation process is considered a complex assignment due to the numerous variables, such as electrostatic interactions between the semiconductor surface, solvent molecules, substrate, and charged radicals, formed during the reaction process. These affect not only the particle charges, but also the size of the aggregates that can be produced, and still allow variations in the positions of the photocatalyst's electronic bands. Changes in pH tend to influence the rate of reaction as a function of the adsorption behavior between the organic molecules and the starting materials, ensuring an important step in the oxidation through simple, cheap, and efficient, optimization. Typically, the ionization state of the molecules can be protonated, or deprotonated, under acidic or alkaline conditions. In the literature, this concept is consolidated by TiO_2, taking into account its value of zero charge potential (PCZ), and its desired pH value, according to the written reactions $TiOH + H^+ - TiOH_2^+$//$TiOH + OH^- - TiO^-$ $+ H_2O$ [98, 100, 103, 104, 106, 107]. In the photocatalytic degradation of orange G dye, Sun et al. [108] showed that nitrogen doped TiO_2 (N: TiO_2) was responsible for the significant increase of dye degradation under visible light. In addition, it was found that lowering the pH of the solution from 6.5 to 2.0 increased the degradation efficiency from 32 to 99% during 150 min under visible light. In acidic conditions, the pH of the solution significantly influences the degradation, with 2.0 being the optimum pH studied.

Maddila et al. [109] reported the influence of pH on the degradation of the bromoxynil pesticide using CS-TiO$_2$ as a photocatalyst. The degradation experiments were performed at pH values of 3, 7 and 11. Their results indicated that the process efficiency was improved with increasing pH from 3 to 11. As the solution became alkaline, the decomposition rate for secondary oxidants, such as the hydroxyl radical, increased, thereby favoring the process. The optimum pH during these experiments was 11, the pH variation study being useful to understand the process mechanism with the highest degree of removal.

5.4   Intensity and Source of Irradiation

The effect of intensity of the light on the rate of degradation of the target molecule also depends on the type of photocatalyst, and it is expected that different behaviors will be observed according to the intensity. In this sense, at low intensities, the rate of degradation increases linearly, as its intensity increases gradually. This relationship is plausible if it is considered that in reactions involving low light intensity, the formation of electrons/holes are in the majority and avoid recombination between charges. However, in increasing light intensity, there is a competition between the separation of the charge carriers and their recombination, thus generating a negative effect on the rate of the photocatalytic reaction [14, 15, 98, 105, 110]. Also in this context, the increase of the irradiation time is expected to favor degradation kinetics, otherwise, there is competition between the substrate and the photogenerated intermediates making the process difficult.

There are other parameters that directly, and/or indirectly, affect the photocatalytic efficiency. When using interfering substances such as chlorides (Cl$^-$), carbonates (CO$_3^{2-}$), nitrates (NO$_3^-$), and phosphates (PO$_4^{3-}$), they may act by blocking or deactivating the oxidative rate during the photocatalytic process due to the interaction and/or adsorption of the ions with the sites of the photocatalysts [96, 97, 105, 106]. Depending on the pH of the solution, these species can compete directly with the model pollutant and influence the overall rate of photocatalytic oxidation. Other substances and parameters may also interfere positively, or negatively, with photocatalytic efficiency (H$_2$O$_2$, Ag$^+$), depending on the reaction medium. The velocity kinetics provides the real interpretation regarding the removal and mineralization of the contaminants. In the same way, parameters derived from the form and type of reactor, such as geometry, light distribution, and flow type, play an important role in the photocatalytic reactions involved.

The literature reports several works involving the variation of these parameters to evaluate the degradation of compounds (pesticide dyes and drugs) [38, 97, 111–115].

These works bring wide versatility and the possibility of their uses against different materials used as photocatalysts.

6 Photocatalytic Degradation of Organic Contaminants for Environmental Treatment

In agreement with the photocatalytic treatment and the use of semiconductive nanoparticles, the photocatalytic degradation of toxic compounds depends on several factors that have already been mentioned in the previous section, and help in the comprehension and interpretation of the photogenerated mechanisms after the process [116].

The most common pollutants in wastewater comprise mainly dyes, pesticides, pharmaceuticals, heavy metal ions, radioactive substances, and polyaromatic hydrocarbons, among many others [117, 118]. These contaminants classed as organic and inorganic, depending on their chemical composition [117, 118]. Table 1 shows the schematic representation of pollutant publications via photocatalytic treatments tested in the last decade. The search only pertains to articles such as the type of document evaluated in the databases.

In analyzing the presented values as showed in Table 1, it was noted how many studies on dyes have been expressly discussed in the past ten years. Considering the articles found, followed by the expressions: (1) dyes, (2) dyes and photocatalytic application, and (3) dyes and photocatalytic treatment, it was verified that the dyes are presented in a number of publications three times greater than that for pesticides,

Table 1 Number of articles found in the Web of Science, and Scopus databases, related to the mentioned pollutants and their usual combinations involved in the photocatalytic process

Description	Web of Science	Scopus
Organic contaminants	15,876	12,776
Organic contaminants and photocatalytic treatment	529	255
Organic contaminants and photocatalytic application	399	196
Pesticides	42,985	45,479
Dyes	119,589	156,005
Pharmaceuticals	78,139	134345
Pesticides and photocatalytic application	110	55
Dyes and photocatalytic application	2,627	2,758
Pharmaceuticals and photocatalytic application	211	118
Pesticides and photocatalytic treatment	237	117
Dyes and photocatalytic treatment	2,559	2,682
Pharmaceuticals and photocatalytic treatment	561	234

Obviously, the number found corresponds to articles concerning diverse applications, not only photocatalytic degradation. The same was observed for the two other pollutants. However, when taken into account for photocatalytic applications, this ratio is larger according to the information from both databases.

Dyes are substances soluble in the medium in which they are present, and are used to attribute permanent color to other materials, absorb light in the visible region of the electromagnetic spectrum, and have a chromophore group in their structure [119]. Synthetic dyes are generally used to color fabrics and textile fibers, and dyes are most commonly found in environmental remediation studies from industrial effluents. Dyes can be classified according to their structure, such as anthraquinones and azo dyes, or they can be classified according to the type of interaction with the textile fiber. From the different interactions, the synthetic dyes can be classified according to Salleh et al. [120]. Reactive dyes have in their structure a reactive group capable of forming a covalent bond with cellulose. They are usually represented by anthraquinones and azo dyes. Direct dyes have interactions with the fiber through Van der Waals forces. A dye class consisting of compounds having more than one azo group in their structure (diazo, triazo), Azo dyes are usually are synthesized on the fiber through a coupling agent. Acid dyes are a group of anion dyes that have three sulfonic groups [2, 14, 120, 121]. Dispersive dyes are generally stabilized in the fibers with the aid of long chain dispersing agents. Several works are found in the literature with the purpose of investigating the removal of these dyes in solution through photochemical reaction using semiconductors as photocatalysts [122–125].

The kinetics of photocatalytic degradation of methylene blue (MB) cationic dye has been reported in the literature. Dariani et al. [126] investigated the photodegradation and mineralization of this dye using TiO_2 under UV light irradiation. The results showed that photocatalysis is an excellent technique for discoloration of the MB dye in solution, reaching up to 90% of complete degradation in 2 h of reaction.

Rhodamine B cationic dye is also present in several studies that address dyes as polluting wastewater models. Fu et al. [127] investigated the photodegradation of Rhodamine B under visible light irradiation using Bi_2WO_6 nanoparticles. The results indicate that the decomposition of the RB by the semiconductor probably does not involve the ˙OH radicals, and in this case, it may be due to the reaction between the directly photogenerated hole with the RB.

An azo dye used in the literature, in research involving photocatalysis for the degradation of emergent pollutants, is methyl orange (MO). Kulkarni et al. [111] used $ZnFe_2O_4@ZnO$ nanoparticles to investigate the photodegradation of MO under visible and UV irradiation. The band gap of the material makes it a good photocatalyst in the visible region, leading to complete degradation of MO in 9 h.

For years, various pesticides are continually being released into the environment via processes from agriculture that, when left untreated, pose risks to biodiversity, terrestrial ecosystems, as well as the life of all living beings [128, 129]. Since then, compliance with strict quality standards is required in the case of toxic substances [130], such as pesticides, which are assessed by international and governmental bodies as to their physicochemical, toxicological and ecotoxicological properties, ensuring compliance with requirements and corrective measures for their use and safety [131, 132].

Pesticides are essential chemical compounds and they are effective tools to control agricultural pests, and/or weeds and insects, that would adversely affect agricultural productivity [132, 133]. Currently, most of the pesticides used are based on semi-volatile organic matter and can react strongly with species such as hydroxyl (OH), ozone (O_3), and nitrate (NO_3) [134]. As these contaminants are non-biodegradable, toxic, and strongly persistent in the environment [130, 135, 136], they become a cumulative problem in the long run. In this sense, treatment by efficient purification technologies is recommended, as well as for the other contaminants highlighted in this chapter, with photocatalytic degradation being the target of studies for the scientific community due to its uses, benefits, and adverse potential.

Contamination by chlorine compounds such as 2,4-dichlorophenoxyacetic acid (2,4-D) and 2,4-dichlorophenoxypropionic acid (2,4-DP), in aqueous solution, were evaluated against photocatalytic activity using purified and pounded clays with TiO_2 [61]. According to the authors, the photocatalysts effectively degrade the selected pesticides, where the activity of the TiO_2 pillarized clay sample is lower than pure TiO_2 due to poor surface accessibility of TiO_2 is the amorphous phase of the TiO_2 pillars. Nevertheless, a titanium intercalation, in the intermediate layer of clays, is one of the most promising methods to improve the photocatalysis, varying the percentages of the oxide present [61].

Atrazine is a herbicide, and/or triazine pesticide, widely used in pest control in agriculture, although it is of great concern because of its persistence and toxicity [133, 137]. Xue et al. [137], visible light-sensitive BiOBr/UiO-66 was synthesized and applied to the degradation of atrazine. Operational parameters, such as pH, presence of anions, inorganic cations, and water type, were investigated extensively in this study. The results showed that BiOBr/UiO-66 exhibited higher and better photocatalytic performance over pure BiOBr. Atrazine was rapidly degraded under acidic conditions (pH = 3.1) and strongly inhibited by HCO_3^- and the SO_4^{2-}, being the inorganic cations few influential in the degradation. Finally, atrazine degradation pathways were investigated via LC-MS/MS [137].

In the case of pharmaceutical micropollutants in the environment, they are seen as a hazard to health, water quality and, above all, food safety, due to the formation of metabolites of unknown nature [138, 139]. The concentrations of drugs released into the sewage system are affected by numerous conditions ranging from their man-ufacture to the seasons, climate, pH, temperature, the type of sewage involved, and other factors [140]. Drugs are substances absorbed by the body. Moreover, they are developed to be persistent, maintaining their chemical properties enough to guarantee their desired purpose [141–143].

On the issue of water decontamination, Bayan et al. [144] synthesized spheres composed of CdS by the simple hydrothermal method and they used them to evaluate the photocatalytic degradation of the drugs, acetaminophen and levofloxacin, under irradiation of visible light. Both drugs were degraded to more than 95% after 4 h of visible light irradiation. Stability was also evaluated by reusing the sample, after four cycles, indicating the high stability of the synthesized sample. Thus, CdS beads are useful, reusable, materials for the photocatalytic degradation of pharmaceuticals.

Ibuprofen is one of the most harmful drugs found in wastewater effluents [145]. In this study, its degradation was evaluated using TiO_2 nanoparticles with different ratios of the anatase/brookite phases. As for the degradation, 100% was achieved with a mixture containing TiO_2 61.8% of anatase and 38.2% of brookite. For the authors, this high oxidation rate is due to the mixture's synergistic effect, high surface area, and mesoporous structure. The photocatalyst, prepared under specific conditions, was considered very stable after eight cycles of reuse, without presenting significant losses to its photocatalytic performance [145].

7 Final Considerations

Due to the growing concern about the presence of these organic contaminants in the environment and their possible effects, researchers are increasingly investigating ways to minimize the disposal of these pollutants, because despite the numerous technologies for their remediation, there is still a deficit about their real effects and affinities in the environment and in the lives of all human beings. In addition, an assessment of the possibility of reducing, minimizing or treating at a lower cost is essential, thus ensuring the required acceptance by environmental standards and requirements. Another persistent approach is the complexity of effluents from which the treatment is discarded in isolation, i.e., if there are no standardized treatment procedures, it is feasible to handle it in isolation or in combination when possible. Tests of treatability, identification, ecotoxicity, antibacterials and others are extremely important in monitoring removal efficiency. Therefore, to know the impacts, establishing concentration standards and acceptable limits for the dismissal of unemployment insurance actions upon payment and maintenance of their health, well-being, health and quality of life for future.

Acknowledgements The authors thank to CAPES, CNPq, FAPEPI for financial support.

References

1. Giannakis, S., Gamarra Vives, F.A., Grandjean, D., Magnet, A., De Alencastro, L.F., Pulgarin, C.: Effect of advanced oxidation processes on the micropollutants and the effluent organic matter contained in municipal wastewater previously treated by three different secondary methods. Water Res. **84**, 295–306 (2015)
2. Brillas, E., Martínez-Huitle, C.A.: Decontamination of wastewaters containing synthetic organic dyes by electrochemical methods. An updated review. Appl. Catal. B: Environ. **166–167**, 603–643 (2015)

3. Moreira, F.C., Boaventura, R.A.R., Brillas, E., Vilar, V.J.P.: Electrochemical advanced oxidation processes: a review on their application to synthetic and real wastewaters. Appl. Catal. B: Environ. **202**, 217–261 (2017)

4. Babuponnusami, A., Muthukumar, K.: A review on Fenton and improvements to the Fenton process for wastewater treatment. J. Environ. Chem. Eng. **2**, 557–572 (2014)

5. Bel Hadjltaief, H., Ben Ameur, S., Da Costa, P., Ben Zina, M., Elena Galvez, M.: Photocatalytic decolorization of cationic and anionic dyes over ZnO nanoparticle immobilized on natural Tunisian clay. Appl. Clay Sci. **152**, 148–157 (2018)

6. Bel Hadjltaief, H., Da Costa, P., Galvez, M.E., Ben Zina, M.: Influence of operational parameters in the heterogeneous photo-Fenton discoloration of wastewaters in the presence of an iron-pillared clay. Ind. Eng. Chem. Res. **52**, 16656–16665 (2013)

7. Puangpetch, T., Sreethawong, T., Yoshikawa, S., Chavadej, S.: Synthesis and photocatalytic activity in methyl orange degradation of mesoporous-assembled $SrTiO_3$ nanocrystals prepared by sol–gel method with the aid of structure-directing surfactant. J. Mol. Catal. A: Chem. **287**, 70–79 (2008)

8. Augugliaro, V., Litter, M., Palmisano, L., Soria, J.: The combination of heterogeneous photocatalysis with chemical and physical operations: a tool for improving the photoprocess performance. J. Photochem. Photobiol. C: Photochem. Rev. **7**, 127–144 (2006)

9. De Araújo, K.S., Antonelli, R., Gaydeczka, B., Granato, A.C., Malpass, G.R.P.: Advanced oxidation processes: a review regarding the fundamentals and applications in wastewater treatment and industrial wastewater. Ambient. Agua – Interdiscip. J. Appl. Sci. **11**, 387 (2016)

10. Ghiasi, M., Malekzadeh, A.: Solar photocatalytic degradation of methyl orange over $La_{0.7}Sr_{0.3}MnO_3$ nano-perovskite. Sep. Purif. Technol. **134**, 12–19 (2014)

11. Bethi, B., Sonawane, S.H., Bhanvase, B.A., Gumfekar, S.P.: Nanomaterials-based advanced oxidation processes for wastewater treatment: a review. Chem. Eng. Process. Process Intensif. **109**, 178–189 (2016)

12. Cheng, M., Zeng, G., Huang, D., Lai, C., Xu, P., Zhang, C., Liu, Y.: Hydroxyl radicals based advanced oxidation processes (AOPs) for remediation of soils contaminated with organic compounds: a review. Chem. Eng. J. **284**, 582–598 (2016)

13. Ribeiro, A.R., Nunes, O.C., Pereira, M.F.R., Silva, A.M.T.: An overview on the advanced oxidation processes applied for the treatment of water pollutants defined in the recently launched Directive 2013/39/EU. Environ. Int. **75**, 33–51 (2015)

14. Rauf, M.A., Meetani, M.A., Hisaindee, S.: An overview on the photocatalytic degradation of azo dyes in the presence of TiO_2 doped with selective transition metals. Desalination **276**, 13–27 (2011)

15. Rauf, M.A., Ashraf, S.S.: Fundamental principles and application of heterogeneous photocatalytic degradation of dyes in solution. Chem. Eng. J. **151**, 10–18 (2009)

16. Byrne, C., Subramanian, G., Pillai, S.C.: Recent advances in photocatalysis for environmental applications. J. Environ. Chem. Eng. **6**, 3531–3555 (2017)

17. Wetchakun, K., Wetchakun, N., Sakulsermsuk, S.: An overview of solar/visible light-driven heterogeneous photocatalysis for water purification: TiO_2-and ZnO-based photocatalysts used in suspension photoreactors. J. Ind. Eng. Chem. **71**, 19–49 (2019)

18. Fujishima, A., Honda, K.: Electrochemical photolysis of water at a semiconductor electrode. Nature **238**, 37–38 (1972)

19. Kabra, K., Chaudhary, R., Sawhney, R.L.: Treatment of hazardous organic and inorganic compounds through aqueous-phase photocatalysis: a review. Ind. Eng. Chem. Res. **43**, 7683–7696 (2004)

20. Schneider, J., Matsuoka, M., Takeuchi, M., Zhang, J., Horiuchi, Y., Anpo, M., Bahnemann, D.W.: Understanding TiO_2 photocatalysis: mechanisms and materials. Chem. Rev. **114**, 9919–9986 (2014)

21. Serpone, N., Emeline, A.V., Horikoshi, S., Kuznetsov, V.N., Ryabchuk, V.K.: On the genesis of heterogeneous photocatalysis: a brief historical perspective in the period 1910 to the mid-1980s. Photochem. Photobiol. Sci. **11**, 1121–1150 (2012)

22. Bora, L.V., Mewada, R.K.: Visible/solar light active photocatalysts for organic effluent treatment: fundamentals, mechanisms and parametric review. Renew. Sustain. Energy Rev. **76**, 1393–1421 (2017)
23. Athanasekou, C.P., Likodimos, V., Falaras, P.: Recent developments of TiO_2 photocatalysis involving advanced oxidation and reduction reactions in water. J. Environ. Chem. Eng. **6**, 7386–7394 (2018)
24. Shen, S., Kronawitter, C., Kiriakidis, G.: An overview of photocatalytic materials. J. Mater. **3**, 1–2 (2017)
25. Yasmina, M., Mourad, K., Mohammed, S.H., Khaoula, C.: Treatment heterogeneous photocatalysis; factors influencing the photocatalytic degradation by TiO_2. Energy Procedia **50**, 559–566 (2014)
26. Qian, R., Zong, H., Schneider, J., Zhou, G., Zhao, T., Li, Y., Yang, J., Bahnemann, D.W., Pan, J.H.: Charge carrier trapping, recombination and transfer during TiO_2 photocatalysis: an overview. Catal. Today **335**, 78–90 (2018)
27. Kumar, S.G., Rao, K.S.R.K.: Comparison of modification strategies towards enhanced charge carrier separation and photocatalytic degradation activity of metal oxide semiconductors (TiO_2, WO_3 and ZnO). Appl. Surf. Sci. **391**, 124–148 (2017)
28. Hitkari, G., Singh, S., Pandey, G.: Photoluminescence behavior and visible light photocatalytic activity of ZnO, ZnO/ZnS and ZnO/ZnS/α-Fe_2O_3 nanocomposites. Trans. Nonferrous Met. Soc. China **28**, 1386–1396 (2018)
29. Senasu, T., Nanan, S.: Photocatalytic performance of CdS nanomaterials for photodegradation of organic azo dyes under artificial visible light and natural solar light irradiation. J. Mater. Sci.: Mater. Electron. **28**, 17421–17441 (2017)
30. Ângelo, J., Andrade, L., Madeira, L.M., Mendes, A.: An overview of photocatalysis phenomena applied to NOx abatement. J. Environ. Manag. **129**, 522–539 (2013)
31. Pawar, R.C., Lee, C.S.: Basics of photocatalysis. In: Heterogeneous Nanocomposite-Photocatalysis for Water Purification, pp. 1–23. Elsevier (2015)
32. Molinari, R., Lavorato, C., Argurio, P.: Recent progress of photocatalytic membrane reactors in water treatment and in synthesis of organic compounds. A review. Catal. Today **281**, 144–164 (2017)
33. Ajmal, A., Majeed, I., Malik, R.N., Idriss, H., Nadeem, M.A.: Principles and mechanisms of photocatalytic dye degradation on TiO_2 based photocatalysts: a comparative overview. RSC Adv. **4**, 37003–37026 (2014)
34. Zhang, L., Mohamed, H.H., Dillert, R., Bahnemann, D.: Kinetics and mechanisms of charge transfer processes in photocatalytic systems: a review. J. Photochem. Photobiol. C: Photochem. Rev. **13**, 263–276 (2012)
35. Kumar, P.S., Lakshmi Prabavathi, S., Indurani, P., Karuthapandian, S., Muthuraj, V.: Light assisted synthesis of hierarchically structured Cu/CdS nanorods with superior photocatalytic activity, stability and photocatalytic mechanism. Sep. Purif. Technol. **172**, 192–201 (2017)
36. Turchi, C., Ollis, D.F.: Photocatalytic degradation of organic water contaminants: mechanisms involving hydroxyl radical attack. J. Catal. **122**, 178–192 (1990)
37. Palominos, R., Freer, J., Mondaca, M.A., Mansilla, H.D.: Evidence for hole participation during the photocatalytic oxidation of the antibiotic flumequine. J. Photochem. Photobiol. A: Chem. **193**, 139–145 (2008)
38. Sanad, M.F., Shalan, A.E., Bazid, S.M., Abdelbasir, S.M.: Pollutant degradation of different organic dyes using the photocatalytic activity of ZnO@ZnS nanocomposite materials. J. Environ. Chem. Eng. **6**, 3981–3990 (2018)
39. Pastrana-Martínez, L.M., Morales-Torres, S., Likodimos, V., Figueiredo, J.L., Faria, J.L., Falaras, P., Silva, A.M.T.: Advanced nanostructured photocatalysts based on reduced graphene oxide-TiO_2 composites for degradation of diphenhydramine pharmaceutical and methyl orange dye. Appl. Catal. B: Environ. **123–124**, 241–256 (2012)
40. Chen, F., Yang, Q., Niu, C., Li, X., Zhang, C., Zhao, J., Xu, Q., Zhong, Y., Deng, Y., Zeng, G.: Enhanced visible light photocatalytic activity and mechanism of $ZnSn(OH)_6$ nanocubes modified with AgI nanoparticles. Catal. Commun. **73**, 1–6 (2016)

41. Chen, F., Yang, Q., Li, X., Zeng, G., Wang, D., Niu, C., Zhao, J., An, H., Xie, T., Deng, Y.: Hierarchical assembly of graphene-bridged $Ag_3PO_4/Ag/BiVO_4(040)$ Z-scheme photocatalyst: an efficient, sustainable and heterogeneous catalyst with enhanced visible-light photoactivity towards tetracycline degradation under visible light irradiation. Appl. Catal. B: Environ. **200**, 330–342 (2017)

42. Liu, R., Ji, Z., Wang, J., Zhang, J.: Solvothermal synthesized Ag-decorated TiO_2/sepiolite composite with enhanced UV–vis and visible light photocatalytic activity. Microporous Mesoporous Mater. **266**, 268–275 (2018)

43. Samsudin, M.F.R., Siang, L.T., Sufian, S., Bashiri, R., Mohamed, N.M., Ramli, R.M.: Exploring the role of electron-hole scavengers on optimizing the photocatalytic performance of $BiVO_4$. Mater. Today Proc. **5**, 21703–21709 (2018)

44. Mohammadi, M., Rezaee Roknabadi, M., Behdani, M., Kompany, A.: Enhancement of visible and UV light photocatalytic activity of $rGO-TiO_2$ nanocomposites: the effect of TiO_2/graphene oxide weight ratio. Ceram. Int. **45**, 12625–12634 (2019)

45. Baeissa, E.S.: Environmental remediation of aqueous methyl orange dye solution via photocatalytic oxidation using $Ag-GdFeO_3$ nanoparticles. J. Alloys Compd. **678**, 267–272 (2016)

46. Inagaki, C.S., Reis, A.E. da S., Oliveira, N.M., Paschoal, V.H., Mazali, Í.O., Alfaya, A.A.S.: Use of SiO_2/TiO_2 nanostructured composited in textiles dyes and their photodegradation. Quim. Nova **38**, 1037–1043 (2015)

47. George, R., Bahadur, N., Singh, N., Singh, R., Verma, A., Shukla, A.K.: Environmentally benign TiO_2 nanomaterials for removal of heavy metal ions with interfering ions present in tap water. Mater. Today Proc. **3**, 162–166 (2016)

48. Sanzone, G., Zimbone, M., Cacciato, G., Ruffino, F., Carles, R., Privitera, V., Grimaldi, M.G.: Ag/TiO_2 nanocomposite for visible light-driven photocatalysis. Superlattices Microstruct. **123**, 394–402 (2018)

49. Litter, M.I.: Heterogeneous photocatalysis: transition metal ions in photocatalytic systems. Appl. Catal. B: Environ. **23**, 89–114 (1999)

50. Ansari, S.A., Khan, M., Ansari, O.: Nitrogen-doped titanium dioxide (N-doped TiO_2) for visible light photocatalysis. New J. Chem. **40**, 3000–3009 (2016)

51. Grabowska, E.: Selected perovskite oxides: characterization, preparation and photocatalytic properties—a review. Appl. Catal. B: Environ. **186**, 97–126 (2016)

52. Song, S., Xu, L., He, Z., Ying, H., Chen, J., Xiao, X., Yan, B.: Photocatalytic degradation of C.I. Direct Red 23 in aqueous solutions under UV irradiation using $SrTiO_3/CeO_2$ composite as the catalyst. J. Hazard. Mater. **152**, 1301–1308 (2008)

53. Hu, R., Li, C., Wang, X., Sun, Y., Jia, H., Su, H., Zhang, Y.: Photocatalytic activities of $LaFeO_3$ and La_2FeTiO_6 in p-chlorophenol degradation under visible light. Catal. Commun. **29**, 35–39 (2012)

54. de Souza, J.K.D., Honório, L.M.C., Ferreira, J.M., Torres, S.M., Santos, I.M.G., Maia, A.S.: Layered niobate KNb_3O_8 synthesized by the polymeric precursor method. Cerâmica **64**, 104–108 (2018)

55. Lucena, G.L., De Lima, L.C., Honório, L.M.C., De Oliveira, A.L.M., Tranquilim, R.L., Longo, E., De Souza, A.G., Maia, A. da S., dos Santos, I.M.G.: $CaSnO_3$ obtained by modified Pechini method applied in the photocatalytic degradation of an azo dye. Ceramica. **63**, 536–541 (2017)

56. Honorio, L.M.C., Santos, M.V.B., da Silva Filho, E.C., Osajima, J.A., Maia, A.S., dos Santos, I.M.G.: Alkaline earth stannates applied in photocatalysis: prospection and review of literature. Cerâmica **64**, 559–569 (2018)

57. Teixeira, A.R.F.A., de Meireles Neris, A., Longo, E., de Carvalho Filho, J.R., Hakki, A., Macphee, D., dos Santos, I.M.G.: $SrSnO_3$ perovskite obtained by the modified Pechini method—insights about its photocatalytic activity. J. Photochem. Photobiol. A: Chem. **369**, 181–188 (2019)

58. Belver, C., Bedia, J., Álvarez-Montero, M.A., Rodriguez, J.J.: Solar photocatalytic purification of water with Ce-doped TiO_2/clay heterostructures. Catal. Today **266**, 36–45 (2016)

59. Belver, C., Bedia, J., Rodriguez, J.J.: Titania–clay heterostructures with solar photocatalytic applications. Appl. Catal. B: Environ. **176–177**, 278–287 (2015)
60. Szczepanik, B.: Photocatalytic degradation of organic contaminants over clay-TiO_2 nanocomposites: a review. Appl. Clay Sci. **141**, 227–239 (2017)
61. Abdennouri, M., Baâlala, M., Galadi, A., El Makhfouk, M., Bensitel, M., Nohair, K., Sadiq, M., Boussaoud, A., Barka, N.: Photocatalytic degradation of pesticides by titanium dioxide and titanium pillared purified clays. Arab. J. Chem. **9**, S313–S318 (2016)
62. Mccabe, R.W., Adams, J.M.: Clay minerals as catalysts. In: Developments in Clay Science, vol. 5, pp. 491–538. Elsevier (2013)
63. Belver, C., Bedia, J., Rodriguez, J.J.: Zr-doped TiO_2 supported on delaminated clay materials for solar photocatalytic treatment of emerging pollutants. J. Hazard. Mater. **322**, 233–242 (2017)
64. Hassani, A., Khataee, A., Karaca, S.: Photocatalytic degradation of ciprofloxacin by synthesized TiO_2 nanoparticles on montmorillonite: effect of operation parameters and artificial neural network modeling. J. Mol. Catal. A: Chem. **409**, 149–161 (2015)
65. Belessi, V., Lambropoulou, D., Konstantinou, I., Katsoulidis, A., Pomonis, P., Petridis, D., Albanis, T.: Structure and photocatalytic performance of TiO_2/clay nanocomposites for the degradation of dimethachlor. Appl. Catal. B: Environ. **73**, 292–299 (2007)
66. Meng, X., Qian, Z., Wang, H., Gao, X., Zhang, S., Yang, M.: Sol-gel immobilization of SiO_2/TiO_2 on hydrophobic clay and its removal of methyl orange from water. J. Sol-Gel Sci. Technol. **46**, 195–200 (2008)
67. Bel Hadjltaief, H., Galvez, M.E., Ben Zina, M., Da Costa, P.: TiO_2/clay as a heterogeneous catalyst in photocatalytic/photochemical oxidation of anionic reactive blue 19. Arab. J. Chem. (2014)
68. González, B., Trujillano, R., Vicente, M.A., Rives, V., Korili, S.A., Gil, A.: Photocatalytic degradation of trimethoprim on doped Ti-pillared montmorillonite. Appl. Clay Sci. **167**, 43–49 (2019)
69. Zhang, Y., Wang, D., Zhang, G.: Photocatalytic degradation of organic contaminants by TiO_2/sepiolite composites prepared at low temperature. Chem. Eng. J. **173**, 1–10 (2011)
70. Lopes, J. da S., Rodrigues, W.V., Oliveira, V.V., Braga, A. do N.S., da Silva, R.T., França, A.A.C., da Paz, E.C., Osajima, J.A., da Silva Filho, E.C.: Modification of kaolinite from Pará/Brazil region applied in the anionic dye photocatalytic discoloration. Appl. Clay Sci. **168**, 295–303 (2019)
71. Wang, B., Zhang, G., Leng, X., Sun, Z., Zheng, S.: Characterization and improved solar light activity of vanadium doped TiO_2/diatomite hybrid catalysts. J. Hazard. Mater. **285**, 212–220 (2015)
72. Sun, Q., Hu, X., Zheng, S., Sun, Z., Liu, S., Li, H.: Influence of calcination temperature on the structural, adsorption and photocatalytic properties of TiO_2 nanoparticles supported on natural zeolite. Powder Technol. **274**, 88–97 (2015)
73. Kun, R., Mogyorósi, K., Dékány, I.: Synthesis and structural and photocatalytic properties of TiO_2/montmorillonite nanocomposites. Appl. Clay Sci. **32**, 99–110 (2006)
74. Wang, L., Wang, X., Cui, S., Fan, X., Zu, B., Wang, C.: TiO_2 supported on silica nanolayers derived from vermiculite for efficient photocatalysis. Catal. Today **216**, 95–103 (2013). https://doi.org/10.1016/j.cattod.2013.06.026
75. Belver, C., Han, C., Rodriguez, J.J., Dionysiou, D.D.: Innovative W-doped titanium dioxide anchored on clay for photocatalytic removal of atrazine. Catal. Today **280**, 21–28 (2017)
76. Vicente, M.A., Gil, A., Bergaya, F.: Pillared clays and clay minerals. Handb. Clay Sci. **10**(5), 523–557 (2013)
77. Galeano, L.A., Vicente, M.Á., Gil, A.: Catalytic degradation of organic pollutants in aqueous streams by mixed Al/M-pillared clays (M = Fe, Cu, Mn). Catal. Rev. Sci. Eng. **56**, 239–287 (2014)
78. Yang, J.-H., Piao, H., Vinu, A., Elzatahry, A.A., Paek, S.-M., Choy, J.-H.: TiO_2-pillared clays with well-ordered porous structure and excellent photocatalytic activity. RSC Adv. **5**, 8210–8215 (2015)

79. Aranda, P., Kun, R., Martín-Luengo, M.A., Letaïef, S., Dékány, I., Ruiz-Hitzky, E.: Titania-sepiolite nanocomposites prepared by a surfactant templating colloidal route. Chem. Mater. **20**, 84–91 (2008)

80. Chen, D., Du, Y., Zhu, H., Deng, Y.: Synthesis and characterization of a microfibrous TiO_2-CdS/palygorskite nanostructured material with enhanced visible-light photocatalytic activity. Appl. Clay Sci. **87**, 285–291 (2014)

81. Bouna, L., Rhouta, B., Amjoud, M., Maury, F., Lafont, M.C., Jada, A., Senocq, F., Daoudi, L.: Synthesis, characterization and photocatalytic activity of TiO_2 supported natural palygorskite microfibers. Appl. Clay Sci. **52**, 301–311 (2011)

82. Shi, Y., Yang, Z., Wang, B., An, H., Chen, Z., Cui, H.: Adsorption and photocatalytic degradation of tetracycline hydrochloride using a palygorskite-supported Cu_2O-TiO_2 composite. Appl. Clay Sci. **119**, 311–320 (2016)

83. Mamulová Kutláková, K., Tokarský, J., Peikertová, P.: Functional and eco-friendly nanocomposite kaolinite/ZnO with high photocatalytic activity. Appl. Catal. B: Environ. **162**, 392–400 (2015)

84. Xu, H., Yu, T., Liu, J.: Photo-degradation of Acid Yellow 11 in aqueous on nano-ZnO/Bentonite under ultraviolet and visible light irradiation. Mater. Lett. **117**, 263–265 (2014)

85. Akkari, M., Aranda, P., Ben Rhaiem, H., Ben Haj Amara, A., Ruiz-Hitzky, E.: ZnO/clay nanoarchitectures: synthesis, characterization and evaluation as photocatalysts. Appl. Clay Sci. **131**, 131–139 (2016)

86. Xiao, J., Peng, T., Ke, D., Zan, L., Peng, Z.: Synthesis, characterization of CdS/rectorite nanocomposites and its photocatalytic activity. Phys. Chem. Miner. **34**, 275–285 (2007)

87. Wu, S., Fang, J., Xu, W., Cen, C.: Bismuth-modified rectorite with high visible light photocatalytic activity. J. Mol. Catal. A: Chem. **373**, 114–120 (2013)

88. Zhang, G., Gao, Y., Zhang, Y., Guo, Y.: Fe_2O_3-pillared rectorite as an efficient and stable fenton-like heterogeneous catalyst for photodegradation of organic contaminants. Environ. Sci. Technol. **44**, 6384–6389 (2010)

89. Zhang, Y., Yang, H.: Co_3O_4 nanoparticles on the surface of halloysite nanotubes. Phys. Chem. Miner. **39**, 789–795 (2012)

90. Kataria, N., Garg, V.K., Jain, M., Kadirvelu, K.: Preparation, characterization and potential use of flower shaped Zinc oxide nanoparticles (ZON) for the adsorption of Victoria Blue B dye from aqueous solution. Adv. Powder Technol. **27**, 1180–1188 (2016)

91. Kaplan, R., Erjavec, B., Dražić, G., Grdadolnik, J., Pintar, A.: Simple synthesis of anatase/rutile/brookite TiO_2 nanocomposite with superior mineralization potential for photocatalytic degradation of water pollutants. Appl. Catal. B: Environ. **181**, 465–474 (2016)

92. Wen, X.J., Niu, C.G., Huang, D.W., Zhang, L., Liang, C., Zeng, G.M.: Study of the photocatalytic degradation pathway of norfloxacin and mineralization activity using a novel ternary Ag/AgCl-CeO_2 photocatalyst. J. Catal. **355**, 73–86 (2017)

93. Wen, X.J., Niu, C.G., Zhang, L., Liang, C., Zeng, G.M.: A novel Ag_2O/CeO_2 heterojunction photocatalysts for photocatalytic degradation of enrofloxacin: possible degradation pathways, mineralization activity and an in depth mechanism insight. Appl. Catal. B: Environ. **221**, 701–714 (2018)

94. Kovalevskiy, N.S., Lyulyukin, M.N., Selishchev, D.S., Kozlov, D.V.: Analysis of air photocatalytic purification using a total hazard index: effect of the composite TiO_2/zeolite photocatalyst. J. Hazard. Mater. **358**, 302–309 (2018)

95. Ma, R., Wang, L., Wang, S., Wang, C., Xiao, F.S.: Eco-friendly photocatalysts achieved by zeolite fixing. Appl. Catal. B: Environ. **212**, 193–200 (2017)

96. Carbonaro, S., Sugihara, M.N., Strathmann, T.J.: Continuous-flow photocatalytic treatment of pharmaceutical micropollutants: activity, inhibition, and deactivation of TiO_2 photocatalysts in wastewater effluent. Appl. Catal. B: Environ. **129**, 1–12 (2013)

97. Wang, C., Xue, Y., Wang, P., Ao, Y.: Effects of water environmental factors on the photocatalytic degradation of sulfamethoxazole by AgI/UiO-66 composite under visible light irradiation. J. Alloys Compd. **748**, 314–322 (2018)

98. Konstantinou, I.K., Albanis, T.A.: TiO$_2$-assisted photocatalytic degradation of azo dyes in aqueous solution: Kinetic and mechanistic investigations: a review. Appl. Catal. B: Environ. **49**, 1–14 (2004)

99. Greco, E., Ciliberto, E., Cirino, A.M.E., Capitani, D., Di Tullio, V.: A new preparation of doped photocatalytic TiO$_2$ anatase nanoparticles: a preliminary study for the removal of pollutants in confined museum areas. Appl. Phys. A: Mater. Sci. Process. **122** (2016)

100. Malato, S., Fernández-Ibáñez, P., Maldonado, M.I., Blanco, J., Gernjak, W.: Decontamination and disinfection of water by solar photocatalysis: recent overview and trends. Catal. Today **147**, 1–59 (2009)

101. Rezaei, M., Habibi-Yangjeh, A.: Simple and large scale refluxing method for preparation of Ce-doped ZnO nanostructures as highly efficient photocatalyst. Appl. Surf. Sci. **265**, 591–596 (2013)

102. Gaya, U.I., Abdullah, A.H.: Heterogeneous photocatalytic degradation of organic contaminants over titanium dioxide: a review of fundamentals, progress and problems. J. Photochem. Photobiol. C: Photochem. Rev. **9**, 1–12 (2008)

103. Akpan, U.G., Hameed, B.H.: Parameters affecting the photocatalytic degradation of dyes using TiO$_2$-based photocatalysts: a review. J. Hazard. Mater. **170**, 520–529 (2009)

104. Haque, M.M., Muneer, M.: Photodegradation of norfloxacin in aqueous suspensions of titanium dioxide. J. Hazard. Mater. **145**, 51–57 (2007)

105. Ahmed, S., Rasul, M.G., Brown, R., Hashib, M.A.: Influence of parameters on the heterogeneous photocatalytic degradation of pesticides and phenolic contaminants in wastewater: a short review. J. Environ. Manag. **92**, 311–330 (2011)

106. Guillard, C., Lachheb, H., Houas, A., Ksibi, M., Elaloui, E., Herrmann, J.-M.: Influence of chemical structure of dyes, of pH and of inorganic salts on their photocatalytic degradation by TiO$_2$ comparison of the efficiency of powder and supported TiO$_2$. J. Photochem. Photobiol. A: Chem. **158**, 27–36 (2003)

107. M.A.Fox, M.T. Dulay, Heterogeneous photocatalysis. In: Chemical Reviews, pp. 341–357. Elsevier (1993)

108. Sun, J., Qiao, L., Sun, S., Wang, G.: Photocatalytic degradation of Orange G on nitrogen-doped TiO$_2$ catalysts under visible light and sunlight irradiation. J. Hazard. Mater. **155**, 312–319 (2008)

109. Maddila, S., Lavanya, P., Jonnalagadda, S.B.: Degradation, mineralization of bromoxynil pesticide by heterogeneous photocatalytic ozonation. J. Ind. Eng. Chem. **24**, 333–341 (2015)

110. Sakthivel, S., Neppolian, B., Shankar, M.V., Arabindoo, B., Palanichamy, M., Murugesan, V.: Solar photocatalytic degradation of azo dye: comparison of photocatalytic efficiency of ZnO and TiO$_2$. Sol. Energy Mater. Sol. Cells **77**, 65–82 (2003)

111. Kulkarni, S.D.D., Kumbar, S., Menon, S.G.G., Choudhari, K.S.S., Santhosh, C.: Magnetically separable core-shell ZnFe$_2$O$_4$@ZnO nanoparticles for visible light photodegradation of methyl orange. Mater. Res. Bull. **77**, 70–77 (2016)

112. Chen, H., Liu, X.Y., Hao, X.D., Zhang, Y.X.: Facile biphasic synthesis of TiO$_2$–MnO$_2$ nanocomposites for photocatalysis. Ceram. Int. **42**, 19425–19428 (2016)

113. Papoulis, D., Panagiotaras, D., Tsigrou, P., Christoforidis, K.C., Petit, C., Apostolopoulou, A., Stathatos, E., Komarneni, S., Koukouvelas, I.: Halloysite and sepiolite –TiO$_2$ nanocomposites: synthesis characterization and photocatalytic activity in three aquatic wastes. Mater. Sci. Semicond. Process. **85**, 1–8 (2018)

114. Szczepanik, B., Rogala, P., Słomkiewicz, P.M., Banaś, D., Kubala-Kukuś, A., Stabrawa, I.: Synthesis, characterization and photocatalytic activity of TiO$_2$-halloysite and Fe$_2$O$_3$-halloysite nanocomposites for photodegradation of chloroanilines in water. Appl. Clay Sci. **149**, 118–126 (2017)

115. Srikanth, B., Goutham, R., Badri Narayan, R., Ramprasath, A., Gopinath, K.P., Sankaranarayanan, A.R.: Recent advancements in supporting materials for immobilised photocatalytic applications in waste water treatment. J. Environ. Manag. **200**, 60–78 (2017)

116. Ayodhya, D., Veerabhadram, G.: A review on recent advances in photodegradation of dyes using doped and heterojunction based semiconductor metal sulfide nanostructures for environmental protection. Mater. Today Energy. **9**, 83–113 (2018)

117. Iranpour, R., Stenstrom, M., Tchobanoglous, G., Miller, D., Wright, J., Vossoughi, M.: Environmental engineering: energy value of replacing waste disposal with resource recovery. Science (80-) **285**, 706–711 (1999)
118. Gao, Q., Xu, J., Bu, X.H.: Recent advances about metal–organic frameworks in the removal of pollutants from wastewater. Coord. Chem. Rev. **378**, 17–31 (2019)
119. Guillermin, D., Debroise, T., Trigueiro, P., de Viguerie, L., Rigaud, B., Morlet-savary, F., Balme, S., Janot, J.-M., Tielens, F., Michot, L., Lalevee, J., Walter, P., Jaber, M., De Viguerie, L., Morlet-savary, F., Balme, S., Tielens, F., Michot, L., Lalevee, J., Walter, P., Jaber, M.: New pigments based on carminic acid and smectites: a molecular investigation. Dye Pigment. **160**, 971–982 (2018)
120. Salleh, M.A.M., Mahmoud, D.K., Karim, W.A.W.A., Idris, A.: Cationic and anionic dye adsorption by agricultural solid wastes: a comprehensive review. Desalination **280**, 1–13 (2011)
121. Zhou, Y., Lu, J., Zhou, Y., Liu, Y.: Recent advances for dyes removal using novel adsorbents: a review. Environ. Pollut. **252**, 352–365 (2019)
122. Fatimah, I., Wang, S., Wulandari, D.: ZnO/montmorillonite for photocatalytic and photochemical degradation of methylene blue. Appl. Clay Sci. **53**, 553–560 (2011)
123. Djellabi, R., Ghorab, M.F., Cerrato, G., Morandi, S., Gatto, S., Oldani, V., Di Michele, A., Bianchi, C.L.: Photoactive TiO_2-montmorillonite composite for degradation of organic dyes in water. J. Photochem. Photobiol. A: Chem. **295**, 57–63 (2015)
124. Khataee, A.R., Kasiri, M.B.: Photocatalytic degradation of organic dyes in the presence of nanostructured titanium dioxide: influence of the chemical structure of dyes. J. Mol. Catal. A: Chem. **328**, 8–26 (2010)
125. Manohar, D.M., Noeline, B.F., Anirudhan, T.S.: Adsorption performance of Al-pillared bentonite clay for the removal of cobalt(II) from aqueous phase. Appl. Clay Sci. **31**, 194–206 (2006)
126. Dariani, R.S., Esmaeili, A., Mortezaali, A., Dehghanpour, S.: Photocatalytic reaction and degradation of methylene blue on TiO_2 nano-sized particles. Optik (Stuttg) **127**, 7143–7154 (2016)
127. Fu, H., Pan, C., Yao, W., Zhu, Y.: Visible-light-induced degradation of rhodamine B by nanosized Bi_2WO_6. J. Phys. Chem. B. **109**, 22432–22439 (2005)
128. Damalas, C.A., Koutroubas, S.D.: Farmers' behaviour in pesticide use: a key concept for improving environmental safety. Curr. Opin. Environ. Sci. Health **4**, 27–30 (2018)
129. Gibbons, D., Morrissey, C., Mineau, P.: A review of the direct and indirect effects of neonicotinoids and fipronil on vertebrate wildlife. Environ. Sci. Pollut. Res. **22**, 103–118 (2015). https://doi.org/10.1007/s11356-014-3180-5
130. Fernández-Alba, A., Hernando, D., Agüera, A., Cáceres, J., Malato, S.: Toxicity assays: a way for evaluating AOPs efficiency. Water Res. **36**, 4255–4262 (2002)
131. Bolognesi, C., Merlo, F.D.: Pesticides: human health effects. Encycl. Environ. Health 438–453 (2011)
132. Jørgensen, L.N., Kudsk, P., Ørum, J.E.: Links between pesticide use pattern and crop production in Denmark with special reference to winter wheat. Crop Prot. **119**, 147–157 (2019)
133. Fernández-Domene, R.M., Sánchez-Tovar, R., Lucas-granados, B., Muñoz-Portero, M.J., García-Antón, J.: Elimination of pesticide atrazine by photoelectrocatalysis using a photoanode based on WO_3 nanosheets. Chem. Eng. J. **350**, 1114–1124 (2018)
134. Mattei, C., Wortham, H., Quivet, E.: Heterogeneous degradation of pesticides by OH radicals in the atmosphere: influence of humidity and particle type on the kinetics. Sci. Total Environ. **664**, 1084–1094 (2019)
135. Evgenidou, E., Fytianos, K., Poulios, I.: Semiconductor-sensitized photodegradation of dichlorvos in water using TiO_2 and ZnO as catalysts. Appl. Catal. B: Environ. **59**, 81–89 (2005)
136. Van der Werf, H.M.G.: Assessing the impact of pesticides on the environment. Agric. Ecosyst. Environ. **60**, 81–96 (1996)

137. Xue, Y., Wang, P., Wang, C., Ao, Y.: Efficient degradation of atrazine by BiOBr/UiO-66 composite photocatalyst under visible light irradiation: environmental factors, mechanisms and degradation pathways. Chemosphere **203**, 497–505 (2018)
138. Mackuľak, T., Marton, M., Radičová, M., Staňová, A.V., Grabic, R., Bírošová, L., Nagyová, K., Vojs, M., Bodík, I., Brandeburová, P., Gál, M.: Monitoring of micropollutants and resistant bacteria in wastewater and their effective removal by boron doped diamond electrode. Mon. Chem. **148**, 539–548 (2017)
139. Rastogi, T., Mahmoud, W.M.M., Kümmerer, K.: Human and veterinary drugs in the environment. In: Encyclopedia of Anthropology, pp. 263–268. Elsevier (2018)
140. Mackuľak, T., Černanský, S., Fehér, M., Birošová, L., Gál, M.: Pharmaceuticals, drugs and resistant microorganisms – environmental impact on population health. Curr. Opin. Environ. Sci., Heal (2019)
141. Stumpf, M., Ternes, T.A., Wilken, R.-D., Rodrigues, Silvana Vianna, Baumann, W.: Polar drug residues in sewage and natural waters in the state of Rio de Janeiro, Brazil. Sci. Total Environ. **225**, 135–141 (1999)
142. Ankley, G.T., Brooks, B.W., Huggett, D.B., Sumpter, J.P.: Repeating history: pharmaceuticals in the environment. Environ. Sci. Technol. **41**, 8211–8217 (2007)
143. López-Serna, R., Pérez, S., Ginebreda, A., Petrović, M., Barceló, D.: Fully automated determination of 74 pharmaceuticals in environmental and waste waters by online solid phase extraction-liquid chromatography-electrospray-tandem mass spectrometry. Talanta **83**, 410–424 (2010)
144. Al Balushi, A.S.M., Al Marzouqi, F., Al Wahaibi, B., Kuvarega, A.T., Al Kindy, S.M.Z., Kim, Y.: Hydrothermal synthesis of CdS sub-microspheres for photocatalytic degradation of pharmaceuticals. Appl. Surf. Sci. **457**, 559–565 (2018)
145. Khedr, T.M., El-Sheikh, S.M., Ismail, A.A., Bahnemann, D.W.: Highly efficient solar light-assisted TiO_2 nanocrystalline for photodegradation of ibuprofen drug. Opt. Mater. (Amst) **88**, 117–127 (2019)

Nanostructured Metallic Oxides for Water Remediation

R. Natividad, L. Hurtado, R. Romero, T. Torres-Blancas, C. E. Barrera-Díaz, G. Santana-Martinez and G. Roa

Abstract This chapter summarizes the application of some nanostructured metallic oxides for water remediation via three main Advanced Oxidation Processes: photocatalysis, photo-Fenton and catalyzed ozonation. Regarding photocatalysis, this section is mainly focused on the synthesis methods, modification, application and mechanism of action of TiO_2. Metallic foams are presented as promising supports for photochemical processes. Within the photo-Fenton section, special attention was given to the synthesis, characterization and application of pillared clays as promising catalysts of the photo-Fenton process. The ozonation section deals with metallic oxides successfully applied to catalyze this oxidation process. It is presented by the first time, results showing the remarkable performance of Fe-Cu-Ni/Liparite as ozonation catalyst. It can be concluded that, albeit the excellent results attained with iron metallic oxides, Cu and Ni oxides are also promising and their assessment is becoming an important task for the scientific community dedicated to develop efficient technologies for water remediation.

1 TiO_2-Based Photocatalysis for Wastewater Treatment

Photocatalysis is an advanced oxidation process (AOP) based on the utilization of light for the occurrence of chemical reactions and is considered as a promising technology for environment remediation purposes. In this AOP, light is used to activate a catalyst to lower the activation energy required for a primary reaction to occur [1].

R. Natividad (✉) · L. Hurtado · R. Romero · T. Torres-Blancas · C. E. Barrera-Díaz · G. Roa
Centro Conjunto de Investigación en Química Sustentable, UAEM-UNAM, Universidad Autónoma del Estado de México, Km 14.5, 50200 Toluca-Atlacomulco Road, Mexico
e-mail: rnatividadr@uaemex.mx

G. Santana-Martinez
Dirección de Ingeniería Industrial y de Sistemas, Universidad Politécnica del Valle de Toluca, Toluca-Almoloya de Juárez Km 5.6, Santiaguito Tlalcilalcali, C.P. 5090 Almoloya de Juárez, México

© Springer Nature Switzerland AG 2019   91
G. A. B. Gonçalves and P. Marques (eds.), *Nanostructured Materials for Treating Aquatic Pollution*, Engineering Materials,
https://doi.org/10.1007/978-3-030-33745-2_4

According to the state of aggregation of the chemicals and photocatalysts it is possible to carry out homogeneous or heterogeneous photocatalytic processes. Generally, for water remediation purposes, heterogeneous processes have been widely studied with exceptional results and in this chapter, we will focus in nanostructured metallic oxides for heterogeneous photocatalysis.

Heterogeneous photocatalysis has been employed in a wide number of applications including air purification, water purification and disinfection and self-cleaning coatings. At the present, due to the growing interest in chemical processes based in renewable sources, some of the developed technologies have been successfully commercialized. The plausible applications of heterogeneous photocatalysis have motivated the vast literature on the subject. Only the topic heterogeneous photocatalysis retrieves 5082 records in the Web of Science System since 1985 until June 2019 and a high percent of the publications in the field has been dedicated to water remediation.

In general terms, in a heterogeneous photocatalytic process a pollutant might be either oxidized at the surface of the photocatalyst (irradiated with light) or by an indirect pathway (later explained in this work) and transformed to carbon dioxide and water as the final products of the process. Among the organic pollutants mineralized through heterogeneous photocatalysis under both, UV and visible light we can find: phenol-derived compounds [2–6], pesticides [7, 8], pharmaceuticals [9, 10], dyes [11, 12], heavy metals [13, 14] and endocrine disrupters [15, 16]. Even though the referred chemicals are successfully degraded by heterogeneous photocatalysis (it means the complete disappearance of the target molecule), the complete mineralization is not reached in many cases. This fact represents a very important area of opportunity for this technology because of the potential toxicity of the residual non-mineralized products.

The central element of a photocatalytic process is the material used as photocatalyst because it plays a very important role in the generation of charge carriers (electron and hole) and the subsequent occurrence of oxidation reactions. The electronic structure in solid materials involves the existence of *energy bands* originated from the localization of electrons in determined orbitals in atoms. In solid materials the conduction band and the valence band are recognized as the most important energy bands. In semiconductors, the gap between them is known as band gap or forbidden band. The valence band in a semiconductor is normally filled with electrons situated at the most external layers of the atom and these electrons are expected to participate in chemical reactions while the conduction band is referred to the lowest energy band, in which the electrons can move easily due to the existence of empty sites in the band [17]. To accomplish the displacement of an electron from the valence band to the conduction band it is necessary to irradiate the semiconductor with light emitting energy larger than the band gap energy.

Despite the large number of semiconductors, Titanium dioxide (TiO_2) was one of the first semiconductors elected as photocatalyst due to its abundance, chemical stability, high photocatalytic performance under UV light and its relatively low toxicity. Since the 70s and until now titanium dioxide is the most assessed material as photocatalyst. Titanium dioxide is classified as an n-type semiconductor indicated by the presence of oxygen vacancies compensated by the presence of Ti^{3+} centers and exists in three different crystalline forms: anatase, rutile and brookite [18]. Anatase and rutile phases have tetragonal unit cell while brookite is the orthorhombic form of TiO_2. For photocatalytic purposes, the use of brookite is scarce and it has been reported that anatase is the structure which shows the highest photocatalytic activity although rutile is the thermodynamically most stable phase [19]. Among the options of commercially available TiO_2, Evonik Aeroxide (formerly Degussa P25®) has become a standard for photocatalytic reactions. This material is composed by a mixture of anatase and rutile phases in an approx. weight ratio of 80/20; it possesses a specific surface area of approximately 50 m^2 g^{-1} and the mean diameter of the primary particles is about 21 nm [20].

The advances in nanotechnology and material science during the last decades has been useful for the engineering of sophisticated nanostructures rather than nanoparticles which enhances the photocatalytic activity through larger specific surface area, better light absorption and inhibition of charge carriers. Table 1 summarizes representative examples of nanosized TiO_2 synthesized with particle, rod, fiber, tubes, wires, sheets and 3D flower-like morphology. It has been included the synthesis method for all the cases giving to the reader a general overview of preparation of TiO_2 nanostructures through different methodologies.

The morphology of TiO_2 nanostructures is dictated by the synthesis method. While TiO_2 nanoparticles can be prepared following a simple procedure like the sol-gel method, the synthesis of ordered TiO_2 nanotubes racks requires a more complex procedure like electrochemical anodization or the utilization of templates. The shown infographic summarizes the most popular TiO_2 nanostructures synthesis methods given its high tunability and high quality of the final products including the sol-gel method [31], the hydrothermal method [32], the solvothermal method [33, 34], electrochemical anodization [35], the template-assisted method [36], electrospinning [24] and the EISA (Evaporation-Induced Self-Assembly) method [4].

Table 1 Representative TiO_2-based nanostructures in heterogeneous photocatalysis

TiO_2-based nanostructure	TiO_2 crystalline phase	Characteristic dimension	Synthesis method	References
TiO_2 nanoparticles	Anatase/brookite	Grain size <20 nm	Sol-gel	[21]
TiO_2 nanoparticles	Anatase	Grain size <20 nm	Evaporation-Induced Self-Assembly	[4]
TiO_2 nanowires	Anatase/rutile	Length: 5 μm	One-step PEG assisted hydrothermal	[22]
	Rutile	Length ~44 μm	Solvothermal	[23]
TiO_2 nanofibers	Anatase/rutile	Diameter: 30–210 nm	Electrospinning	[24]
TiO_2 nanotubes	Anatase/rutile	Diameter: 26–106 nm	Electrochemical anodization	[25]
MoS_2 quantum dots@TiO_2 nanotubes	Anatase	Length: 3 μm Diameter: 70 nm	Electrochemical anodization	[26]
TiO_2 nanosheets	Anatase	Thickness ~5 nm	C_3N_4 templating	[27]
Carbon dots-TiO_2 nanosheets	Anatase	Thickness: 5 nm	Hydrothermal	[28]
TiO_2 nanorods	Anatase	Length: 76 nm	Hydrothermal	[29]
	Rutile	Length: 29.48 nm		
	Brookite	Length: 98.98 nm		
3D flowerlike $α–Fe_2O_3$@TiO_2	Anatase	Cubic morphology Side length: 400 nm	Hydrothermal	[30]

TiO$_2$ Nanostructures
Synthesis methods

Sol-gel

- Offers simplicity and high purity of the products.
- Stages: hydrolysis, condensation, aging and annealing.

Electrospinning

Innovative, highly tunable and scalable process for the synthesis of long nanostructures

Hydrothermal

- Utilizes aqueous solutions at high-temperature and high water vapor pressure
- Steps: crystal growth, crystal transformation, phase equilibrium, and final ultrafine crystals formation

Solvothermal

- Similar to hydrothermal method, but here organic solvents are employed.
- Extended possibilities of tuning synthesis conditions for a better control of the prepared nanostructures

Electrochemical anodization

Usually based on the anodization of a Ti plate in a fluoride-based electrolyte followed by the annealing of the plate in presence of oxygen.

Templated-assisted

- Method highly recommended for the synthesis of well-defined nanostructures
- Encompasses a sacrificial template easily removed by combustion or chemical etching

EISA

- Spontaneous association of individual elements into an organised pattern induced by the evaporation of a solvent.
- Synthesis of mesoporous structures with high surface areas.

The success of TiO$_2$ nanometric materials in photocatalysis is given by the advantage of increasing the surface available to participate in the photo-assisted process in addition to the reduction of the intra-particle diffusive pathway to conduct the chemical reaction. Also, in a photocatalytic process once electron and holes are formed, they need to be transferred to the electron and hole acceptors, respectively. The efficiency of this step is dependent of the charge-transfer rate at the interface, the charge recombination rate within the particle and of the transit time of the charges carriers to the surface. The charge transfer time is closely related to size particle, for example, Warman et al. [37], reported that electron transfer time increases from 19 to 65 ps when particle size increases from 5 to 50 nm. In this sense, TiO_2 nanorods and nanowires are composed by nanocrystals where intraparticle charge carrier occurs promoting the separation of photogenerated charge carriers. Furthermore, the TiO_2 two-dimensional nanostructures (nanosheets) offers specific exposed facets and a large fraction of unsaturated atoms on the surface. Decoration of nanosheets with metallic nanoparticles is another common way for enhancing the photocatalytic process. Regarding to TiO_2 nanotubes, the preparation of highly ordered racks delivers an efficient illumination through the channels. TiO_2 hierarchical nanostructures have raised as an interesting option since they offer a large surface-to-volume ratio and because of the delocalization of electrons in the 3D nanostructures, the separation of charge carriers is enhanced.

Within the context of oxidation of hazardous organic contaminants in water, two mechanisms are widely recognized: the direct oxidation mechanism, where the totality of the phenomena occurs on the surface of the catalyst; and the indirect oxidation mechanism via the oxidant species produced on the surface of the catalyst and eventually transported to the bulk solution. In a reaction system both mechanisms might be present but only one of them is dominant. In general terms, the most accepted and reported mechanism of heterogeneous photocatalysis based on TiO_2 involves the following primary steps [38],

i.   Formation of charge carriers in the catalytic particle by photon absorption.
ii. Charge carriers recombination in catalytic particle.
iii. Trapping of a conduction band electron at Ti (IV) to yield Ti (III).
iv. Trapping of a valence-band hole at a surficial titanol group.
v.   Initiation of an oxidative pathway by a conduction-band electron.
vi. Further thermal and photocatalytic reactions to yield mineralization products.

These steps are represented by the following reactions [39–41],

$$TiO_2 + h\upsilon \xrightarrow{\lambda < 385nm} e_{cb}^- + h_{vb}^+ \tag{1}$$

$$e_{cb}^- + h_{vb}^+ \rightarrow heat \tag{2}$$

$$h_{vb}^+ + H_2O \rightarrow \ ^{\cdot}OH + H^+ \tag{3}$$

$$h_{vb}^+ + OH^- \rightarrow \ ^{\cdot}OH \tag{4}$$

$$e_{cb}^- + O_2 \rightarrow O_2^- \tag{5}$$

$$O_2^{\cdot-} + O_2^{\cdot-} + 2H^+ \rightarrow H_2O_2 + O_2 \tag{6}$$

$$O_2^{\cdot-} + H^+ \rightarrow HO_2^{\cdot} \tag{7}$$

$$HO_2^{\cdot} + H^+ + e_{cb}^- \rightarrow H_2O_2 \tag{8}$$

$$H_2O_2 + h\upsilon \rightarrow 2\,^{\cdot}OH \tag{9}$$

$$H_2O_2 + e_{cb}^- \rightarrow {}^{\cdot}OH + OH^- \tag{10}$$

Potential organic compound (OC) degradation pathways:

$$OC + e_{cb}^- \rightarrow products \tag{11}$$

$$OC + h_{vb}^+ \rightarrow products \tag{12}$$

$$OC + {}^{\cdot}OH \rightarrow products \tag{13}$$

$$OC + HO_2^{\cdot}/O_2^- \rightarrow products \tag{14}$$

$$OC + H_2O_2 \quad \rightarrow \quad products \tag{15}$$

The direct reduction and oxidation of an organic compound (OC) proceeds via reactions 11 and 12, respectively. By the other side, the indirect oxidation is expected to occur via reactions 13–15 with the oxidant species produced during reactions 3–10. Thereby the resulting kinetic equation for the oxidation of an organic compound is the result of combining the reaction rates of Eqs. 1–15. Furthermore, it is important to recall that the products of reactions 11–15 can also undergo oxidation in a similar way than the parent compound. The indirect process can occur either near the surface or in the bulk solution. In both cases, an effective mass transport in the homogeneous phase and near the catalyst must be ensured.

According to the aforementioned reactions, the photocatalytic efficiency is usually limited by the fast recombination of the electron-holes pairs. Traditionally, it has been assumed that TiO_2 electron-hole pairs recombination is given by a non-irradiative pathway which leads to the release of heat (R2). In their investigations, Leytnar et al. [42], report that approximately 60% of all trapped electron-hole pairs recombine on the time scale of about 25 ns releasing 154 kJ mol^{-1} of energy as heat. It has been stated that charge recombination can be affected by factors such as sample preparation, reaction temperature, charge carriers trapping, interfacial charge transfer and light intensity and wavelength.

TiO$_2$ heterogeneous photocatalysis has proved to be an efficient alternative for the degradation of organic contaminants in water; however, this technology is limited to the utilization of UV light due to large band gap of anatase (3.2 eV) and rutile (3.0 eV) phases. The low conversion efficiency of nanosized TiO$_2$ under visible light reduces the attractiveness of a photocatalytic processes for large scale applications and this fact has been the main motivation for the continuous study of novel photocatalysts as well as the modification of pristine TiO$_2$ properties to provide superior photocatalytic performance and light response in the visible region. Extensive research has been conducted for the development of photocatalysts based on TiO$_2$ able to take advantage of visible light and the he following approaches have shown promising results in terms of visible light absorption and efficient electron-hole pairs recombination.

Cation-doped TiO$_2$. The doping of TiO$_2$ with metallic elements such as rare earth metals, noble metals and transition metals has been reported in numerous publications. It has been established that such groups of materials promote in a photocatalytic process an extended light absorption interval, the increase of the redox potential of the photogenerated radicals and the enhancement of the quantum efficiency by inhibiting the recombination of electron/holes pairs [43, 44]. As an example let us examine the case of TiO$_2$ mono and co-doped with 1 wt% W and 1 wt% Mo, synthesized by the EISA method [4]. Such works present a complete structural, morphological and textural characterization and comparison of the mono and co-doped material with bare TiO$_2$ synthesized by EISA method and the typical Degussa P25. Table 2 summarizes the main established differences and the results obtained in the oxidation of 4-chlorophenol. It is worth mentioning that anatase was obtained at all cases.

Figure 1 presents representative STEM images of the pristine and doped TiO$_2$ nanoparticles. The superior photocatalytic performance of this material was

Table 2 Effect of cation-doping TiO$_2$ (data abstracted from [45] and [46])

Material/catalytic parameter	Material				
	Degussa P25	TiO$_2$ (EISA)	W(1%)–TiO$_2$	Mo(1%)–TiO$_2$	W(1%)–Mo(1%)/TiO$_2$
Crystal size (nm)	–	8.6	7.4	8.1	8.3
Specific surface area (m^2 g^{-1})	50	144	151	151	172
Average Pore diameter (nm)	–	6.1	6.1	6.1	6.6
Band Gap (eV)	3.2	3.16	3.1	3.1	2.87
$-r_{4CP_0} \times 10^5$ (mol g^{-1} min^{-1})	0.69	1.44	2.8	2.8	2.14
4-CP removal (%)	50	68	89	94	97
TOC removal (%)	30	40	70	65	74

$-r_{4CP,o}$ = initial oxidation rate of 4-Chlorophenol

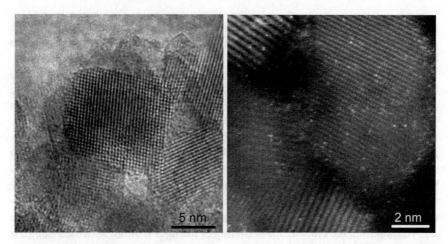

Fig. 1 STEM images of TiO_2 nanoparticles (left) and TiO_2 doped with W and Mo cations (right) obtained through the EISA (Evaporation-Induced Self-Assembly) method

attributed to the synergistic effect of both dopant cations reducing the recombination of photogenerated charges and by enhancing the harvesting of light to promote the generation of electron-hole pairs.

Anion-doped TiO_2. The incorporation of anionic non-metals to the TiO_2 structure such as carbon, nitrogen, sulfur, fluorine and iodine has also been studied. The higher photocatalytic activity under visible light has been found to be ensured by doping of TiO_2 with carbon or nitrogen anions [47]. The main effect of C-doped TiO_2 is the increase in surface area and the better absorption of visible light. Elsewhere, N-doped TiO_2 results in modified refraction index, hardness, electrical conductivity and the absorption of visible light [48].

Coupled semiconductors based on TiO_2. Zhang and coworkers have reported that the good matching of the conduction and valence bands of two semiconductors could ensure an efficient transfer of the charge carriers between semiconductors [44]. Among the numerous reports in the literature, the coupling of TiO_2 with CdS, SnO_2, WO_3, FeO_3 and Bi_2S_3, have helped to enhance the photocatalytic performance of TiO_2 under visible irradiation [49].

Semiconductor heterojunctions. The purpose of the synthesis of semiconductor heterostructures (or also called heterojunctions) from two or even more nanocomposites or chemical elements with potential photocatalytic properties is to originate an electric field able to push the carriers (electron and holes) to opposite sides of the heterojunction resulting in long lifetime of the charge carriers. Normally, the synthesis of the heterojunction favors the reduction of the band gap energy [50]. Examples of heterojunctions based on TiO_2 with remarkable activity in solar applications are the g-C_3N_4/TiO_2 heterojunction [51], the TiO_2–In_2O_3 nanocrystals heterojunction [52] and the $CH_3NH_3PbI_3$/TiO_2 heterojunction [53].

Photocatalyst properties of nanosized materials (chemical composition, morphology, crystalline structure and textural and optical properties) exert a strong influence on the pollutant degradation processes but additionally, a good photoreactor design and operation of the reactor under kinetic regime, are essential aiming to maximize the photocatalytic activity of a nanosized material. At the present, besides the maturity reached in lab conditions regarding novel photocatalysis synthesis and innovative reactor designs for the effective mineralization of pollutants, important efforts are ongoing for the implementation of heterogeneous photocatalysis to real effluents owing to the complexity of a heterogeneous photocatalytic process at large scale. For the treatment of real wastewater effluents, heterogeneous photocatalysis normally is employed as a secondary step given the high concentrations of pollutants and impurities in those sewages. During the initial pre-treatment, commonly the turbidity of the samples is reduced and in consequence the photo-generation of charge carriers is more effective in comparison to the original sample because the light irradiated towards the reactor may have access to any point and, without solids of large size coming from the effluent, the scattering of the light is diminished increasing the probability to be absorbed by the photocatalyst.

It is possible to find in the literature numerous publications regarding the photocatalytic degradation of pollutants in synthetic aqueous solutions, but our discussion will be focused on real effluents treatment because of the promising results reported until date contributing to the development of efficient solutions for purification of water bodies. To do so, there are summarized in Table 3 representative TiO_2-based photocatalytic processes for the remediation of effluent coming from industries like petrochemical, pulp and paper, textile and tannery from 2015 onwards.

Although desirable, a straightforward comparison of the photocatalytic performance of the listed experiments may not be possible due to the differences in process variables like reaction volume, initial COD concentration, catalyst loading and light intensity. Nevertheless, with the data provided in each reference, we estimated the removal rate of Chemical Oxygen Demand (COD) achieved by each process as the parameter to compare its efficiency on a normalized basis (column 7, Table 3). It is worth noticing, and is not a coincidence, that the processes with the highest removal rate are those that were conducted with TiO_2 and the presence of Fe (rows 1 and 6, Table 3). Given the inherent complex composition of a real effluent, it is rather difficult the individual study of each hazardous compound and COD is widely accepted as parameter of water quality. COD concentration represents the requirements of oxygen for the complete oxidation of organic compounds to CO_2 and water. According to the EPA, the maximum permissible limit of COD in wastewaters is 120 mg L^{-1} [54]. From Table 2, most of the wastewater effluents exceeds considerably this value before being treated and through photocatalytic processes, it was possible to reduce up to 84% of the initial COD. Furthermore, among the summarized processes it was found that the highest COD removal rate achieved was 2548 (mg L^{-1}) g_{cat}^{-1} h^{-1} corresponding to the photocatalytic degradation of the Kraft pulp mill effluent where commercial nanosized TiO_2 was exposed to UV light [55]. Although this outstanding COD removal rate was achieved, it should be considered the small volume treated (50 mL) which contrast with a COD removal rate lower than 50 (mg L^{-1}) g_{cat}^{-1} h^{-1} when 5 L of effluent were treated [56, 57].

Table 3 TiO_2-based heterogeneous photocatalytic processes in the treatment of wastewater (real effluents)

Treated wastewater effluent	COD concentration (mg L^{-1})	Catalyst/catalyst concentration	Particle size (nm)	Reaction volume (L)	% COD removal	COD removal rate (mg L^{-1}) g_{cat}^{-1} h^{-1}	References
Petroleum refinery	602	TiO_2/Fe-ZSM-5 $3 g L^{-1}$	–	0.5	80	193	[58]
Petroleum refinery[a]	240	Commercial TiO_2 $1.2 g L^{-1}$	25	5	41	8	[56]
Tannery industry[a b]	1428	TiO_2 (anatase) $1 g L^{-1}$	75	5	83	47	[57]
Olive oil mill	7	Commercial TiO_2 $2 g L^{-1}$	<25	0.05	5	1.75	[59]
Kraft pulp mill[a]	391	Commercial TiO_2 $1 g L^{-1}$	<25	0.05	65	2548	[55]
Industrial pulp and paper mill[a]	648	Fe_2O_3–TiO_2 $1.5 g L^{-1}$	22.99	0.05	62	1814	[60]
Textile industry	34.2	TiO_2-ZY60 $0.6 g L^{-1}$	6	0.75	84	6.45	[61]

[a]Pre-treated effluent
[b]Conducted under sunlight illumination

The utilization of zeolites like ZY-60 (CBV 760) and ZM5 was found to be a feasible option to support TiO_2 looking for the high dispersion of the active sites for an enhanced performance of the catalyst. Elsewhere, in most of the studies highly energetic light of wavelength in the UV region was used to excite the photocatalyst and for this reason the COD removal rate achieved by Goutam and coworkers [57] in the treatment of kraft pulp mill effluent is remarkable. In such study, natural sunlight is utilized thus demonstrating that the remediation of wastewater utilizing renewable sources of energy is plausible. Interestingly, despite the advances in the engineering of a wide variety of TiO_2 nanostructures as presented before, the utilization of TiO_2 nanoparticles of variable size is the most extended morphology used in real wastewater treatment applications.

In a similar way to lab approaches, TiO_2 is by far the preferred photocatalyst for the treatment of real effluents and the incorporation of Iron species is commonly used since it offers an extended light absorption interval of the catalyst and inhibition of the recombination of electron/holes pairs in addition to its accessible cost in comparison to other metals. The enhanced photocatalytic performance of iron-doped TiO_2 regarding to pristine TiO_2 has been discussed in several reports and is generally stated that the incorporation of Fe in the TiO_2 lattice is expected to diminish the band gap energy because of the introduction of two additional energy levels located between the valence band and the conduction band. According to Crisan and coworkers [62], when light irradiates the surface of Fe^{3+}-doped TiO_2, electrons and holes are generated in the valence and conduction bands, respectively (R16). Fe^{3+} centres can serve as electron and hole traps leading to the formation of Fe^{2+} and Fe^{4+} species according to reactions 17 and 18. In the intermediate energy levels created by the incorporation of Fe in the TiO_2 lattice, occurs the oxidation process on the Fe^{3+}/Fe^{4+} level above the valence band while a reduction process takes place on the Fe^{3+}/Fe^{2+} level located below the conduction band. Elsewhere, Fe^{2+} ions are oxidized to Fe^{3+} (19) through the transfer of electrons to adsorbed molecular oxygen to form highly reactive superoxide species (O_2^-). In a similar way, Fe^{4+} ions are reduced to Fe^{3+} (20) by accepting electrons leading to the generation of OH. radicals. Either, hydroxyl radicals (OH.) and superoxide anions (O_2^-) promote the oxidation of the organic pollutant molecules through reactions 13 and 14, respectively.

$$TiO_2 + h\upsilon \rightarrow e_{CB}^- + h_{VB}^+ \tag{16}$$

$$Fe^{3+} + h_{VB}^+ \rightarrow Fe^{4+}(holetrap) \tag{17}$$

$$Fe^{3+} + e_{CB}^- \rightarrow Fe^{2+} \quad (electrontrap) \tag{18}$$

$$Fe^{2+} + O_2 \rightarrow Fe^{3+} + O_2^{\cdot-} \tag{19}$$

$$Fe^{4+} + OH^- \rightarrow Fe^{3+} + OH^{\cdot} \tag{20}$$

The works summarized in Table 3 are proof of the feasibility of the contribution of heterogeneous photocatalysis to the remediation of effluents coming from activities such as the textile industry in Ethiopia [61], the petroleum refinery in Iran [58] or the olive oil extraction in the Mediterranean region [59] and undoubtedly in the upcoming years, the research in this field will contribute to the solution of the limited access to clean water around the world taking advantage of the sunlight.

The implementation of photocatalytic processes at large scale is at the present, in its early stages. One of the challenges in the area, is to extrapolate the remarkable photocatalytic performance of TiO_2-based photocatalysis achieved in lab conditions to at least, pilot scale. Elsewhere, the utilization of catalysts slurries is impractical since it is necessary to separate the catalyst particles at the end of the process firstly, to reutilize the catalyst for numerous cycles and secondly, because of the possible toxicity of TiO_2-based nanomaterials to aquatic organisms. Several technological solutions have been proposed and nowadays photocatalytic films are the most accepted supported TiO_2 form leading to an efficient light absorption with easy handling of the catalyst in the reactor. Furthermore, commercially available porous materials like foams monoliths are another interesting solution for the deposit of a photocatalyst aiming

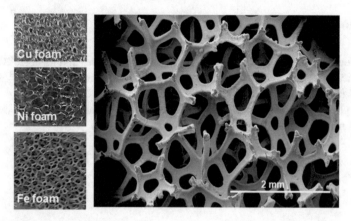

Fig. 2 Metallic foams commercially available manufactured in different materials and SEM image of the interconnected 3D structure

for an efficient light utilization. A foam monolith is a reticulated structure (usually metallic or ceramic) which offers a high ratio of active surface area to reactor volume. Representative images of metallic foams monoliths are presented below in Fig. 2. During the last years, foams have been receiving increased attention as prospective material for supporting photocatalysts since its high porosity may improve light penetration through the reactor as demonstrated by Ochuma and coworkers [63]. In this work, alumina foam monoliths were coated with TiO_2 and installed in the reactor for processing 15 L of an aqueous solution of 1, 8-diazabicyclo [5.4.0] undec-7-ene. It was found that the photonic efficiency of the process employing a TiO_2 slurry was 2-fold improved when the catalyst was supported on foam monoliths. The promising results in water purification attained by means of coated foams (achieved in lab and pilot scale conditions) are highly motivating for increasing the research in this area in the upcoming years.

Despite the countless options to be explored for improving the photocatalytic performance aiming for successful implementation to large scale, the design of hybrid processes based on advanced oxidation processes deserves attention since none of them (photocatalysis, ozonation, photo-Fenton, electro-oxidation, etc.) demands extreme operational conditions (i.e.: high temperature and/or pressure) and they have been proved to be an effective alternative for the abatement of organic pollutants in both, residential and industrial effluents during the last years. Thus, in the following sections, the use of nanostructured metallic oxides in other advanced oxidation processes like photo-Fenton and ozonation will be introduced.

2 Photo-Fenton

Photo-fenton is a very important advanced oxidation process and its objective is the generation of hydroxyl radicals, based on the combined action of H_2O_2, Fe^{2+} or Fe^{3+} and UV radiation.

The photoreduction of ferric ions to ferrous ions and the formation of hydroxyl radicals may be expressed by the following reactions,

$$Fe^{3+} + H_2O + hv \rightarrow Fe^{2+} + OH^{\cdot} + H^{+} \tag{21}$$

$$H_2O_2 + hv \rightarrow 2OH^{\cdot} \tag{22}$$

$$Fe^{2+} + H_2O_2 \rightarrow Fe^{3+} + OH^{\cdot} + OH^{-} \tag{23}$$

The photo-fenton can be either homogeneous or heterogeneous. Although, the former exhibits higher pollutant oxidation rates, the latter presents the following advantages,

- Relatively easy separation of the catalyst.
- Catalyst can be reused.
- Working pH can be other than acid.
- Sludge is not formed during the process.
- Fe^{3+} species cannot easily transform into less photoactive species such as $Fe(OH)_3$.

Therefore, there has been a growing interest on the synthesis of solids that can be used as a heterogenous source of iron or copper to efficiently conduct the heterogeneous photo-fenton process. The challenge is to obtain stable catalysts that do not present metal leaching. That is, Fe ions leached out the catalyst during the reaction, which causes loss of catalytic activity and consequent contamination.

By using simple techniques, iron species or iron oxides have been immobilized on various solid supports as hematite, magnetite, active carbon, silica, zeolites, resins, clays, organic supports, etc. [64, 65]. In this sense, clays are especially attractive because of its original iron content. Nevertheless, the initial iron content in natural clays is low and for this reason, several methods have been used to increase the iron content, the process of pillaring being one of the most promising methods. Some advantages of these materials are the low cost of natural clays, its easy process of preparation, a high specific surface area and environmental safety [66]. In the following section clays, their pillaring process and applications will be discussed in more detail.

2.1 Pillared Clays: Synthesis and Characterization

Mineral clays of the smectite family are generally used to prepare pillared clays due to their low density and swelling capacity. Each layer of such clay minerals consists of two tetrahedral (T) Si-O sheets sandwiching a central octahedral (O) Al-O/Al-OH sheet (TOT type) [66]. A small quantity of the tetrahedral Si atoms is isomorphically substituted by Al and a fraction of the octahedral atoms (Al, Mg or Fe) is substituted by atoms of lower oxidation number such as Li^+. An octahedral layer between two tetrahedrals, strongly linked by covalent bonds, forms the basic structural unit of smectites.

Pillared clays (PILCs) are obtained by ion exchange of the interlayer cations of smectite clays by inorganic hydrated polyoxocations as pillaring agent providing thermal stability to pillared clays and giving place to increased interlayer spacing, specific surface area (200–500 m^2/g) and pore volume [67, 68]. Different oxocations containing Al, Zr, Ti, La, Ga, Fe, Mn, Ta, Cr and Ce have been used to prepare pillared clays [65, 69, 70]. These polyoxocations introduced into the interlaminar region are then calcined and transformed into the corresponding metal oxides that fix the clay sheets (pillars) preventing them collapse. A stable structure is obtained by careful dehydration, converting the hydroxide pillars to stable oxides into the clay layer. A schematic representation of the process is given in Fig. 3.

The aforementioned is due to swelling properties of mineral clays showing an enlargeable cage structure that allows ion exchange with voluminous cations, causing up to a 5-fold increase in basal spacing (d_{001}) of the clay mineral [31].

The pore size of a pillared clay is determined by the interlayer distance (height) and density of the pillars within the layer (width). The interlayer distance of a PILC depends upon the size of the pillaring agent, while the lateral distance is regulated by the charge density of the layers and effective charge on the pillars [71]. According to the calculations made by Clearfield and Kuchenmeister [72], the distance between pillars must be 17–29 Å. When subtracting the width of the pillar that is approximately 7 Å, the free distance would be 10–22 Å.

Tzou and Pinnavaia [36] reported for the first time the preparation of Fe pillared clays (Fe-PILCs). An interesting aspect of Fe-PILCs clays is its magnetic characteristic given by iron oxide pillars [73] that facilitates their recovery even more.

Generally speaking, the method of preparation of pillared clays consists of the use of a suspension containing mineral clay (bentonite or montmorillonite) that is mixed with a solution containing a polyoxocation. A reproducible method is that reported by Valverde [74]: Fe-PILC is prepared using the necessary amount of 0.2 M $FeCl_3 \cdot 6H_2O$ and 0.2 M NaOH solutions to obtain the required OH/Fe molar ratio of 2.0 (this value can be modified depending on the required final Fe content). The mixture is aged for 4 h under stirring at room temperature. After that, the pillaring solution is dropwise added to 0.1%wt aqueous bentonite suspension. In order to avoid precipitation of iron species, the pH is kept constant at 1.78–1.8 (this value is crucial) adjusted by HCl 5 M. The mixture is kept under vigorous stirring for 12 h at room temperature. The next

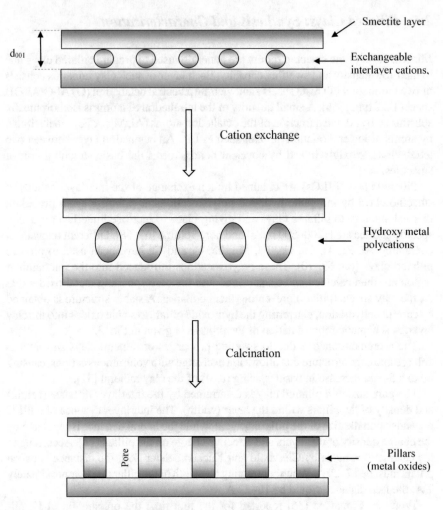

Fig. 3 Pillaring process of mineral clays

step is the recovery of the prepared powder by centrifugation for further chloride ions elimination by washing with deionized water until conductivity <10 μS/cm. Finally, the solid is dried overnight at 70 °C and calcined for 2 h at 400 °C. The surface area after calcination at 400 °C is 283 m²/g. With this procedure the final amount of iron is 16.72% wt and this can be determined by atomic absorption. The pillaring process can be verified by XRD analysis [75–77]. In order to establish the iron species in the Fe-PILC, X-ray photoelectronic spectroscopy (XPS) analysis can be performed. In a solid prepared as above mentioned, by XPS three peaks can identified: at 710.4 eV corresponding to Fe_3O_4 (magnetite, Fe^{2+}/Fe^{3+}), at 711 eV the characteristic signal of Fe_2O_3 (Fe^{3+}) and at 709.5 eV that corresponds to FeO (Fe^{2+}), according to the National Institute of Standards and Technology (NIST) [36,

38]. A recent publication, Minella et al. [78] reported magnetite as a material with high stability under a wide interval of pH. In addition, the Fe^{2+}/Fe^{3+} ratio available was found to favor the production of hydroxyl radicals.

2.2 Use of Fe-PILC on Water Purification

Sum et al. [79] used a modified laponite clay-based Fe nanocomposites for photo-Fenton degradation of acid black 1. In this report, the degradation experiments were conducted at pH close to neutral observing the Fe leaching is negligible in the treated solution after 120 min reaction.

Feng et al. [80] synthesized a novel bentonite clay-based Fe-nanocomposite as a heterogeneous catalyst for photo-Fenton discoloration and mineralization of an azo-dye Orange II. The result indicates that the nanocomposite catalyst exhibits a high catalytic activity not only in the photo-Fenton discoloration of Orange II but also in the mineralization of Orange II. The reactions started by the photoreduction of Fe^{3+} on the surface of pillared clay to Fe^{2+}. Then the Fe^{2+} formed accelerates the decomposition of H_2O_2 in solution, generating highly oxidative OH radicals while the Fe^{2+} is oxidized by H_2O_2 into Fe^{3+}. Thus, the OH radicals attack Orange II adsorbed on the surface of nanocomposite, resulting in reaction intermediates that are colorless.

De León et al. [65], synthesized Fe-PILCs using two different-sized particle fractions of montmorillonite (<250 μm and within the range of 250–450 μm) and were tested for catalytic performance in photo-Fenton discoloration of methylene blue aqueous solutions. To study the influence of pH, they conducted experiments in a range of 3.0–6.0 and observed that their influence may be attenuated in the heterogeneous system since immobilized Fe(III) species could not be transformed into less photoactive materials: $Fe^{3+}(H_2O)_6$ at lower pH and $Fe(OH)_3$ at higher pH. Thus, the stabilization of iron oxide pillars may also be attributable to interactions with the clay sheet. From the results of the photo Fenton process, they were found that a smaller-sized particle fraction of the mineral improved the catalytic activity in methylene blue discoloration.

Martín el Campo et al. [75] examined an iron pillared inter-layered clay as catalyst of the photo-Fenton process to oxidize phenolic compounds, achieving 94% elimination of 4-chlorophenol without adjusting the pH at any time during the experiment.

Bel Hadjltaief et al. [81] prepared Fe pillared clays and used as the catalyst in the oxidation of Red Congo and Malachite Green in aqueous solution. With pH in the range 2.5–3.0 and an amount of catalyst of 0.3 g/L were the best conditions for achieving complete discoloration, either for Red Congo or Malachite Green although oxidation for Red Congo was slower. Also, the catalyst presented a low leaching of iron ions.

Hurtado et al. [77] used Fe-PILCs ion-exchanged with copper in the mineralization of paracetamol through photo-Fenton process at near to neutral pH. The reaction did not show precipitation of Fe complexes. This catalyst allowed to conduct the mineralization of paracetamol at least twice faster than applying only UV light and H_2O_2.

From all the aforementioned studies, it can be concluded that the most important variables affecting the efficiency, i.e. mineralization extent and/or product distribution, of a photo-fenton process in the context of water purification are,

- Mass of catalyst.
- Fe content.
- Radiation wavelength.
- pH.

It can also be concluded that the use of Fe-PILC is practically unlimited since catalyzes the H_2O_2 dissociation. It could only be deactivated by some organic pollutants chemisorbing onto the catalytic surface then limiting the accessibility of the active sites for H_2O_2.

For long time, the use of UV lamps for water remediation was questioned because of energy consumption. Nowadays, however, solar panels to provide electricity are a reality and can be used to power UV lamps. Nevertheless, homogeneous solar photo-fenton has already been proven to be effective and the solar application of heterogeneous catalysts is an ongoing research all around the globe.

3 Ozonation

Ozone (O_3) is a weak polar (0.53 D) and highly reactive molecule that exhibits a high redox potential (2.07 V vs. NHE) only below atomic oxygen (2.42 V vs. NHE), hydroxyl radical (2.80 V vs. NHE) and fluorine (3.06 V vs. NHE). Its high redox potential allows O_3 to effectively participate in reduction oxidation reactions. Furthermore, because of its electronic configuration, ozone can also participate in dipolar cycloaddition, nucleophilic and electrophilic substitution reactions [82].

Regarding redox reactions, the role of O_3 is not only the transference of electrons as in any typical reaction as follows [83],

$$O_3 + HO_2^- \rightarrow O_3^- \cdot + HO_2^\cdot \tag{1}$$

but also and mainly O_3 transfers oxygen and this is the way the other compound is oxidized (direct oxidation) and mainly occur under acidic conditions. Indirect oxidation, however, proceeds either under alkaline conditions or under the presence of initiators like hydrogen peroxide and under such condition ozonation could be recognized as an Advanced Oxidation Process (AOP) since indirect oxidation is based on the production of hydroxyl radicals rather than on an direct attack by ozone molecule. Nevertheless, ozonation has been proven to not being able to mineralize humic substances nor oxalic acid [84]. This fact along with the imminent necessity of lowering costs has motivated the search of alternatives to intensify the ozonation process. As an attempt to achieve so, ozonation has been coupled to advanced oxidation processes like photocatalysis, fenton, photo-fenton or has been intensified by the addition of initiators like hydrogen peroxide or catalysts or the use of a radiation source. In the context of catalyzed ozonation, this can be homogeneous or heterogeneous. As expected, one of the main disadvantages of the homogeneous catalyzed ozonation is the recovery and reuse of the catalyst. This has motivated the search and assessment of heterogeneous catalysts. An ozonation process can be said that has been heterogeneously catalyzed when one of the following reactions steps is demonstrated [84].

- Chemisorption of ozone onto the catalytic surface.
- Chemisorption of organic molecule onto the catalytic surface.
- Chemisorption of both, ozone and organic molecules onto the catalyst.

To demonstrate any of the aforementioned phenomena one can apply mechanisms based on such steps to obtain Eley-Rideal or Langmuir-Hinshelwood mechanistic models. Similar to the homogeneous catalysis, when only ozone is adsorbed then this decomposes thus enhancing the production of hydroxyl radicals via reaction 24,

$$M^{(n-1)+} + O_3 + H^+ \rightarrow M^{n+} + HO^{\cdot} + O_2 \tag{24}$$

In reaction 24, M represents the transition metal and n the oxidation state. According to this reaction, the metal ions catalyzes ozone decomposition into hydroxyl radicals under acidic conditions. This highlights once again the importance of pH during the ozonation process. However, if the organic molecule is the one adsorbed then the resulting complex is more viable to be oxidized by direct ozonation than the original precursor albeit the concomitant production of hydroxyl radicals. An analysis of some results from literature leads to conclude that is the pH, the catalyst and the substrate nature that dictate which of above mentioned reaction steps is the predominant one during a catalyzed ozonation treatment. It is also worth pointing out that a catalytic effect can be established as such if the effect of catalyst addition is higher than the observed effect when combining adsorption and ozonation without a catalyst at the same pH [85]. Therefore, in order to establish such a catalytic effect all experimentation aiming to assess the activity of a catalyst in an ozonation process should always include adsorption and not catalyzed ozonation experiments. Ozonation products adsorption tests should also be conducted [85].

Unlike industrial wastewater catalyzed ozonation, the existing literature regarding the synthetic effluents catalyzed ozonation is vast and the tested catalysts can be classified within metal oxides, supported metals, minerals and activated carbon [86]. The main identified catalysts and/or supports are TiO_2 [87], Al_2O_3 [87, 88], Fe_2O_3 [89–92], MnO_2 [93] [94] [95] [96] [97] [98], NiO [99–103], CuO [103–105], CeO_2 [106, 107], pillared clays [108], Fe/pimenta dioica [109] and supported oxides on activated carbon [110, 111]. Actually, a comprehensive review on the catalyzed ozonation by metal oxides of synthetic solutions is presented in [85, 112]. From this kind of catalysts only a reduced group of them has been applied to wastewater treatment, this is mainly due to economical reasons.

Table 4 Ozonation catalysts of synthetic solutions

Catalyst	Organic compound	References
MnO_2, TiO_2, CoO	Oxalic acid	[113]
TiO_2/Al_2O_3	Dimethyl phthalate	[114]
α- Al_2O_3	2,4-dimethylphenol	[115]
RuO_2/Al_2O_3	Dimethyl phthalate	[116]
Fe_2O_3/Al_2O_3@SBA-15	Ibuprofen	[89]
Fe_3O_4/Al_2O_3, Fe_3O_4/SnO_2	2,4-dichlorophenoxyacetic acid and para-chlorobenzoic acid	[92]
MnO_2/Graphene	Gaseous toluene	[95]
CuO/SBA-15	Mesoxalic and oxalic acids	[103]
CuO/γ-Al_2O_3	Azo dyes	[105]
NiO	2,4-Dichlorophenoxyacetic acid	[99]
β-FeOOH/NiO	4-Chlorophenol	[100]
Fe/pimenta dioica, Fe/Pillared clays	Indigo carmine	[108, 109]

Some metallic oxides employed for the ozonation of synthetic solutions are summarized in Table 4 which is based on the last 9 years publications.

When using a supported metal as catalyst of the ozonation process, two mechanisms might be distinguished [117]. In the first one, for instance, an organic molecule that is not readily oxidized by ozone alone, i.e. oxalic acid, forms a complex with the catalyst (i.e. oxalate, reaction 25) and then this complex becomes more plausible to be directly oxidized by ozone or by a hydroxyl radical than the parent compound (reaction 26).

Mechanism 1

$$Me - OH + AH \rightarrow Me - A + H_2O \tag{25}$$

$$Me - A + O_3 \text{ or } HO^. \rightarrow P' + Me - P \tag{26}$$

$$Me - P + O_3 \text{ or } HO^. \rightarrow R' + Me - R \tag{27}$$

$$Me - R + H_2O \rightarrow R + Me - OH \tag{28}$$

Mechanism 2

$$Me_{red} - OH + O_3 + H^+ \rightarrow Me_{ox} - OH + HO_3^. \quad HO_3^. \rightarrow HO^. + O_2 \tag{29}$$

$$Me_{ox} - OH + AH \rightarrow Me_{ox} - A + H_2O \tag{30}$$

$$Me_{ox} - A \rightarrow Me_{red} - A^{\cdot} \tag{31}$$

$$Me_{red} - A^{\cdot} + H_2O \rightarrow Me_{red} - OH + A^{\cdot} \tag{32}$$

Table 5 presents the results of assessing nanoparticles oxides systems for water purification by ozonation. The elected model reacting molecule was indigo carmine (a dye widely used in the textile industry). All the results were obtained in a 1 L up-flow glass bubble column reactor at pH 3. Ozone was constantly fed at the same rate at all experiments. Indigo carmine is an excellent model molecule to assess the oxidation capacity of a system since its quantification can be easily performed by UV/Vis spectroscopy. It absorbs energy with a wavelength of 610 nm and its main oxidation product, Isatin 5-sulfonic acid, absorbs at 320 nm. It is observed in Table 5, that Indigo Carmine was only partially oxidized by ozonation and its mineralization significantly increased by the use of nanoparticles, especially Cu. The effect of the support is also of paramount importance. Although the attained TOC removal is very similar with supported and unsupported metallic nanoparticles (MNP), it is worth noticing that the mass of catalyst is about 4 times less when Liparite is used as support.

The mentioned in Table 4 nanoparticles oxides were synthesized by reduction with sodium borohydride in aqueous medium (Eq. 22) at room temperature [109] with yields of 80%.

$$2Fe^{2+} + 2Cu^{2+} + 2H_2O + BH_4^- + 4e^- \rightarrow 2Fe(Cu)_{(s)} + BO_2^- + 4H^+ + 2H_{2(g)} \tag{33}$$

Table 5 Effect of nanoestructured metallic oxides, unsupported and supported, on the TOC and COD removal of indigo carmine solutions. Reaction conditions: 60 min [108, 109, 118]

Treatments		TOC removed (%)	COD removed (%)
O₃	Concentration of solid (mg/L) and pH	65	10
Fe	1000, pH 3	61	20
Cu		80	50
Fe-Cu		85	60
Liparite		20	50
Fe/Liparite	250, pH 3	70	60
Cu/Liparite		50	70
Fe-Cu/Liparite		80	80
Fe-pillared clays	100, pH 3	not reported	30
De-oiled allspice	2500, pH 4		10
De-oiled allspice husk xanthate			70

The MNP were studied by EPR before and after the treatment. It was found that the oxidation state of Cu was, before and after the ozonation treatment, was 2+. For the iron MNPs, it is observed under the EPR studies that before the reaction the oxidation state of the iron is 2 + because there is not paramagnetic signal.

Figure 4 shows the concentration profiles evolution with time of Indigo carmine (IC) and its main oxidation product, Isatin sulfonic acid (ISA). The results shown in Fig. 5 were obtained during 60 min of treatment by ozonation catalyzed with different nanostructured metallic oxide systems (Fe, Cu, Ni and Fe-Cu-Ni) supported on Liparite. The initial catalyst loading was 250 mg/L at all cases. It can be observed that oxidation of IC readily occurs with all systems. It is worth noticing, however, that when Ni is added to the mixture Fe-Cu, IC removal is practically instantaneous. ISA oxidation is also faster when Ni is added.

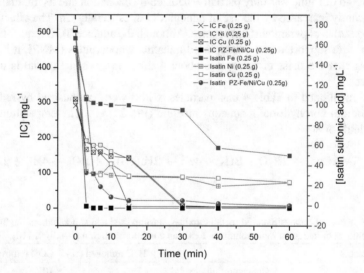

Fig. 4 Concentration of Indigo carmine and Isatin sulfonic acid at 60 min of treatment, pH3 for nanoparticulate metal oxides supported on Liparite

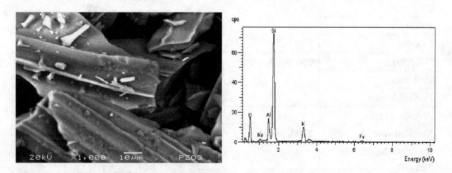

Fig. 5 Scanning electron microscopy (SEM/EDS) of raw Liparite

Figure 5 shows SEM/EDS studies on Liparite and it can be observed that only iron is in the raw Liparite.

By HR-SEM/EDS of the synthesized Fe, Cu and Fe/Cu particles, a spherical morphology was established (size 50–90 × 10^{-3} μm). These spheres were connected together forming chains. This is likely to be due to the magnetic interaction between particles, which causes the rapid reduction rate of metal ions, in all cases.

References

1. Chou, H.L., Hwang, B.J., Sun, C.L.: Catalysis in fuel cells and hydrogen production. New Futur. Dev Catal Batter. Hydrog Storage Fuel Cells 217–270 (2013). https://doi.org/10.1016/B978-0-444-53880-2.00014-4
2. Ji, K., Dai, H., Deng, J., et al.: 3DOM BiVO < inf > 4</inf > supported silver bromide and noble metals: High-performance photocatalysts for the visible-light-driven degradation of 4-chlorophenol. Appl. Catal. B Environ. 168–169:274–282 (2015). https://doi.org/10.1016/j.apcatb.2014.12.045
3. Hurtado, L., Solís-Casados, D.A., Escobar-Alarcón, L., et al.: Multiphase photo-capillary reactors coated with TiO < inf > 2</inf > films: Preparation, characterization and photocatalytic performance. Chem. Eng. J. 304.(2016). https://doi.org/10.1016/j.cej.2016.06.003
4. Avilés-García, O., Espino-Valencia, J., Romero-Romero, R., et al.: Enhanced photocatalytic activity of titania by co-doping with Mo and W. Catalysts 8:631 (2018). https://doi.org/10.3390/catal8120631
5. Avilés-García, O., Espino-Valencia, J., Romero, R., et al.: Oxidation of 4-chlorophenol by mesoporous titania: Effect of surface morphological characteristics. Int. J. Photoenergy 2014 (2014). https://doi.org/10.1155/2014/210751
6. Avilés-García, O., Espino-Valencia, J., Romero, R., et al.: W and Mo doped TiO2: synthesis, characterization and photocatalytic activity. Fuel. (2017). https://doi.org/10.1016/j.fuel.2016.10.005
7. Affam, A.C., Chaudhuri, M.: Degradation of pesticides chlorpyrifos, cypermethrin and chlorothalonil in aqueous solution by TiO2 photocatalysis. J. Environ. Manage. 130:160–165 (2013) . https://doi.org/10.1016/J.JENVMAN.2013.08.058
8. Verma, A., Prakash, N.T., Toor, A.P.: Photocatalytic degradation of herbicide isoproturon in TiO 2 Aqueous Suspensions: study of reaction intermediates and degradation pathways. Environ. Prog. Sustain. Energy. 33, 402–409 (2014). https://doi.org/10.1002/ep.11799
9. Moctezuma, E., Leyva, E., Aguilar, C.A., et al.: Photocatalytic degradation of paracetamol: Intermediates and total reaction mechanism. J. Hazard. Mater 243:130–138. (2012) https://doi.org/10.1016/j.jhazmat.2012.10.010
10. Kanakaraju, D., Glass, B.D., Oelgemöller, M.: Titanium dioxide photocatalysis for pharmaceutical wastewater treatment. Environ. Chem. Lett. 12, 27–47 (2014). https://doi.org/10.1007/s10311-013-0428-0
11. Ji, K., Deng, J., Zang, H., et al.: Fabrication and high photocatalytic performance of noble metal nanoparticles supported on 3DOM InVO4–BiVO4 for the visible-light-driven degradation of rhodamine B and methylene blue. Appl. Catal. B Environ. 165, 285–295 (2015). https://doi.org/10.1016/J.APCATB.2014.10.005
12. Vaiano, V., Sacco, O., Sannino, D., et al.: N-doped TiO2/s-PS aerogels for photocatalytic degradation of organic dyes in wastewater under visible light irradiation. J. Chem. Technol. Biotechnol. 89, 1175–1181 (2014). https://doi.org/10.1002/jctb.4372
13. Litter, M.I., Quici, N.: New advances in heterogeneous photocatalysis for treatment of toxic metals and arsenic. Nanomaterials for Environmental Protection, pp. 143–167. Wiley, Hoboken, NJ (2014)

14. Alanis, C., Natividad, R., Barrera-Diaz, C., et al.: Photocatalytically enhanced Cr(VI) removal by mixed oxides derived from MeAl (Me: Mg and/or Zn) layered double hydroxides. Appl. Catal. B Environ. **140–141**, 546–551 (2013)

15. Khadgi, N., Li, Y., Upreti, A.R., et al.: Enhanced photocatalytic degradation of 17 α -ethinylestradiol exhibited by multifunctional $ZnFe_2O_4$-Ag/rGO nanocomposite under visible light. Photochem. Photobiol. **92**, 238–246 (2016). https://doi.org/10.1111/php.12565

16. Orozco-Hernández, L., Gómez-Oliván, L.M., Elizalde-Velázquez, A., et al.: 17-β-estradiol: significant reduction of its toxicity in water treated by photocatalysis. Sci. Total Environ. **669**, 955–963 (2019). https://doi.org/10.1016/J.SCITOTENV.2019.03.190

17. Trefil, J.: The Nature of Science : An A-Z Guide to the Laws and Principles Governing our Universe. Houghton Mifflin (2003)

18. Hernández-Alonso, M.D., Fresno, F., Suárez, S., et al.: Development of alternative photocatalysts to TiO_2: Challenges and opportunities. Energy Environ. Sci. **2**:1231 (2009). https://doi.org/10.1039/b907933e

19. Dunnill, C.W., Kafizas, A., Parkin, I.P.: CVD production of doped titanium dioxide thin films. Chem. Vap. Depos. **18**, 89–101 (2012). https://doi.org/10.1002/cvde.201200048

20. Industries, E.: AEROXIDE®, AERODISP® and AEROPERL®—Titanium Dioxide as Photocatalyst (2018)

21. Mutuma, B.K., Shao, G.N., Kim, W.D., Kim, H.T.: Sol–gel synthesis of mesoporous anatase–brookite and anatase–brookite–rutile TiO_2 nanoparticles and their photocatalytic properties. J. Colloid Interface Sci. **442**, 1–7 (2015). https://doi.org/10.1016/J.JCIS.2014.11.060

22. Zhang, Y., Han, C., Zhang, G., et al.: PEG-assisted synthesis of crystal TiO_2 nanowires with high specific surface area for enhanced photocatalytic degradation of atrazine. Chem. Eng. J. **268**, 170–179 (2015). https://doi.org/10.1016/J.CEJ.2015.01.006

23. Li, H., Yu, Q., Huang, Y., et al.: Ultralong rutile TiO_2 nanowire arrays for highly efficient dye-sensitized solar cells. ACS Appl. Mater. Interfaces. **8**, 13384–13391 (2016). https://doi.org/10.1021/acsami.6b01508

24. Nalbandian, M.J., Greenstein, K.E., Shuai, D., et al.: Tailored synthesis of photoactive TiO_2 nanofibers and Au/TiO_2 nanofiber composites: structure and reactivity optimization for water treatment applications. Environ. Sci. Technol. **49**, 1654–1663 (2015). https://doi.org/10.1021/es502963t

25. Ye, Y., Feng, Y., Bruning, H., et al.: Photocatalytic degradation of metoprolol by TiO_2 nanotube arrays and UV-LED: effects of catalyst properties, operational parameters, commonly present water constituents, and photo-induced reactive species. Appl. Catal. B Environ. **220**, 171–181 (2018). https://doi.org/10.1016/J.APCATB.2017.08.040

26. Wang, Q., Huang, J., Sun, H., et al.: MoS_2 quantum dots@TiO_2 nanotube arrays: an extended-spectrum-driven photocatalyst for solar hydrogen evolution. Chemsuschem **11**, 1708–1721 (2018). https://doi.org/10.1002/cssc.201800379

27. Pan, X., Chen, X., Yi, Z.: Defective, porous TiO_2 nanosheets with Pt decoration as an efficient photocatalyst for ethylene oxidation synthesized by a C_3N_4 templating method. ACS Appl. Mater. Interfaces. **8**, 10104–10108 (2016). https://doi.org/10.1021/acsami.6b02725

28. Li, Y., Liu, Z., Wu, Y., et al.: Carbon dots-TiO_2 nanosheets composites for photoreduction of Cr(VI) under sunlight illumination: favorable role of carbon dots. Appl. Catal. B Environ. **224**, 508–517 (2018). https://doi.org/10.1016/J.APCATB.2017.10.023

29. Yang, Z., Wang, B., Cui, H., et al.: Synthesis of crystal-controlled TiO_2 nanorods by a hydrothermal method: rutile and brookite as highly active photocatalysts. J. Phys. Chem. C **119**, 16905–16912 (2015). https://doi.org/10.1021/acs.jpcc.5b02485

30. Liu, J., Yang, S., Wu, W., et al.: 3D Flowerlike α-Fe_2O_3 @TiO_2 core-shell nanostructures: general synthesis and enhanced photocatalytic performance. ACS Sustain. Chem. Eng. **3**, 2975–2984 (2015). https://doi.org/10.1021/acssuschemeng.5b00956

31. Macwan, D.P., Dave, P.N., Chaturvedi, S.: A review on nano-TiO_2 sol–gel type syntheses and its applications. J. Mater. Sci. **46**, 3669–3686 (2011). https://doi.org/10.1007/s10853-011-5378-y

32. Byrappa, K., Adschiri, T.: Hydrothermal technology for nanotechnology. Prog. Cryst. Growth Charact. Mater. **53**, 117–166 (2007). https://doi.org/10.1016/J.PCRYSGROW.2007.04.001
33. Li, G., Hong, Z., Yang, H., Li, D.: Phase composition controllable preparation of zirconia nanocrystals via solvothermal method. J. Alloys. Compd. **532**, 98–101 (2012). https://doi.org/10.1016/J.JALLCOM.2012.03.050
34. Chen, X., Mao, S.S.: Titanium dioxide nanomaterials: synthesis, properties, modifications and applications. Chem. Rev. **107**, 2891–2959 (2007). https://doi.org/10.1021/cr0500535
35. Wender, H., Feil, A.F., Diaz, L.B., et al.: Self-organized TiO_2 nanotube arrays: synthesis by anodization in an ionic liquid and assessment of photocatalytic properties. ACS Appl. Mater. Interfaces. **3**, 1359–1365 (2011). https://doi.org/10.1021/am200156d
36. Ge, M., Li, Q., Cao, C., et al.: One-dimensional TiO_2 nanotube photocatalysts for solar water splitting. Adv. Sci. **4**, 1600152 (2017). https://doi.org/10.1002/advs.201600152
37. Warman, J.M., de Matthijs, P.H., Pierre, P., et al.: Electronic processes in semiconductor materials studied by nanosecond time-resolved microwave conductivity—III. Al_2O_3, MgO and TiO_2 powders. Int. J. Radiat. Appl. Instrumentation Part C Radiat. Phys. Chem. **37**, 433–442 (1991). https://doi.org/10.1016/1359-0197(91)90015-T
38. Schneider, J., Matsuoka, M., Takeuchi, M., et al.: Understanding TiO_2 photocatalysis: mechanisms and materials. Chem. Rev. **114**, 9919–9986 (2014). https://doi.org/10.1021/cr5001892
39. Rahim Pouran, S., Abdul Aziz, A.R., Wan Daud, W.M.A.: Review on the main advances in photo-Fenton oxidation system for recalcitrant wastewaters. J. Ind. Eng. Chem. **21**, 53–69 (2015). https://doi.org/10.1016/j.jiec.2014.05.005
40. Hirakawa, T., Yawata, K., Nosaka, Y.: Photocatalytic reactivity for O_2- and OH radical formation in anatase and rutile TiO_2 suspension as the effect of H_2O_2 addition. Appl. Catal. A Gen. **325**, 105–111 (2007). https://doi.org/10.1016/J.APCATA.2007.03.015
41. Hirakawa, T., Nosaka, Y.: Properties of $O_2.-$ and OH. Formed in TiO_2 aqueous suspensions by photocatalytic reaction and the influence of H_2O_2 and some ions. Langmuir **18**, 3247–3254 (2002). https://doi.org/10.1021/la015685a
42. Leytner, S., Hupp, J.T.: Evaluation of the energetics of electron trap states at the nanocrystalline titanium dioxide/aqueous solution interface via time-resolved photoacoustic spectroscopy. Chem. Phys. Lett. **330**, 231–236 (2000). https://doi.org/10.1016/S0009-2614(00)01112-X
43. Teh, C.M., Mohamed, A.R.: Roles of titanium dioxide and ion-doped titanium dioxide on photocatalytic degradation of organic pollutants (phenolic compounds and dyes) in aqueous solutions: a review. J. Alloys. Compd. **509**, 1648–1660 (2011). https://doi.org/10.1016/j.jallcom.2010.10.181
44. Zhang, H., Chen, G., Bahnemann, D.W., et al.: Photoelectrocatalytic materials for environmental applications. J. Mater. Chem. **19**, 5089 (2009). https://doi.org/10.1039/b821991e
45. Avilés-García, O., Espino-Valencia, J., Romero, R., et al.: W and Mo doped TiO_2: synthesis, characterization and photocatalytic activity. Fuel **198**, 31–41 (2017). https://doi.org/10.1016/j.fuel.2016.10.005
46. Avilés-García, O., Espino-Valencia, J., Romero-Romero, R., et al.: Enhanced photocatalytic activity of titania by co-doping with Mo and W. Catalysts **8**, 631 (2018). https://doi.org/10.3390/catal8120631
47. Chen, D., Jiang, Z., Geng, J. et al.: Carbon and nitrogen co-doped TiO_2 with enhanced visible-light photocatalytic activity (2007). https://doi.org/10.1021/IE061491K
48. Peng, F., Cai, L., Huang, L., et al.: Preparation of nitrogen-doped titanium dioxide with visible-light photocatalytic activity using a facile hydrothermal method. J. Phys. Chem. Solids **69**, 1657–1664 (2008). https://doi.org/10.1016/j.jpcs.2007.12.003
49. Daghrir, R., Drogui, P., Robert, D.: Modified TiO_2 for Environmental photocatalytic applications: a review. Ind. Eng. Chem. Res. **52**, 3581–3599 (2013). https://doi.org/10.1021/ie303468t
50. Wang, S., Yun, J.-H., Luo, B., et al.: Recent progress on visible light responsive heterojunctions for photocatalytic applications. J. Mater. Sci. Technol. **33**, 1–22 (2017). https://doi.org/10.1016/j.jmst.2016.11.017

51. Hao, R., Wang, G., Tang, H., et al.: Template-free preparation of macro/mesoporous g-C_3N_4/TiO_2 heterojunction photocatalysts with enhanced visible light photocatalytic activity. Appl. Catal. B Environ. **187**, 47–58 (2016). https://doi.org/10.1016/j.apcatb.2016.01.026

52. Jiang, Z., Jiang, D., Yan, Z., et al.: A new visible light active multifunctional ternary composite based on TiO_2-In_2O_3 nanocrystals heterojunction decorated porous graphitic carbon nitride for photocatalytic treatment of hazardous pollutant and H_2 evolution. Appl. Catal. B Environ. **170–171**, 195–205 (2015). https://doi.org/10.1016/j.apcatb.2015.01.041

53. Etgar, L., Gao, P., Xue, Z., et al.: Mesoscopic $CH_3NH_3PbI_3$/TiO_2 heterojunction solar cells. J. Am. Chem. Soc. **134**, 17396–17399 (2012). https://doi.org/10.1021/ja307789s

54. U.S. Environmental Protection Agency.: National pollutant discharge elimination system (2018)

55. Nogueira, V., Lopes, I., Rocha-Santos, T.A.P., et al.: Treatment of real industrial wastewaters through nano-TiO_2 and nano-Fe_2O_3 photocatalysis: case study of mining and kraft pulp mill effluents. Environ. Technol. (U. K.) **39**, 1586–1596 (2018). https://doi.org/10.1080/09593330.2017.1334093

56. Khan, W.Z., Najeeb, I., Tuiyebayeva, M., Makhtayeva, Z.: Refinery wastewater degradation with titanium dioxide, zinc oxide, and hydrogen peroxide in a photocatalytic reactor. Process. Saf. Environ. Prot. **94**, 479–486 (2015). https://doi.org/10.1016/J.PSEP.2014.10.007

57. Goutam, S.P., Saxena, G., Singh, V., et al.: Green synthesis of TiO_2 nanoparticles using leaf extract of *Jatropha curcas* L. for photocatalytic degradation of tannery wastewater. Chem. Eng. J. **336**, 386–396 (2018). https://doi.org/10.1016/J.CEJ.2017.12.029

58. Ghasemi, Z., Younesi, H., Zinatizadeh, A.A.: Kinetics and thermodynamics of photocatalytic degradation of organic pollutants in petroleum refinery wastewater over nano-TiO_2 supported on Fe-ZSM-5. J. Taiwan Inst. Chem. Eng. **65**, 357–366 (2016). https://doi.org/10.1016/J.JTICE.2016.05.039

59. Nogueira, V., Lopes, I., Rocha-Santos, T.A.P., et al.: Photocatalytic treatment of olive oil mill wastewater using TiO_2 and Fe_2O_3 nanomaterials. Water Air Soil Pollut. **227**, 88 (2016). https://doi.org/10.1007/s11270-016-2787-1

60. Subramonian, W., Wu, T.Y., Chai, S.P.: Photocatalytic degradation of industrial pulp and paper mill effluent using synthesized magnetic Fe_2O_3–TiO_2: Treatment efficiency and characterizations of reused photocatalyst. J. Environ. Manage. **187**, 298–310 (2017). https://doi.org/10.1016/j.jenvman.2016.10.024

61. Guesh, K., Mayoral, Á., Márquez-Álvarez, C., et al.: Enhanced photocatalytic activity of TiO_2 supported on zeolites tested in real wastewaters from the textile industry of Ethiopia. Microporous Mesoporous Mater. **225**, 88–97 (2016). https://doi.org/10.1016/J.MICROMESO.2015.12.001

62. Crişan, M., Mardare, D., Ianculescu, A., et al.: Iron doped TiO_2 films and their photoactivity in nitrobenzene removal from water. Appl. Surf. Sci. **455**, 201–215 (2018). https://doi.org/10.1016/J.APSUSC.2018.05.124

63. Ochuma, I.J., Osibo, O.O., Fishwick, R.P., et al.: Three-phase photocatalysis using suspended titania and titania supported on a reticulated foam monolith for water purification. Catal. Today **128**, 100–107 (2007). https://doi.org/10.1016/J.CATTOD.2007.05.015

64. Ameta, R., Chohadia, A.K., Jain, A., Punjabi, P.B.: Fenton and photo-fenton processes. In: Advanced Oxidation Processes for Wastewater Treatment: Emerging Green Chemical Technology (2018)

65. De León, M.A., Castiglioni, J., Bussi, J., Sergio, M.: Catalytic activity of an iron-pillared montmorillonitic clay mineral in heterogeneous photo-Fenton process. Catal. Today (2008). https://doi.org/10.1016/j.cattod.2007.12.130

66. Chen, L., Zhou, C.H., Fiore, S., et al.: Functional magnetic nanoparticle/clay mineral nanocomposites: preparation, magnetism and versatile applications. Appl. Clay Sci (2016)

67. De León, M.A., Rodríguez, M., Marchetti, S.G., et al.: Raw montmorillonite modified with iron for photo-Fenton processes: Influence of iron content on textural, structural and catalytic properties. J. Environ. Chem. Eng. (2017). https://doi.org/10.1016/j.jece.2017.09.014

68. Kloprogge, J.T.: Synthesis of smectites and porous pillared clay catalysts: a review. J. Porous Mater. (1998). https://doi.org/10.1023/A:1009625913781
69. Cañizares, P., Valverde, J.L., Sun Kou, M.R., Molina, C.B.: Synthesis and characterization of PILCs with single and mixed oxide pillars prepared from two different bentonites. A comparative study. Microporous Mesoporous Mater (1999). https://doi.org/10.1016/S1387-1811(98)00295-9
70. Gil, A., Gandía, L.M., Vicente, M.A.: Recent advances in the synthesis and catalytic applications of pillared clays. Catal. Rev. - Sci. Eng. (2000). https://doi.org/10.1081/CR-100100261
71. Clearfield, A., Perry, H.P., Gagnon, K.J.: Porous pillared clays and layered phosphates. In: Comprehensive Inorganic Chemistry II, 2nd edn. From Elements to Applications (2013)
72. Clearfield, A., Kuchenmeister, M.: Pillared layered materials, pp. 128–144 (1992)
73. Doff, D.H., Gangas, N.H.J., Allan, J.E.M., Coey, J.M.D.: Preparation and characterization of iron oxide pillared montmorillonite. Clay Miner. (1988). https://doi.org/10.1180/claymin.1988.023.4.04
74. Valverde, J.L., Romero, A., Romero, R., et al.: Preparation and characterization of Fe-PILCS. Influence of the synthesis parameters. Clays Clay Miner. (2005). https://doi.org/10.1346/CCMN.2005.0530607
75. Martin Del Campo, E., Romero, R., Roa, G., et al.: Photo-Fenton oxidation of phenolic compounds catalyzed by iron-PILC. In: Fuel (2014)
76. Bernal, M., Romero, R., Roa, G., et al.: Ozonation of indigo carmine catalyzed with Fe-pillared clay. Int. J. Photoenergy (2013). https://doi.org/10.1155/2013/918025
77. Hurtado, L., Romero, R., Mendoza, A., et al.: Paracetamol mineralization by Photo Fenton process catalyzed by a Cu/Fe-PILC under circumneutral pH conditions. J. Photochem. Photobiol. A Chem. (2019). https://doi.org/10.1016/j.jphotochem.2019.01.012
78. Minella, M., Marchetti, G., De Laurentiis, E., et al.: Photo-Fenton oxidation of phenol with magnetite as iron source. Appl. Catal. B Environ. (2014). https://doi.org/10.1016/j.apcatb.2014.02.006
79. Sum, O.S.N., Feng, J., Hu, X., Yue, P.L.: Pillared laponite clay-based Fe nanocomposites as heterogeneous catalysts for photo-Fenton degradation of acid black. Chem. Eng. Sci. (2004)
80. Feng, J., Hu, X., Yue, P.L.: Novel Bentonite clay-based Fe - Nanocomposite as a heterogeneous catalyst for photo-fenton discoloration and mineralization of orange II. Environ. Sci. Technol. (2004). https://doi.org/10.1021/es034515c
81. Bel Hadjltaief, H., Da Costa, P., Galvez, M.E., Ben Zina, M.: Influence of operational parameters in the heterogeneous photo-fenton discoloration of wastewaters in the presence of an iron-pillared clay. Ind. Eng. Chem. Res. (2013). https://doi.org/10.1021/ie4018258
82. Beltrán, F.J.: Ozone reaction kinetics for water and wastewater systems. 1st ed. New York (2004)
83. Hoigné, J.: Chemistry of aqueous ozone and transformation of pollutants by ozonation and advanced oxidation processes. Quality and Treatment of Drinking Water II, pp. 83–141. Springer, Berlin Heidelberg (1998)
84. Martínez-Huitle, C.A., Rodrigo, M.A., Sirés, I., Scialdone, O.: Single and coupled electrochemical processes and reactors for the abatement of organic water pollutants: a critical review. Chem. Rev. 115, 13362–13407 (2015). https://doi.org/10.1021/acs.chemrev.5b00361
85. Nawrocki, J., Kasprzyk-hordern, B.: Applied catalysis B: environmental the efficiency and mechanisms of catalytic ozonation. Appl. Catal B Environ. 99, 27–42 (2010). https://doi.org/10.1016/j.apcatb.2010.06.033
86. Nawrocki, J.: Applied catalysis B : environmental catalytic ozonation in water: controversies and questions. discussion paper. Appl. Catal B Environ. 142–143, 465–471 (2013). https://doi.org/10.1016/j.apcatb.2013.05.061
87. Biard, P., Werghi, B., Soutrel, I., et al.: Efficient catalytic ozonation by ruthenium nanoparticles supported on SiO_2 or TiO_2 : towards the use of a non-woven fiber paper as original support 289:374–381 (2016). https://doi.org/10.1016/j.cej.2015.12.051.

88. Álvarez, P.M., Beltrán, F., Pocostales, P., Masa, F.J.: Preparation and structural characterization of Co/Al 2O 3 catalysts for the ozonation of pyruvic acid. Appl. Catal B-Environ. Appl Catal B-Environ. **72**, 322–330 (2007). https://doi.org/10.1016/j.apcatb.2006.11.009

89. Bing, J., Hu, C., Nie, Y., et al.: Mechanism of catalytic ozonation in Fe2O3/Al2O3@ SBA-15 aqueous suspension for destruction of ibuprofen. Environ. Sci. Technol. **49**, 1690–1697 (2015)

90. Trapido, M., Veressinina, Y., Munter, R., Kallas, J.: Catalytic ozonation of m-dinitrobenzene. Ozone Sci. Eng. **27**, 359–363 (2005). https://doi.org/10.1080/01919510500250630

91. Cooper, C., Burch, R.: An investigation of catalytic ozonation for the oxidation of halocarbons in drinking water preparation. The research was performed in the Catalysis Research Group, Department of Chemistry, University of Reading, Berkshire RG6 6AD, U.K. Water Res **33**, 3695–3700 (1999). https://doi.org/10.1016/S0043-1354(99)00091-3

92. Yang, Z., Lv, A., Nie, Y., Hu, C.: Catalytic ozonation performance and surface property of supported Fe3O4 catalysts dispersions. Front Environ. Sci. Eng. **7**, 451–456 (2013). https://doi.org/10.1007/s11783-013-0509-0

93. Sánchez-Polo, M., Rivera-Utrilla, J.: (2004) Ozonation of 1,3,6-naphthalenetrisulfonic acid in presence of heavy metals. J. Chem. Technol. Biotechnol. **79**, 902–909 (2004)

94. Li, G., Lu, Y., Lu, C., et al.: Efficient catalytic ozonation of bisphenol-A over reduced graphene oxide modified sea urchin-like α-MnO_2 architectures. J. Hazard Mater. **294**, 201–208 (2015). https://doi.org/10.1016/j.jhazmat.2015.03.045

95. Hu, M., Hui, K.S., Hui, K.: Role of graphene in MnO_2/graphene composite for catalytic ozonation of gaseous toluene. Chem. Eng. J. **254**, 237–244 (2014). https://doi.org/10.1016/j.cej.2014.05.099

96. Dong, Y., Yang, H., He, K., et al.: β-MnO_2 nanowires: a novel ozonation catalyst for water treatment. Appl. Catal B Environ. **85**, 155–161. https://doi.org/10.1016/j.apcatb.2008.07.007 (2009)

97. Alsheyab, M.A., Muñoz, A.H.: Comparative study of ozone and MnO_2/O_3 effects on the elimination of TOC and COD of raw water at the Valmayor station. Desalination **207**, 179–183 (2007). https://doi.org/10.1016/j.desal.2006.07.010

98. Ma, J., Graham, N.J.D.: Degradation of atrazine by manganese-catalysed ozonation: influence of humic substances. Water Res. **33**, 785–793 (1999). https://doi.org/10.1016/S0043-1354(98)00266-8

99. Rodríguez, J.L., Valenzuela, M.A., Poznyak, T., et al.: Reactivity of NiO for 2,4-D degradation with ozone: XPS studies. J. Hazard. Mater. **262**, 472–481 (2013). https://doi.org/10.1016/j.jhazmat.2013.08.041

100. Chowdhury, M., Oputu, O., Nyamayaro, K., et al.: Novel β-FeOOH/NiO composite material as a potential catalyst for catalytic ozonation degradation of 4-chlorophenol. RSC Adv. **5** (2015). https://doi.org/10.1039/C5RA09177B

101. Avramescu, S.M., Bradu, C., Udrea, I., et al.: Degradation of oxalic acid from aqueous solutions by ozonation in presence of Ni/Al_2O_3 catalysts. Catal. Commun. **9**, 2386–2391 (2008). https://doi.org/10.1016/j.catcom.2008.06.001

102. Qin, W., Li, X., Qi, J.: Experimental and theoretical investigation of the catalytic ozonation on the surface of NiO−CuO Nanoparticles. Langmuir **25**, 8001–8011 (2009). https://doi.org/10.1021/la900476m

103. Petre, A.L., Carbajo, J.B., Rosal, R., et al.: CuO/SBA-15 catalyst for the catalytic ozonation of mesoxalic and oxalic acids. Water matrix effects. Chem. Eng. J. **225**, 164–173 (2013). https://doi.org/10.1016/j.cej.2013.03.071

104. Turkay, O., Inan, H., Dimoglo, A.: Experimental and theoretical investigations of CuO-catalyzed ozonation of humic acid. Sep. Purif. Technol. **134**, 110–116 (2014). https://doi.org/10.1016/j.seppur.2014.07.040

105. Hua, L., Ma, H., Zhang, L.: Degradation process analysis of the azo dyes by catalytic wet air oxidation with catalyst CuO/gamma-Al2O3. Chemosphere **90**, 143–149 (2013). https://doi.org/10.1016/j.chemosphere.2012.06.018

106. Dai, Q., Wang, J., Yu, J., et al.: Catalytic ozonation for the degradation of acetylsalicylic acid in aqueous solution by magnetic CeO_2 nanometer catalyst particles. Appl. Catal. B Environ. **144**, 686–693 (2014). https://doi.org/10.1016/j.apcatb.2013.05.072
107. Chen, C., Yoza, B.A., Chen, H., et al.: Manganese sand ore is an economical and effective catalyst for ozonation of organic contaminants in petrochemical wastewater (2015). https://doi.org/10.1007/s11270-015-2446-y
108. Romero, R., Bernal, M., Barrera-Díaz, C., et al.: Ozonation of indigo carmine catalyzed with fe-pillared clay. Int. J. Photoenergy (2013). https://doi.org/10.1155/2013/918025
109. Torres-Blancas, T., Roa-Morales, G., Ureña-Núñez, F., et al.: Ozonation enhancement by Fe–Cu biometallic particles. J. Taiwan Inst. Chem. Eng. **74** (2017). https://doi.org/10.1016/j.jtice.2017.02.025
110. Faria, P.C.C., Monteiro, D.C.M., Órfão, J.J.M., Pereira, M.F.R.: Cerium, manganese and cobalt oxides as catalysts for the ozonation of selected organic compounds. Chemosphere **74**, 818–824 (2009). https://doi.org/10.1016/j.chemosphere.2008.10.016
111. Pocostales, J.P., Álvarez, P., Beltrán, F.J.: Kinetic modeling of granular activated carbon promoted ozonation of a food-processing secondary effluent. Chem. Eng. J. **183**, 395–401 (2012). https://doi.org/10.1016/j.cej.2011.12.020
112. Kasprzyk-hordern, B., Ziółek, M., Nawrocki, J.: Catalytic ozonation and methods of enhancing molecular ozone reactions in water treatment **46**, 639–669 (2003). https://doi.org/10.1016/S0926-3373(03)00326-6
113. Madiha, O., Monia, G., Abdelmottaleb, O.: Comparison between MnO_2, TiO_2 and CoO for the Ozonation of oxalic acid **7**, 683–688 (2014)
114. Chen, Y.-H., Hsieh, D.-C., Shang, N.-C.: Efficient mineralization of dimethyl phthalate by catalytic ozonation using $TiO_2/Al2O3$ catalyst. J. Hazard. Mater. **192**, 1017–1025 (2011). https://doi.org/10.1016/j.jhazmat.2011.06.005
115. Vittenet, J., Rodriguez, J., Petit, E., et al.: Microporous and mesoporous materials removal of 2, 4-dimethylphenol pollutant in water by ozonation catalyzed by SOD, LTA, FAU-X zeolites particles obtained by pseudomorphic transformation (binderless). Microporous Mesoporous Mater. **189**, 200–209 (2014). https://doi.org/10.1016/j.micromeso.2013.09.042
116. Wang, J., Cheng, J., Wang, C., et al.: Catalytic ozonation of dimethyl phthalate with RuO_2/Al_2O_3 catalysts prepared by microwave irradiation. Catal. Commun. **41**, 1–5 (2013). https://doi.org/10.1016/J.CATCOM.2013.06.030
117. Legube, B., Leitner, N.K.V.: Catalytic ozonation: a promising advanced oxidation technology for water treatment. Catalysis Today **53**, 61–72 (1999)
118. Palma-Anaya, E., Fall, C., Torres-Blancas, T., et al.: Pb(II) removal process in a packed column system with xanthation-modified deoiled allspice husk. J. Chem. (2017). https://doi.org/10.1155/2017/4296515

Treating of Aquatic Pollution by Carbon Quantum Dots

Z. M. Marković and B. M. Todorović Marković

Abstract In this chapter, structural, optical, antibacterial and photocatalytic properties of carbon quantum dots are described. Carbon quantum dots are a new class of carbon based nanomaterials with extraordinary properties and because of that they can be used in different fields. The special attention is devoted to methods for eliminating various pathogens, organic dyes, chemicals and pesticides from water by using carbon quantum dots. Water pollution is one of the greatest problems worldwide and successful water cleaning from various pollutants has the highest priority.

Keywords Carbon quantum dots · Reactive oxygen species · Antibacterial activity · Photocatalytic activity

1 Introduction

Nowadays water pollution is a great problem due to different factors: increase of urban population, industrialization, huge changes in our environment followed by the presence of various toxic substances as well as uncontrolled usage of antibiotics which contributes to development of drug-resistance pathogens [1]. If some substances are present in water to certain degree that it cannot be used for drinking, cooking or bathing, water is considered polluted. Oceans, rivers, lakes can self-clean to a certain level from various toxic substances but if they are present in a huge amount self-cleaning are not enough. Thus water pollution affects the health of all plants, animals and humans [2].

Water pollutants can be divided into a few groups: organic pollutants (oxygen demanding wastes, synthetic organic pollutants, oil, pathogens, nutrients, thermal pollutants, radioactive pollutants, suspended solids and inorganic pollutants) [3]. Organic pollutants are also responsible for endangering water birds and coastal plants whereas pathogens can cause many very dangerous diseases such as cholera, typhoid,

Z. M. Marković · B. M. Todorović Marković (✉)
Vinča Institute of Nuclear Sciences, University of Belgrade,
Mike Alasa 12-14, 11001 Belgrade, Serbia
e-mail: biljatod@vin.bg.ac.rs

© Springer Nature Switzerland AG 2019 121
G. A. B. Gonçalves and P. Marques (eds.), *Nanostructured Materials for Treating Aquatic Pollution*, Engineering Materials,
https://doi.org/10.1007/978-3-030-33745-2_5

dysentery, polio and infectious hepatitis in humans. The presence of nutrients in water can lead to growth of undesirable aquatic life whereas organic compounds result in thermal stratification. Radioactive pollutants accumulate in the bones, teeth and can cause serious disorders [4]. Suspended solids and sediments can block the sun penetration in water and thus prevent very important processes such as photosynthesis. Inorganic pollutants (mineral acids, inorganic salts, metals, metals compounds) have adverse effect on aquatic flora and fauna [5].

One of the critical points of water pollution related to human health is the pollution of tap water. Bacteria are a natural component of water. But some of them can be very harmful to human health. Namely, according to US Centers for Disease Control and Prevention about 1.7 million infections and 99.000 deaths annually occur in American hospitals due to healthcare infections [6]. It is assumed that the most common source of these infections is tap water. It is figured out that almost 1400 deaths occur each year as a result of nosocomial pneumonias attributable to *Pseudomonas aeruginosa* alone [7]. But, although there is a big risk for transmission of various pathogens from tap water to humans, very little attention is devoted to this problem. Waterborne pathogens can exist to certain degree both in hot and cold water. Whereas cold water is delivered directly to the point of use, hot water is supplied via a recirculation loop, which contains nutrients to nourish waterborne microbes, maintains favourable temperatures for microbial growth, and promotes the formation of biofilm on internal surfaces of pipes and fixtures. The most common pathogens existing in tap water are *Escherichia coli, Campylobacter jejuni, Hepatitis A, Giardia lamblia, Salmonella, Legionella pneumophila* and *Cryptosporidium* [8]. Harmful bacteria such as mentioned above can be very dangerous for human health in the course of cause of very serious illnesses.

How water disinfection can be conducted? It is sure that applied methods must be safe, affordable, robust and sustained. Seven water treatment strategies excluding point-of-use (POU) filters are applied to prevent aquatic pollution: hot water flushing of the plumbing system, chlorination, chlorine dioxide, monochloramine, copper-silver ionization, UV light and ozonation. Each of mentioned strategies has advantages and disadvantages related to different parameters: ease of implementation, cost, maintenance as well as efficiency. Hot water flushing is the easiest to implement but its main drawback is the demand that whole system must be exposed to high-temperature water. Chlorination is simple to implement but the highest priority demand is maintenances of adequate levels of chlorine throughout the system and microbe free environment [9]. It has been indicated that the most effective water disinfection regime for *Legionella* is chlorine dioxide [10]. But main disadvantage of this system is the high cost of installment. Performance studies of chloramines showed that their usage alone or in combination with free chlorine is complete neither as a disinfect nor as an antimicrobial agent [11]. Copper-silver ionization can be effective only in combination with other disinfection technologies [12]. UV light which can be used as disinfection agent has poor penetrating power. It is only effective as the source of radiation. The only benefit of UV light is the usage of light-emitting diodes to deliver UV-A radiation which has been shown as bactericidal [13]. All mentioned above strategies has level of uncertainty related to toxic by-products. Only

POU filtration offers the potential benefit of immediate and complete effectiveness against waterborne bacteria, fungi, and protozoa. This technology has been used in Europe for the last ten years [14].

Organic dyes as very toxic substances can be remediated from water by applying different methods: physical (electrokinetic coagulation, irradiation, ion exchange, membrane filtration, adsorption), chemical (oxidation process, Fenton method, ozonation, photochemical, sodium hypochloride, reverse osmosis) and biological (decolorization by white-rot fungi, adsorption by living/dead microbial biomass, anaerobic textile dye bioremedition systems, other microbial culture) [15, 16].

Due to large problems caused by contaminated water, many various materials are used for water disinfection. Most commonly used materials for that purpose are chlorine, chlorine dioxide, ozone and monochloramine as mentioned above [17]. As barrier materials which should prevent microbiological contamination, coagulants such as aluminium and iron salts are often used. Other materials used for water purification especially for removal different bacterial pathogens and organic dyes are metal/metal oxide nanoparticles, polypeptides, semiconductor nanoparticles, polymeric nanostructures, and carbon based nanomaterials [18–30].

Carbon based nanomaterials (porous carbon, fullerene, nanodiamonds, carbon nanotubes, carbon quantum dots, graphene quantum dots, graphene, graphene oxide, reduced graphene oxide) can be used for removal of different contaminants due to their properties: structural, photoluminescent and photocatalytic; low cytotoxicity, high antimicrobial activity [31, 32].

Porous carbon as carbonaceous material exhibits a high degree of porosity and extended particulate surface area. Thus it can be used as adsorbent of choice for the removal of organic pollutants [33]. The adsorption capacity of porous carbon is changed by chemical treatments (varying of surface functional groups and porous structure). The acidity of the porous carbon is varied along with the textural characteristics depending on the nature of mineral acids. While hydrochloric acid treatment decreases the active acidic groups, thereby enhancing the adsorption of larger molecules on the porous carbon, nitric acid treatment produces more active acidic surface groups such as carboxyl and lactone, and generates a more homogeneous porous size, finally resulting in a reduced adsorption of basic dyes [33, 34].

Fullerenes are classified as the third allotrope of carbon and due to their unique properties can be used in different fields i.e. as a sorbent of pollutants from water [33]. C_{60} molecules consist of 60 carbon atoms arranged as 12 pentagons and 20 hexagons as the basis of icosohedral symmetry closed cage structure. Each carbon atom is bonded to three others and is sp^2 hybridized [35]. Fullerenes are powerful antioxidants, reacting readily and at a high rate with free radicals, which are often the cause of cell damage or death. They are known as "free-radical sponge" because they can sponge-up more than 20 radicals per fullerene molecule [36, 37]. Compared to porous carbon fullerenes have higher adsorption capacity against organic pollutants. It is found out that fullerenes remove pollutants mainly by physical adsorption through dispersive interaction forces [33]. Thus fullerenes can be used for removal various organic compounds, covering dyes, antibiotics, polycyclic aromatic hydrocarbons, phenolic compounds and pesticides [38].

Carbon nanotubes (CNTs) represent enrolled cylindrical graphitic sheet (called graphene) rolled up in a seamless cylinder with nanometer sized diameter and micrometer sized length. In CNTs carbon atoms are arranged in hexagons as in graphite [35, 39]. CNTs belong to fullerene structural family but they have cylindrical shape whereas fullerenes have spherical shape. CNTs can be used for removal heavy metals. The electrostatic attraction and chemical bonding such as ion exchange dominated the adsorption process where the surface functional groups of oxidized CNTs provide the major adsorption sites to heavy metals [38]. CNTs in combination with catalyst nanoparticles (i.e. TiO_2) can be used for removal of toxic substances and non-biodegradable pollutants from water by photocatalysis process [38]. Membrane filters can be made from CNTs and polymers. These nanocomposites or thin film nanocomposites can be used for water filtration.

Graphene based nanomaterials (graphene oxide-GO, reduced graphene oxide-RGO) including graphene nanocomposites (different types of nanocomposites can be prepared from GO or RGO and polymers) represent materials with graphene like structure. GO is an atom thick layer of carbon atoms arranged in a honeycomb structure and bonded together by σ bonds. Apart from intrinsic corrugations and topological defects (i.e. pentagons, heptagons or their combination), GO can have other types of defects such as vacancies, adatoms, edges/cracks, adsorbed impurities [40]. These materials can be used for removal of organic pollutants, pathogens or as membrane filters. Their surface properties can be tuned successfully by chemical treatments (i.e. doping). These materials have very strong adsorption capacity due to large surface area, porous structure, and feasible surface properties [38].

Figure 1 shows schematic view of potential application of carbon based nanomaterials in wastewater treatment. They can be used as adsorbents, photocatalysis and disinfection agents in the form of filters, hydrogels, nanocomposites, fibers and thin films composites.

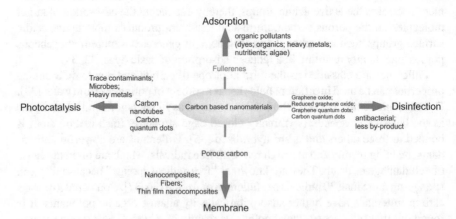

Fig. 1 Schematic view of potential application of carbon based nanomaterials in wastewater treatment

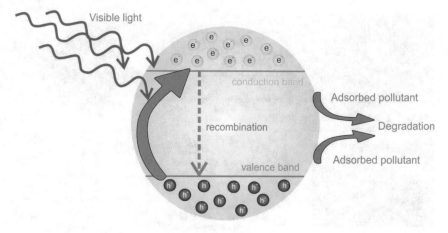

Fig. 2 Basic principles of photocatalytic process

In this paper we discuss the usage of carbon quantum dots as a potential water disinfection agent especially for photocatalytic degradation of organic molecules and antibacterial water disinfection. Photocatalysis is a process which can be described by the following way: absorbed light generates electron-hole pairs which are followed by separation of excited charges. Electrons and holes transfer to the photocatalytic surface and due to charges, redox reaction occurs on the surface. During transfer to the surface many electron-hole pairs recombine or they recombine on the surface sites themselves [41]. Figure 2 shows the main processes involved in the photocatalytic process. Photocatalysis can be classified as green technology because the same materials can be used many times.

2 Carbon Quantum Dots: Structure, Synthesis and Characterization

Carbon quantum dots (CQDs) are a new class of carbon based nanomaterials with lateral dimension smaller than 10 nm and quasi-spherical shape—Fig. 3a–d [42]. Carbon network of CQDs are predominantly composed of amorphous and crystalline cores with either graphitic or turbostratic carbon or with graphene and graphene oxide sheets mixed by sp^3 carbon bonds (Fig. 4) [43]. Oxygen content through epoxy, carbonyl and carboxyl groups presented on the surface of CQDs can be changed by various synthetic routes [42]. Due to oxygen surface functionalization CQDs possess very good dispersibility in water which enables their usage in biomedicine. By surface functionalization and passivation with different materials physical and chemical properties of CQDs can be modified significantly. Doping of CQDs with different atoms (N, S) can affect very significantly the properties of CQDs (structural, photoluminescent, antibacterial) [44]. CQDs have very good chemical and

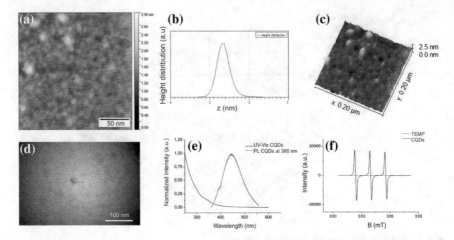

Fig. 3 **a** Top view AFM image of CQDs; **b** height profile of CQDs; **c** 3D view AFM image of CQDs; **d** TEM micrograph of CQDs; **e** Photoluminescence of CQDs; **f** EPR spectra of CQDs. CQDs presented in (**a**)–(**d**) are synthesized by thermal decomposition of citric acid at 210 °C for 1 h as described in Ref. [58] in detail

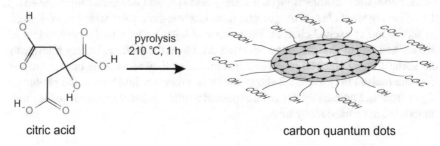

citric acid   carbon quantum dots

Fig. 4 Structure of CQDs synthesized from citric acid by pyrolysis at 210 °C for 1 h. Detailed description of synthesis procedure can be found in Ref. [58]

thermal stability, photoinduced electron transfer, resistance to photobleaching, high photoluminescence [43, 45].

Methods for CQDs synthesis are commonly grouped into two categories: bottom-up and top-down [42]. Schematic view of two common approaches of CQDs synthesis is presented in Fig. 5. Bottom-up methods involve CQDs synthesis in microwave or hydrothermal reactors as well as solvothermal treatment and pyrolysis [42, 45]. During synthesis of CQDs in microwave reactor, as carbon precursors can be used glucose, sucrose, succinic acid, catechol, resorcinol, hydroquinone) [46–48]. As reaction media are used polyethylene glycol, NH₄OH, tris(2-aminoethyl)amine. Synthesis of various materials in microwave reactor is applied due to simultaneous, homogeneous, efficient heating, fast reaction rates and uniform size distribution of nanoparticles. At the same time this method is time-saving and low-cost [47]. CQDs synthesis in hydrothermal reactor is considered as green procedure, eco-friendly and non-toxic

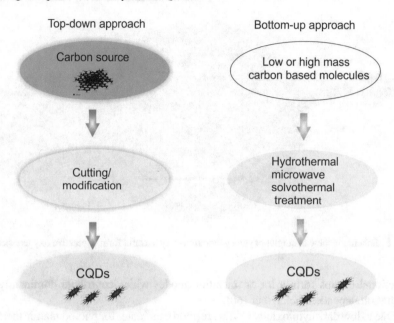

Fig. 5 Schematic view of two common approaches for CQDs synthesis

[42]. As carbon precursors are used citric acid, carbon hydrates or glucose, chitosan, banana or orange juice, cabbage, broccoli etc. [49–54]. In solvothermal treatment, carbon based compounds are heated in high boiling point organic solvents followed by further purification process including filtration and dialysis [42, 55]. Top-down methods comprise usage of nano-diamonds, carbon nanotubes, graphite, carbon soots, activate carbon or GO to produce CQDs [42, 45].

Photoluminescence is one of the very important features of CQDs—Fig. 3e. Nowadays, there are a few theories concerning photoluminescence but the main feature of all theories is: photoluminescence of CQDs is due to quantum confinement effect or conjugated π-domains, determined by the carbon core, zig-zag edges and surface defects i.e. surface states. Surface states are induced by hybridization of the carbon backbone and the connected chemical groups [56–58]. Commonly CQDs emit blue to yellow photoluminescence. But the highest intensity of luminescence is blue or green light. Photoluminescence colour of CQDs cannot be tuned by the size of CQDs. Namely, it is related to the surface groups presented on the basal plane of CQDs rather than the size [56]. Photoluminescence of CQDs is a laser excitation wavelength dependent i.e. there is a shift (blue or red) between excitation and emission wavelength. Blue photoluminescence of CQDs is a feature of quantum size effect and zig-zag edges whereas the red shifted photoluminescence is predominantly due to surface defects presented on the basal plane and edges of sp^2 domain inside sp^3 matrix and the increase of size of aromatic π-conjugated domains [59]. Surface

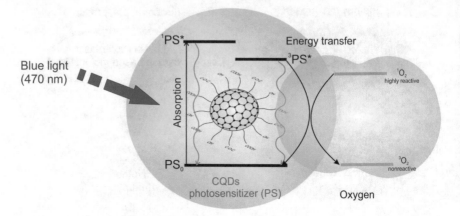

Fig. 6 Schematic view of singlet oxygen generation-very harmful form of reactive oxygen species

functional groups form a lot of transition modes which contribute dominantly to excitation dependent behaviour [60].

Due to low dark cytotoxicity CQDs are good candidates for photodynamic therapy because of their very good antibacterial properties toward different bacterial strains [61, 62]. Photodynamic therapy is based on the usage of light in combination with materials called photosensitizers to cure different types of disease including cancer especially skin or eye cancer [63]. Under blue light irradiation CQDs become photoactive, i.e. they generate reactive oxygen species (ROS)—Fig. 3f [58]. ROS (singlet oxygen, superoxide anion or hydroxyl radicals) are generated by the following way: photoactive materials (PS) are excited under visible light from ground singlet state to excited state and further to long-lived excited triplet state—Fig. 6. During reaction with molecular oxygen, PSs transfer their energy to it and molecular oxygen transfers from ground triplet state to excited triplet state and further to excited singlet state ($^1O_2$). Besides fullerenes, CQDs and graphene quantum dots (GQDs) also generate singlet oxygen under blue light irradiation at 470 nm [58, 64]. GQDs differ from CQDs in terms of morphology, chemical composition of sp^2 and sp^3 bonds and photoluminescence. Ge et al. claim that energy transfer, not electron transfer, is responsible for singlet oxygen generation of GQDs [65]. In our previous research we got similar results [58]. CQDs, GQDs and N-doped CQDs generate singlet oxygen only under blue light irradiation. Under ambient light conditions there is no singlet oxygen production in significant quantities. Compared to GQDs and N-doped CQDs, CQDs produced singlet oxygen in the highest quantities only under blue light irradiation. By applying electron paramagnetic resonance technique (EPR) we established that none of these quantum dots had produced hydroxyl or superoxide anion radicals. Chong et al. reported that photoexcited GQDs produced singlet oxygen both via energy and electron transfer. At the same time they produce hydroxyl and superoxide anion radicals due to formation of electron-hole pairs and charge separation [66]. Furthermore, GQDs accelerate the oxidation of ordinarily nonenzymatic antioxidant biomolecules and thus lipid peroxidation under photoexcitation. In this way, GQDs can produce or quench ROS depend on light conditions (the presence or absence of light).

3 Removal of Pathogens by Carbon Quantum Dots

Nowadays, one of the main subjects in research field is the development of new types of antibacterials materials. Antibacterial surfaces can be designed in different ways, i.e. as repelling surfaces made from either neutral polymers such as poly(ethylene glycol) or from charged anionic polymers [67, 68]. These surfaces do not allow bacteria to adhere the surfaces at all by steric hidrance. Besides repelling surfaces, there are contact killing surfaces, i.e. polymer surfaces embedding with different photoactive materials [69–72]. First antibacterial surfaces were based on leaching biocides (i.e. essential oils) and releasing of metallic ions. Later new types of antibacterial surfaces were developed: superhydrophobic surfaces and surfaces encapsulated with photoactive materials which produce ROS [73]. Surface energy, wettability, surface charges, surface roughness affect the antibacterial activity of designed antibacterial surfaces as well [73].

CQDs belong to novel types of materials with good antibacterial properties. They generate very toxic form of molecular oxygen-singlet oxygen under blue light irradiation. Thus they could be used as antibacterial agent against many bacterial strains. Hazarika et al. claim that CQDs/TiO$_2$ containing waterborne hyperbranched polyester nanocomposite have very strong antibacterial activity especially toward *Staphylococcus aureus (S. aureus), Bacillus subtilis (B. subtilis), Klebsiella pneumoniae (K. pneumoniae)* and *Pseudomonas aeruginosa (P. aeruginosa)* [74]. Duarah et al. reported that hyperbranched polyurethane CQDs/Ag nanocomposites showed very high antibacterial potency against *Escherichia coli MTCC 40 (E. coli)* and *S. aureus MTCC 3160* [75]. Kováčová et al. showed that CQDs polyurethane based nanocomposites can be used as self-cleaning surface for removing of various bacterial pathogens such as *E. coli* and *S. aureus* [69]. CQDs encapsulated in polydimethylsiloxane have induced high mortality of *E. coli, S. aureus* and *K. pneumonie* [73]. CQDs deposited in the form of thin films by Langmuir-Blodgett method on SiO$_2$/Si substrates have very good antibacterial activity against *E. coli* and *S. aureus* [76]. Antibiofouling testing of these thin films showed moderate effect on biofilms of *Bacillus cereus (B. cereus)* and *P. aeruginosa*. Park et al. used polyethylene glycol as a precursor to synthesize CQDs in atmospheric plasma involving reactive gases such as nitrogen and oxygen [62]. These dots show low cytotoxicity and good antibacterial activity against *E. coli* and *Acinetobacter baumannii (A. Baumanni)*. Meziani et al. synthesized CQDs functionalized with 2,2′-(Ethylenedioxy)bis(ethylamine) (EDA) [61]. These dots inhibited the growth of *E. coli* bacterial strains under household LED lighting or ambient laboratory light conditions. Travlou and co-authors reported that N, S doped CQDs synthesized by a simple hydrothermal method had

showed antibacterial activity toward *E. coli* and *B. subtilis* [44]. Their antibacterial activity is affected by amides and amines presented on the plane surface of
N-doped CQDs. They cause bacterial death due to the electrostatic interactions
between their protonated forms and the lipids of the bacterial cell membrane. On
the contrary, S-doped CQDs containing mainly a negatively charged surface showed
a size dependent rather than a chemistry dependent (electrostatic interactions) inhibition of *B. subtilis* growth. Dong et al. claim that the CQDs functionalized with
EDA and mixed with H_2O_2 show synergistic effect and could reach the goal of
inhibiting bacteria growth by using lower concentration of each individual chemical in the combination than using one chemical treatment alone [77]. By using
other chemicals such as Na_2CO_3 and acetic acid for mixing with EDA functionalized CQDs did not show any synergistic effect. Awak and co-workers reported
that antibacterial action of CQDs functionalized with EDA was correlated with their
fluorescence quantum yields [78]. CQDs-EDA with higher fluorescence quantum
yield show higher antibacterial activity compared to dots with lower one under the
same conditions both for Gram-positive and Gram-negative bacterial strains. Yang
et al. prepared quaternized CQDs which had simultaneous antibacterial and bacterial differentiation capabilities [79]. These authors used simple carboxyl-amine reaction between lauryl betaine and amine-functionalized CQDs. Besides prolong and
polarity-sensitive fluorescence emission, these dots have the capability to selectively
attach to Gram-positive bacteria such as *S. aureus* due to presence of hydrophobic hydrocarbon chains and positively charged quaternary ammonium groups on
their surface. Roy et al. described that silver nanoparticles decorated with CQDs
had showed very good antibacterial activity toward *S. aureus* and *E. coli* [80]. Otis
et al. investigated the effect of CQDs synthesized from aminoguanidine and citric
acid precursors on the growth of *P. aeruginosa* [81]. *P. aeruginosa* is a prominent
pathogen associated with pneumonia, urinary tract infections, and surgical wound
infections [82]. The authors claim that CQDs synthesized from aminoguanidine and
citric acid can selectively stain and inhibit the growth of this bacteria. Kuang and
co-workers investigated antibacterial action of ZnO nanorods embedded by CQDs
prepared by in-situ sol-gel chemistry [83]. They established that these modified ZnO
nanorods had strong antibacterial activity. Namely, at concentration of 0.1 mg/L was
able to kill more than 96% of bacteria. Thakur et al. established that CQDs prepared from gum arabic and conjugated with ciprofloxacin had strong antibacterial
activity against both Gram-positive and Gram-negative bacteria [84]. Marković et al.
claim that N-CQDs are an excellent photodynamic antibacterial agent for treatment
of bacterial infections induced by *Enterobacter aerogenes (E. aerogenes), Proteus
mirabilis (P. mirabilis), Staphylococcus saprophyticus (S. saprophyticus), Listeria
monocytogenes (L. monocytogenes), Salmonella typhimurium (S. typhimurium)* and
K. pneumoniae [58]. Moradlou et al. recently reported that CQDs@hematite nanostructures deposited on titanium were bactericidal against *E. coli* and *S. aureus* [85].
These samples were tested against bacteria strains in dark and visible light conditions.
Better results were obtained during irradiation with tungsten lamp.

4   Mechanism of Antibacterial Activity of Carbon Quantum Dots

Mechanism of antibacterial activity of CQDs is a complex phenomenon due to many parameters which can affect it. One of the parameters improving the antibacterial activity of CQDs is doping of CQDs, i.e. doping with nitrogen atoms contributes to formation of amine and amino groups which enhance antibacterial activity of CQDs. In that case, the electrostatic interaction between protonated forms of amines and amides and the lipids of bacterial membrane induces bacterial dead. Travlou et al. emphasize that the bactericidal activity of the CQDS was linked to their specific surface chemistry, and their sizes in the range of nanometers [44]. But under blue light irradiation CQDs become photoactive, i.e. generate ROS (especially singlet oxygen)—Fig. 6. Generated singlet oxygen attacks bacterial membrane wall, increases its porosity and contributes to lipid peroxidation. In this way, CQDs induce bacterial death—Fig. 7a–c. Figure 7a–c shows how CQDs in different forms (colloidal sample, polymer sample and thin film sample) kill bacteria under blue light. Lifetime of generated singlet oxygen contributes to bactericidal efficiency of CQDs/polymer (polyurethane, polydimethylsiloxane) surfaces. Namely, the lifetime of singlet oxygen in CQDs/polydymethylsiloxane samples is almost 4 times higher than in CQDs/polyurethane samples and the former samples kill *S. aures* and *E. coli* completely only after 15 min compared to CQDs/polyurethane samples which kill the same bacteria strains for 60 min [69, 73]. In the case of N-doped CQDs there is a synergetic effect of ROS and amines and amides on bacterial death. Marković et al. established that amino groups are adsorbed onto the bacterial membrane wall [58]. Thus absorption provides the molecules bearing these functional groups are diffusing into the cell interior, where the disruption of the cytoplasmic membrane finally leads to the bacterial cell destruction [44, 86].

Furthermore, the action of N-doped CQDs on bacterial membrane wall can be twofold: generated singlet oxygen penetrates membrane wall whereas amino groups adsorb on membrane wall thus enabling N-doped CQDs to enter bacteria cell and cause oxidative stress [58].

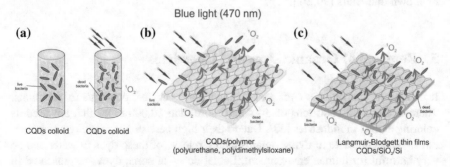

Fig. 7 CQDs as a bactericidal agent under blue light irradiation

When bacterial strains (*E. coli* and *S. aureus*) are deposited on the surface of CQDs@hematite nanostructures, they can be inactivated by the following way: under visible light electron-hole pairs are generated on the surface of these nanostructures. The electrons with high reducing power in the conduction band can react with molecular oxygen to produce reactive superoxide anion radicals, and the holes in the valence band of semiconductor can abstract electrons from water to generate reactive hydroxyl radicals through an oxidative process [85]. In the case of CQDs@TiO$_2$ nanocomposites, Hazarika et al. claim that TiO$_2$ generates ROS [74]. Namely, upon exposure to light, electrons of TiO$_2$ transfer from valence band to conduction band and thus form holes in the valence band. At the same time, CQDs absorb the visible light, emit shorter wavelength and TiO$_2$ nanoparticles are excited again. In this way, CQDs help forming electron/hole pairs which subsequently react with O$_2$ through a reductive process and H$_2$O through an oxidative process to produce active oxygen radicals like superoxide anion and hydroxyl radicals.

By producing ROS, CQDs contributes to bacterial death. But in numerous investigations it was established that CQDs have low dark cytotoxicity [69, 76, 87]. In our previous research it was found that cancer cells and normal cells might be more resistant to phototoxicity of GQDs than bacteria [64]. The reason for that might be the level of isocitrate dehydrogenase (IDPc) in the cells. Kim et al. reported that the level of IDPc in the cells is determined by singlet oxygen [88]. The cells with low level of IDPc are more sensitive to death by singlet oxygen. Thus bacteria strains (*S. aureus*, *E. coli*) and cells (adenocarcinomic human alveolar basal epithelial cells and mouse fibroplast embryonic cells) have different mechanisms to cope with oxidative stress [89].

Apart from ROS generation and specific role of surface functional groups present on the basal plane and edges of CQDs, the very important parameters which contribute significantly to bacterial death are surface wettability and roughness. Surfaces with moderate wettability are more favorable compared to high hydrophobic or hydrophilic surfaces whereas the effect of surface roughness is limited due to the shape and size of bacteria [73]. Materials with low surface energy adhere bacteria weaker than materials with higher surface energy. Katsikogianni and Missirlis reported that bacteria dominantly adhere to surface topography features as large as their own diameters [90, 91].

5   Removal of Organic Dyes and Chemicals

It is thought that organic dyes and chemicals are very serious aquatic pollutants. There is about 700,000 tons of dyes and more than 10,000 types that are used as coloring agents in industries [38]. Due to their high persistence, toxicity and potential to bioaccumulate in living organisms the release of these dyes in water can be very harmful for human environment. Lethal doses of some dyes are presented in ref. [15]. Different materials and methods should be applied to remove these pollutants from water [38, 71]. One of the most used techniques for removal organic dyes from water is absorption technique due to its low cost and flexibility in design. Any harmful substances do not remain after destroying the target compound. CQDs and their composites with other materials can be used for effective removal of organic dyes and chemicals. Prekodravac et al. reported that N-doped CQDs produced in microwave reactor for only 1 min degraded almost 93% Rose Bengal (RB) dye for only 30 min under visible light [92]. Kováčová et al. showed that hydrophobic CQDs encapsulated into polyurethane foil had very good photocatalytic property under blue light irradiation [69]. Namely, after 90 min blue light irradiation organic dye RB was degraded about 90%. Budimir et al. investigated the effect of gamma rays on structural and photocatalytic properties of CQDs/polyurethane nanocomposites [93]. Obtained results indicate dose dependent photocatalytic activity of gamma rays modified CQDs/polyurethane nanocomposites. The best results are obtained for 200 kGy irradiated sample which degraded RB dye almost 97% for 180 min [93]. Recently, Zhou et al. studied the size dependent photocatalytic activity of CQDs against organic dyes (RhB and MB) [94]. Authors found out that the smallest size CQDs with narrow band gap of 2.04 eV have best photocatalytic capability and excellent stability. The authors showed that 2 nm size CQDs were capable to photo-degrade p-nitrophenol (70% for 150 min). Saud et al. reported synthesis of CQDs/TiO_2 composite nanofibers and their usage as photocatalytic and antibacterial agents [95]. TiO_2 is well known catalyst under UV light. But anchoring of CQDs to TiO_2 fibers contributes to spreading of catalyst acting to visible light region. These composite nanofibers have been shown as very good photocatalytic agents in visible light region for methylene blue (MB) organic dye degradation (almost completely degraded MB for 95 min). CQDs/TiO_2 composite nanofibers have shown very high antibacterial activity toward *E. coli* as well. Zhang et al. claim that N doped CQDs/TiO_2 hybrid composite has very good photocatalytic activity especially toward MB under visible light [96]. They reported that these hybrid composites had degraded MB for 86.9% within 420 min. Elkodous et al. prepared macro-mesoporous TiO_2 nanospheres (MMPT) doped with CQDs to enhance photocatalytic activity toward MB organic dye [97]. Obtained results indicate that CQDs@MMPT 30% sample shows the highest photocatalytic activity among all samples due to the relatively higher surface area than CQDs@MMPT 32% sample and the better particle sizes than C-dots@MMPT 29% and CQDs@MMPT 28% samples, respectively. Thus it promotes the formation of relatively increased pores (macropores and mesopores) which facilitated the diffusion of reactants, provided a large number of readily accessible active sites and fostered the light transfer into the inner surface of the photocatalyst. Martins et al. reported better photocatalytic activity of CQDs/TiO_2 nanocomposites toward MB compared to TiO_2 itself (91% for 60 min under UV) [98]. This nanocomposite exhibits a NO conversion (27.0%) more than two times higher of that observed for TiO_2 (10%) under visible light. The selectivity of the process is increased from 37.4 to 49.3%. Zhang et al. claim that N-CQDs/TiO_2 nanocomposites have better photocatalytic activity toward Rhodamine B (RhB) than N-CQDs itself under visible light (more than 95% for 30 min) [99]. Miao et al. achieved to enhance photocatalytic activity of mesoporous TiO_2 prepared by sol-gel method by incorporating of CQDs in it [100]. The CQDs are randomly packed in inverse surfactant micelles and mesopores are formed by interconnected intraparticles. The 5% CQDs/meso-TiO$_2$ can mostly remove MB (98%) under visible light compared to commercial TiO$_2$, which is capable of removing 10% MB under the same conditions. Tian et al. reported the synthesis of novel CQDs/hydrogenated TiO$_2$ nanobelts with improved photocatalytic activity toward MO organic dye in UV/visible/NIR regions [101]. Authors claim that oxygen vacancies and Ti^{3+} ions created by hydrogenation and CQDs play key roles in improving the activity of the TiO$_2$ nanobelts photocatalysts. Oxygen vacancies on the surface of TiO$_2$ nanobelts, associating with Ti^{3+} sites, can enhance UV and visible light photocatalytic activity through improved optical absorption, charge carrier trapping, and prevention of electron-hole recombination. The up-converted CQDs enable the CQDs/H-TiO$_2$ heterostructures to make use of NIR light. Muthulingam et al. prepared N-CQDs/ZnO composites with very good photocatalytic activity toward organic dyes (malachite green, MB and fluorescein) [102]. It was found that the CQD/N-ZnO photocatalyst established a high compatibility to degrade all three commercial dyes within 30–45 min, under daylight irradiation. Also, it can be reused for repeated photocatalytic performance due to anti-photocorrosion offered by CQDs. Bozetine et al. showed that CQDs/ZnO nanocomposites had very good photocatalytic activity toward RhB under visible light irradiation [103]. But the photocatalytic activity of this nanocomposite depends on the preparation conditions. Namely, the highest photocatalytic activity has samples of CQDs/ZnO annealed at 200 °C (94% degradation of RhB for 105 min) compared to samples of CQDs/ZnO annealed at 80 °C (83% degradation of RhB for 105 min) and ZnO nanoparticles themselves (70% under the same conditions). Zhang et al. reported the synthesis of CQDs/ZnO nanoflowers composites with enhanced photocatalytic activity toward RhB organic dye under visible light condition [104]. Namely, ZnO nanoflowers were prepared by combined electrospinning and hydrothermal methods and dispersed in CQDs solution. Results show that the photocatalytic efficiency of ZnO nanoflowers increases gradually from 10 to 98% with the increasing concentration of CQDs from 0.5 to 1 mg/mL. This fact indicates that introducing a suitable amount of CQDs can effectively improve the visible light photocatalytic performance. The authors assumed that the enhanced photocatalytic activity of these composites origin from upconverted photoluminescence behaviour of the CQDs and the novel 3-D structure of the ZnO. Ding et al. prepared ZnO foam/CQDs nanocomposites [105]. They established that ZnO foam consists of a large number of ZnO nanoparticles filling in irregular, hierarchical pores. This nanocomposite showed very good photocatalytic activity against MB, methyl orange (MO) and RhB (MB > RhB > MO) due to synergistic effect of several factors, including the light-trapping effect of ZnO foam, the up-converted photoluminescence behavior and photo-induced electron transfer property of CQDs. Li et al. prepared ZnO/CQDs heterostructure by a sol–gel approach combined with a spin-coating processing [106]. Authors found that the degree of photocatalytic activity enhancement of this heterostructure strongly depends on the layers of CQDs coating on the surface of ZnO layer. The heterostructure with 4 layers of CQDs shows the highest photocatalytic activity (3 times higher compared to pristine ZnO). Zhang et al.

shown very good photocatalytic property of CQDs/AgPO$_4$ nanocomposites. Furthermore, the highest photocatalytic activity has shown CQDs/Ag/AgPO$_4$ nanocomposites (degrade MO completely only after 10 min irradiation by visible light) [107]. Zhang et al. claimed that Fe$_2$O$_3$/CQDs nanocomposites had very high photocatalytic activity toward benzene and methanol. Benzene can be eliminated by Fe$_2$O$_3$/CQDs nanocomposites about 80% for 24 h visible light irradiation whereas methanol is degraded for 70% after 24 h. Without CQDs, Fe$_2$O$_3$ showed very weak photocatalytic activity [108]. Tadesse et al. claimed that Fe$_2$O$_3$/NCQDs effectively degraded MB in 20 min till 90.84% [109]. Hazarika et al. showed very good efficiency of CQDs/TiO$_2$ containing waterborne hyperbranched polyester nanocomposite toward MO, MB, phenol, B-phenol [110]. MO is degraded after 60 min of sunlight exposure in the presence of this nanocomposite whereas MB is degraded after 120 min of sunlight exposure. Lower degradation percentage of MB in the mixture is mainly related with the occurrence of the oxidizing species as well as presence of N = N linkage that makes MO more reactive, while the presence of –CH$_3$ groups in MB makes it resistance to photodegradation [110]. The rate of degradation of this nanocomposite could be enhanced by increasing the loading of nanocomposite. In this way, the number of absorbed photons and consequently the number of the adsorbed contaminants is increased. Zhang et al. established that CQDs/Bi$_2$MoO$_6$ had very good photocatalytic potential especially toward RhB under simulated solar light irradiation [111]. RhB is degraded almost 97.1% after 50 min of irradiation. The content of CQDs in this composite affect its photocatalytic activity. The higher content of CQDs in the composites can compete for light absorption, and the availability of light for RhB degradation is decreased. The same group of author claim that CQDs/Bi$_2$MoO$_6$ due to up-converting property of CQDs has photocatalytic property in the near infrared (NIR) region. Namely, the absorption spectra of RhB is changed during irradiation (absorption peak at 552 nm decreases and blue shifts). Thus CQDs/Bi$_2$MoO$_6$ composite can harvest NIR light to realize pollutant degradation. Wang et al. prepared CQDs/Bi$_2$WO$_6$ hybrid with good photocatalytic property toward MO and bisphenol A [112]. Kannan et al. prepared cerium oxide heteroatom doped CQDs/RGO (HDCQD@RGO nanohybrid catalyst for the degradation of MO and MB [113]. As heteroatoms are used sulfur and nitrogen. The catalytic activity of cerium oxide was improved significantly by incorporating heteroatoms in CQDs/RGO catalyst. In this hybrid, CQDs improves the interaction between cerium oxide and reduced graphene oxide. At the same time CQDs serve as a sensitizer for the electron-transfer process with cerium oxide. The nanostructured cerium oxide possesses multiple oxygen vacancies. Thus it generates singlet oxygen and hydroxyl radicals. Generation of these radicals contributes to the photodegradation of MO (about 95% for 60 min visible light irradiation) and MB (about 81% for 60 min visible light irradiation [113]. Liu et al. reported that CQDs doped CdS (C/CdS) showed enhanced photocatalytic activity toward degradation of RhB under the solar-simulated light irradiation [114]. Furthermore, under visible light irradiation, a higher and more stable photocurrent is generated at the C/CdS electrode. Results showed that degradation of RhB was about 50% after 60 min irradiation of CdS microspheres under visible light. But in the presence of 1% CQDs in CdS microspheres, degradation of RhB was 90%.

The stability of this catalyst for degradation of RhB for three cycles is very good indicating that C/CdS composites are well-reproductive catalysts.

Based on reports mentioned above we assume that CQDs themselves are capable to absorb long wavelength photons in the visible region and emit the light of shorter wavelength in UV region and vice versa [115]. At the same time they can inhibit the recombination of the electron-hole pairs generated during photocatalysis. By anchoring to different catalysts (i.e. TiO_2, ZnO) the photocatalytic property of these catalysts are improved significantly against different organic dyes (i.e. RB, MB, MO or RhB) [115].

CQDs can effectively remove radionuclides from water as well. Co-immobilization of cationic and anionic radionuclides is highly desirable for total remediation of radioactive wastewater. $MgAl-NO_3$ layered double hydroxide (LDH) modified with CQDs can be used as an effective adsorbent for total remediation of anionic and cationic radioactive nuclides (Sr^{2+} and SeO_4^{2-}) from wastewater [116]. Adsorption of SeO_4^{2-} and Sr^{2+} on $MgAl-NO_3-LDH/CQDs$ composites showed that the Sr^{2+} immobilization capacities increased with an increase in the amount of CQDs. The mechanism of Sr^{2+} adsorption on these composites occurs via coordination with the $-COO-$ group of CQDs, whereas that of SeO_4^{2-} occurs through ion exchange with NO_3^- in the interlayer galleries of LDH.

6 Mechanism of Photocatalytic Activity of Carbon Quantum Dots

To be a good photocatalytic agent CQDs must be excellent acceptors and electron donors whereas photocatalytic activity itself strongly depends on transport and separation efficiencies of electrons and holes produced under light. Photocatalytic activity of CQDs is affected by band structure and interfacial interaction. These parameters can be tuned by CQDs size, doping level, attached functional groups [87, 94]. N doping of CQDs affects the work function whereas hydrogen bonding to CQDs surface bends the energy band upward [99, 117].

Figure 8 presents the mechanism of photocatalytic activity of CQDs. In general, under ambient light irradiation of CQDs, electron-hole pairs are generated. The holes in the valence band of CQDs abstract electrons from water and generate reactive hydroxyl radicals which degrade organic dyes (for example RB) [92].

In CQDs@TiO_2 nanocomposites, CQDs reduce the high recombination rate of electron-hole pairs. Thus CQDs with TiO_2 could prevent the charge recombination and expand the range of light wavelengths to be adsorbed [110]. In CQDs/Bi_2WO_6 hybrids photocatalytic activity can be achieved by generation of hydroxyl and superoxide anion radicals, and holes [112]. Namely, CQDs absorb NIR light and after that emit light with shorter wavelength (due to their up-converting properties). In this way they excite m-Bi_2WO_6 to form electron-hole pairs and consequently this nanohybrid can absorb the full spectrum light. At the same time, CQDs can trap

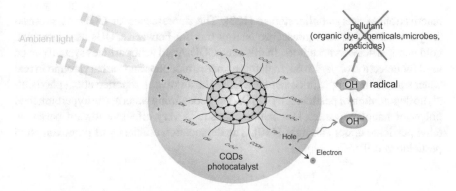

Fig. 8 CQDs as a photocatalytic agent under ambient light irradiation

electron emitted by m-Bi$_2$WO$_6$ due to visible light irradiation, inhibiting the recombination of electron-hole pairs. Superoxide anions can be formed by the combination of electrons with O$_2$ on the surfaces of CQDs thus degrading the bisphenol A. The holes from the valence zone of m-Bi$_2$WO$_6$ oxidize H$_2$O to form hydroxyl radicals which degrade MO [112].

In CeOx-HDCQD@RGO nanohybrids after visible light irradiation the CeOx causes an electron from the conduction band and a hole from the valence band to diffuse to the CeOx. The generated active oxygen molecule is a free radical and possesses the strong oxidizing ability, thereby effectively degrading the dye molecule (MO and MB) [113]. In C/CdS composites, electrons transfer from CdS surface to CQDs. The photogenerated electron-hole pairs then react with the adsorbed oxidants/reductants (O$_2$/OH$^-$) to produce active superoxide anion and hydroxyl radicals which subsequently cause degradation of RhB [114]. Prekodravac et al. showed that bare CQDs degraded RB due to generation of hydroxyl radicals under visible light conditions whereas Zhou et al. claim that photocatalytic degradation of organic compounds is mainly due to the hole in valence bands and the generated superoxide anion radicals [92, 94].

7 Removal of Pesticides by Carbon Quantum Dots

The development of sensors for successful monitoring of different pesticides is very important due to overusing of pesticides in different areas. There are a few reports referring to CQDs as sensing platform for determination of pesticides such as paraoxon, malathion, methyl parathion, and paraoxon-ethyl, and one commercial product, MAPA MALATHION EC 57 (MM57), flumioxazin [118–120]. Lin et al. used CQDs for detection of chlorpyrifos (pesticide) based on the fact that the fluorescence intensity of CQDs was proportional to pesticide concentration [121]. Korram et al. used CQDs/gold nanoparticle system as a probe for the inhibition and reactivation of acetylcholinesterase [122]. The fluorescence of CQDs is successfully quenched due to fluorescence energy transfer between CQDs as donors and gold nanoparticles as acceptors. In this way, CQDs/gold nanoparticles system can be used for detection of paraoxon, malathion, methamidophos and carbaryl found in real water samples and apple juice samples. Only Hazarika et al. reported about photocatalytic degradation of pesticides by CQDs/TiO$_2$ containing waterborne hyperbranched polyester nanocomposite. This nanocomposite is very effective toward paraoxon ethyl pesticide under sun light [110]. The degradation efficiency of paraoxon ethyl pesticide was 91%.

8   Conclusions

In this chapter we discussed possible usage of CQDs as photocatalytic and antibacterial agents in treating aquatic pollution. CQDs are novel class of carbon based nanomaterials with lateral dimension smaller than 10 nm, high photoluminescence, good thermal and chemical stability and resistance to photobleaching. Due to these unique properties, CQDs can be used in different fields. As photoactive materials, i.e. ability to generate ROS under blue light irradiation, CQDs can be used as bactericidal agent especially for destroying bacteria which cause nosocomial infection. Under visible light irradiation, CQDs can be used as photocatalytic agent due to production of hydroxyl radicals which degrade chemicals, pesticides and organic dyes. Thus CQDs have a promising future as antibacterial and photocatalytic agents.

Acknowledgements Authors thank for support to the Ministry of Education, Science and Technological Development of the Republic of Serbia via project no. 172003.

References

1. Lumb, G., Clare, A.S.: The problems of water pollution: an overview. Pa Med **95**, 28–32 (1992)
2. Inyinbor Adejumoke, A., Adebesin Babatunde, O., Oluyori Abimbola, P., Adelani Akande Tabitha, A., Dada Adewumi, O., Oreofe Toyin, A.: A water pollution: effects, prevention, and climatic impact. In: Glavan, M. (ed.) Water Challenges of an Urbanizing World, pp. 34–35. Intechopen (2018)
3. Owa, F.W.: Water pollution: sources, effects, control and management. Int. Lett. Nat. Sci. **3**, 1–6 (2014)
4. Konstantinova, E., Shalaumova, Y., Maslakova, T., Varaksin, A., Zhivoderov, A.: Effects of environmental radioactive pollution on the cardiovascular systems of Ural region residents: a comparative study. Int. J. Med. Res. Health Sci. **7**, 1–7 (2018)
5. Jarup, L.: Hazards of heavy metal contamination. Brit. Med. Bull. **68**, 167–182 (2003)
6. Centers for Disease Control and Prevention: Estimates of Healthcare-associated infections. Available at: http://www.cdc.gov/ncidod/dhqp/hai.html. Accessed 6 Aug 2019
7. Anaissie, E.J., Penzak, S.R., Dignani, C.: The hospital water supply as a source of nosocomial infections—a plea for action. Arch. Int. Med. **162**, 1483Y1492 (2002)

8. https://www.novatx.com/antimicrobial-testing/7-examples-of-harmful-bacteria-found-in-water/visited 7/26/2019
9. Clevenger, T, Wu, Y, Degruson, E, Brazos, B, Banerji, S.: Comparison of the inactivation of *Bacillus subtilis* spores and MS2 bacteriophage by MIOX, ClorTec and hypochlorite. J. Appl. Microbiol. **103**, 2285–2290 (2007)
10. Loret, J.F., Robert, S., Thomas, V., Cooper, A.J., McCoy, W.F., Lévi, Y.: Comparison of disinfectants for biofilm, protozoa and *Legionella* control. J. Water Health **3**, 423–433 (2005)
11. Chen, C., Zhang, X.J., He, W.J., Han, H.D.: Simultaneous control of microorganisms and disinfection by-products by sequential chlorination. Biomed. Environ. Sci. **20**, 119–125 (2007)
12. Cachafeiro, S.P., Naveira, I.M., Garcia, I.G.: Is copper-silver ionisation safe and effective in controlling *Legionella*? J. Hosp. Infect. **67**, 209–216 (2007)
13. Nessim, Y, Gehr, R.: Fouling mechanisms in a laboratory-scale UV disinfection system. Water Environ. Res. **78**, 2311–2323 (2006)
14. Cervia, J.S., Ortolano, G.A., Canonica, F.P.: Hospital tap water: a reservoir of risk for health care-associated infection. Infect. Dis. Clin. Practice **16**, 349–353 (2008)
15. Kaykhaii, M., Sasani, M., Marghzari, S.: Removal of dyes from the environment by adsorption process. Chem. Mater. Eng. **6**, 31–35 (2018)
16. Shanker, U., Rani, M., Jassal, V.: Degradation of hazardous organic dyes in water by nanomaterials. Environ. Chem. Lett. **15**, 623–642 (2017)
17. Richardson, S.D., Thruston Jr., A.D., Caughran, T.V., Chen, P.H., Collette, T.W., Schenck, K.M., Lykins Jr., B.W., Rav-Acha, C., Glezer, V.: Identification of new drinking water disinfection by - products from ozone, chlorine dioxide, chloramine, and chlorine. Water Air Soil Poll. **123**, 95–102 (2000)
18. Heinlaan, M., Ivask, A., Blinova, I., Dubourguier, H.C., Kahru, A.: Toxicity of nanosized and bulk ZnO, CuO and TiO_2 to bacteria *Vibrio fischeri* and crustaceans *Daphnia magna* and *Thamnocephalus platyurus*. Chemosphere **71**, 1308–1316 (2008)
19. Mabey, T., Cristaldi, D.A., Oyston, P., Lymer, K.P., Stulz, E., Wilks, S., Keevil, C.W., Zhang, X.L.: Bacteria and nanosilver: the quest for optimal production. Crit. Rev. Biotechnol. **39**, 272–287 (2019)
20. Benetti, G., Cavaliere, E., Brescia, R., Salassi, S., Ferrando, R., Vantomme, A., Pallecchi, L., Pollini, S., Boncompagni, S., Fortuni, B., Van Bael, M.J., Banfi, F., Gavioli, L.: Tailored Ag–Cu–Mg multielemental nanoparticles for wide-spectrum antibacterial coating. Nanoscale **11**, 1626–1635 (2019)
21. Liu, Y.H., Kuo, S.C., Yao, B.Y., Fang, Z.S., Lee, Y.T., Chang, Y.C., Chen, T.L., Hu, C.M.L.: Colistin nanoparticle assembly by coacervate complexation with polyanionic peptides for treating drug-resistant gram-negative bacteria. Acta Biomater. **82**, 133–142 (2018)
22. Xi, Y.W., Ge, J., Guo, Y., Lei, B., Ma, P.X.: Biomimetic elastomeric polypeptide-based nanofibrous matrix for overcoming multidrug-resistant bacteria and enhancing full-thickness wound healing/skin regeneration. ACS Nano **12**, 10772–10784 (2018)
23. Shen, W., He, P., Xiao, C.S., Chen, X.S.: From antimicrobial peptides to antimicrobial poly(α-amino acid)s. Adv. Healthcare Mater. **7**, 1800354 (2018)
24. Wang, Q.W., Zhou, H.X., Liu, X.L., Li, T., Jiang, C.J., Song, W.H., Chen, W.: Facet-dependent generation of superoxide radical anions by ZnO nanomaterials under simulated solar light. Environ. Sci. Nano **5**, 2864–2875 (2018)
25. Premanathan, M., Karthikeyan, K., Jeyasubramanian, K., Manivannan, G.: Selective toxicity of ZnO nanoparticles toward Gram-positive bacteria and cancer cells by apoptosis through lipid peroxidation. Nanomed. Nanotechnol. Biol. Med. **7**, 184–192 (2011)
26. Noimark, S., Weiner, J., Noor, N., Allan, E., Williams, C.K., Shaffer, M.S.P., Parkin, I.P.: Dual-mechanism antimicrobial polymer–ZnO nanoparticle and crystal violet-encapsulated silicone. Adv. Funct. Mater. **25**, 1367–1373 (2015)
27. Chen, J., Wang, F.Y.K., Liu, Q.M., Du, J.Z.: Antibacterial polymeric nanostructures for biomedical applications. Chem. Commun. **50**, 14482–14493 (2014)

28. Wang, C., Cui, Q.L., Wang, X.Y., Li, L.D.: Preparation of hybrid gold/polymer nanocomposites and their application in a controlled antibacterial assay. ACS Appl. Mater. Interfaces **8**, 29101–29109 (2016)
29. Da Silva, F.A.G., Queiroz, J.C., Macedo, E.R., Fernandes, A.W.C., Freire, N.B., Da Costa, M.M., De Oliveira, H.P.: Antibacterial behavior of polypyrrole: the influence of morphology and additives incorporation. Mater. Sci. Eng. C **62**, 317–322 (2016)
30. Lee, I., Roh, J., Lee, J., Song, J., Jang, J.: Antibacterial performance of various amine functional polymers coated silica nanoparticles. Polymer **83**, 223–229 (2016)
31. Wang, L., Yuan, Z., Karahan, H.E., Wang, Y., Sui, X., Liu, F., Chen, Y.: Nanocarbon materials in water disinfection: state-of-the-art and future directions. Nanoscale **11**, 9819–9839 (2019)
32. Perathoner, S., Ampelli, C., Chen, S., Passalacqua, R., Su, D., Centi, G.: Photoactive materials based on semiconducting nanocarbons—a challenge opening new possibilities for photocatalysis. J. Energy Chem. **26**, 207–218 (2017)
33. Gupta, V.K., Saleh, T.A.: Sorption of pollutants by porous carbon, carbon nanotubes and fullerene—an overview. Environ. Sci. Pollut. Res. **20**, 2828–2843 (2013)
34. Wang, S., Zhu, Z.H.: Effects of acidic treatment of activated carbons on dye adsorption. Dyes Pigments **75**, 306–314 (2007)
35. Dresselhaus, M.S., Dresselhaus, G., Eklund, P.C.: Science of Fullerenes and Carbon Nanotubes. Academic Press, New York (1996)
36. Yadav, B.C., Kumar, R.: Structure, properties and applications of fullerenes. Int. J. Nanotechnol. Appl. **2**, 15–24 (2008)
37. Marković, Z., Trajković, V.: Biomedical potential of the reactive oxygen species generation and quenching by fullerenes (C_{60}). Biomaterials **29**, 3561–3573 (2008)
38. Shan, S.J., Zhao, Y., Tang, H., Cui, F.Y.: A Mini-review of carbonaceous nanomaterials for removal of contaminants from wastewater. IOP Conf. Ser. Earth Environ. Sci. **68**, 012003 (2017)
39. Kholoud, A.A., Abou, M.M., Reda, E.N., Ammar, A.A., Warthan, A.A.: Carbon nanotubes, science and technology part (I) structure, synthesis and characterisation. Arab. J. Chem. **5**, 1–23 (2012)
40. Zhu, Y., Murali, S., Cai, W., Li, X., Suk, J.W., Potts, J.R., Ruoff, R.S.: Graphene and graphene oxide: synthesis, properties, and applications. Adv. Mater. **22**, 3906–3924 (2010)
41. Zhu, S., Wang, D.: Photocatalysis: basic principles, diverse forms of implementations and emerging scientific opportunities. Adv. Energy Mater. 1700841 (2017)
42. Wang, Y., Hu, A.: Carbon quantum dots: synthesis, properties and applications. J. Mater. Chem. C **2**, 6921–6939 (2014)
43. Lim, S.Y., Shen, W., Gao, Z.: Carbon quantum dots and their applications. Chem. Soc. Rev. **44**, 362–381 (2015)
44. Travlou, N.A., Giannakoudakis, D.A., Algarra, M., Labella, M.A., Rodríguez-Castellón, E., Bandosz, T.J.: S- and N-doped carbon quantum dots: surface chemistry dependent antibacterial activity. Carbon **135**, 104–111 (2018)
45. Dong, Y., Wang, R., Li, H., Shao, J., Chi, Y., Lin, X., Chen, G.: Polyamine-functionalized carbon quantum dots for chemical sensing. Carbon **50**, 2810–2815 (2012)
46. Liu, Y., Xiao, N., Gong, N., Wang, H., Shi, X., Gu, W., Ye, L.: One-step microwave-assisted polyol synthesis of green luminescent carbon dots as optical nanoprobes. Carbon **68**, 258–264 (2014)
47. Chae, A., Choi, Y., Jo, S., Nur'aeni, Paoprasert P., Park, S.Y., In, I.: Microwave-assisted synthesis of fluorescent carbon quantum dots from an A2/B3 monomer set. RSC Adv. **7**, 12663–12669 (2017)
48. Wang, J., Cheng, C., Huang, Y., Zheng, B., Yuan, H., Bo, L., Zheng, M.W., Yang, S.Y., Guo, Y., Xiao, D.: A facile large-scale microwave synthesis of highly fluorescent carbon dots from benzenediol isomers. J. Mater. Chem. C **2**, 5028–5035 (2014)
49. Zhu, S., Meng, Q., Wang, L., Zhang, J., Song, Y., Jin, H., Zhang, K., Sun, H., Wang, H., Yang, B.: Highly photoluminescent carbon dots for multicolor patterning, sensors, and bioimaging. Angew. Chem. Int. Ed. **125**, 3953–3957 (2013)

50. Yang, Z.C., Wang, M., Yong, A.M., Wong, S.Y., Zhang, X.H., Tan, H., Chang, A.Y., Li, X., Wang, J.: Intrinsically fluorescent carbon dots with tunable emission derived from hydrothermal treatment of glucose in the presence of monopotassium phosphate. Chem. Commun. **47**, 11615–11617 (2011)

51. Yang, Y., Cui, J., Zheng, M., Hu, C., Tan, S., Xiao, Y., Yang, Q., Liu, Y.: One-step synthesis of amino-functionalized fluorescent carbon nanoparticles by hydrothermal carbonization of chitosan. Chem. Commun. **48**, 380–382 (2012)

52. De, B., Karak, N.: A green and facile approach for the synthesis of water soluble fluorescent carbon dots from banana juice. RSC Adv. **3**, 8286–8290 (2013)

53. Alam, A.M., Park, B.Y., Ghouri, Z.K., Park, M., Kim, H.Y.: Synthesis of carbon quantum dot from cabbage with down- and up-conversion photoluminescence properties: excellent imaging agent for biomedical application. Green Chem. **17**, 3791–3797 (2015)

54. Arumugam, N., Kim, J.: Synthesis of carbon quantum dots from Broccoli and their ability to detect silver ions. Mater. Lett. **219**, 37–40 (2019)

55. Bhunia, S.K., Saha, A., Maity, A.R., Ray, S.C., Jana, N.R.: Carbon nanoparticle-based fluorescent bioimaging probes. Sci. Rep. **3**, 1473 (2013)

56. Zhu, S., Song, Y., Zhao, X., Shao, J., Zhang, J., Yang, B.: The photoluminescence mechanism in carbon dots (graphene quantum dots, carbon nanodots, and polymer dots): current state and future perspective. Nano Res. **8**, 355–381 (2015)

57. Jovanović, S.P., Marković, Z.M., Syrgiannis, Z., Dramićanin, M.D., Arcudi, F., La Parola, V., Budimir, M., Todorović Marković, B.M.: Enhancing photoluminescence of graphene quantum dots by thermal annealing of the graphite precursor. Mater. Res. Bull. **93**, 183–193 (2017)

58. Marković, Z.M., Jovanović, S.P., Mašković, P.Z., Danko, M., Mičušik, M., Pavlović, V.B., Milivojević, D.D., Kleinova, A., Špitalsky, Z., Todorović Marković, B.M.: Photo-induced antibacterial activity of four graphene based nanomaterials on a wide range of bacteria. RSC Adv. **8**, 31337 (2018)

59. Yoon, H., Chang, Y.H., Song, S.H., Lee, E.S., Jin, S.H., Park, C., Lee, J., Kim, B.H., Kang, H.J., Kim, Y.H., Jeon, S.: Intrinsic photoluminescence emission from subdomained graphene quantum dots. Adv. Mater. **28**, 5255–5261 (2016)

60. Sun, Z., Li, X., Wu, Y., Wei, C., Zeng, H.: Green luminescence origin of carbon quantum dots: specific luminescence bands originate from oxidized carbon groups. New J. Chem. **42**, 4603–4611 (2018)

61. Meziani, M.J., Dong, X., Zhu, L., Jones, L.P., LeCroy, G.E., Yang, F., Wang, S., Wang, P., Zhao, Y., Yang, L., Tripp, R.A., Sun, Y.P.: Visible-light-activated bactericidal functions of carbon "quantum" dots. ACS Appl. Mater. Interfaces **8**, 10761–10766 (2016)

62. Park, S.O., Lee, C.Y., An, H.R., Kim, H., Chul Lee, Y., Changkyun Park, E., Chun, H.S., Yang, H.S., Choi, S.H., Kim, H.S., Kang, K.S., Park, H.G., Kim, J.P., Choi, Y., Lee, J., Lee, H.U.: Advanced carbon dots via plasma-induced surface functionalization for fluorescent and bio-medical applications. Nanoscale **9**, 9210–9217 (2017)

63. Dolmans, D.E., Fukumura, D., Jain, R.K.: Photodynamic therapy for cancer. Nat. Rev. Cancer **3**, 381–387 (2003)

64. Marković, Z., Ristić, B., Arsikin, K., Klisić, D., Harhaji-Trajković, L., Todorović-Marković, B., Kepić, D., Kravić-Stevović, T., Jovanović, S., Milenković, M., Milivojević, D., Bumbaširević, V., Dramićanin, M., Trajković, V.: Graphene quantum dots as autophagy-inducing photodynamic agents. Biomaterials **33**, 7084–7092 (2012)

65. Ge, J., Lan, M., Zhou, B., Liu, W., Guo, L., Wang, H., Jia, Q., Niu, G., Huang, X., Zhou, H., Meng, X., Wang, P., Lee, C.S., Zhang, W., Han, X.A.: Graphene quantum dot photodynamic therapy agent with high singlet oxygen generation. Nat. Commun. **5**, 4596 (2014)

66. Chong, Y., Ge, C., Fang, G., Tian, X., Ma, X., Wen, T., Wamer, W.G., Chen, C., Chai, Z., Yin, J.J.: Crossover between anti- and pro-oxidant activities of graphene quantum dots in the absence or presence of light. ACS Nano **10**, 8690–8699 (2016)

67. Gour, N., Ngo, K.X., Vebert-Nardin, C.: Anti-infectious surfaces achieved by polymer modification. Macromol. Mater. Eng. **299**, 648–668 (2014)

68. Lin, L., Zhang, H., Cui, H., Xu, M., Cao, S., Zheng, G., Dong, M.: Preparation and antibacterial activities of hollow silica − Ag spheres. Colloids Surf. B **101**, 97–100 (2013)
69. Kováčová, M., Marković, Z.M., Humpolíček, P., Mičušík, M., Švajdlenková, H., Kleinová, A., Danko, M., Kubát, P., Vajďák, J., Capáková, Z., Lehocký, M., Münster, L., Todorović Marković, B., Špitalský, Z.: Carbon quantum dots modified polyurethane nanocomposites as effective photocatalytic and antibacterial agents. ACS Biomater. Sci. Eng. **4**, 3983–3993 (2018)
70. Sehmi, S.K., Noimark, S., Weiner, J., Allan, E., MacRobert, A.J., Parkin, I.P.: Potent antibacterial activity of copper embedded into silicone and polyurethane. ACS Appl. Mater. Interfaces. **7**, 22807–22813 (2015)
71. Bovis, M.J., Noimark, S., Woodhams, J.H., Kay, C.W.M., Weiner, J., Peveler, W.J., Correia, A., Wilson, M., Allan, E., Parkin, I.P., MacRobert, A.J.: Photosensitisation studies of silicone polymer doped with methylene blue and nanogold for antimicrobial applications. RSC Adv. **5**, 54830–54842 (2015)
72. Felgenträger, A., Maisch, T., Spath, A., Schroder, J.A., Baumler, W.: Singlet oxygen generation in porphyrin-doped polymeric surface coating enables antimicrobial effects on *Staphylococcus aureus*. Phys. Chem. Chem. Phys. **16**, 20598 (2014)
73. Marković, Z.M., Kováčová, M., Humpolíček, P., Budimir, M.D., Vajďák, J., Kubát, P., Mičušík, M., Švajdlenková, H., Danko, M., Capáková, Z., Lehocký, M., Todorović Marković, B.M., Špitalský, Z.: Antibacterial photodynamic activity of carbon quantum dots/polydimethylsiloxane nanocomposites against *Staphylococcus aureus*, *Escherichia coli* and *Klebsiella pneumoniae*. Photodiagn Photodyn **26**, 342–349 (2019)
74. Hazarika, D., Karak, N.: Photocatalytic degradation of organic contaminants under solar light using carbon dot/titanium dioxide nanohybrid, obtained through a facile approach. Appl. Surf. Sci. **376**, 276–285 (2016)
75. Duarah, R., Singh, Y.P., Gupta, P., Mandal, B.M., Karak, N.: High performance biobased hyperbranched polyurethane/carbon dot-silver nanocomposite: a rapid self-expandable stent. Biofabrication **8**, 045013 (2016)
76. Stanković, N.K., Bodik, M., Šiffalovič, P., Kotlar, M., Mičušik, M., Špitalsky, Z., Danko, M., Milivojević, D.D., Kleinova, A., Kubat, P., Capakova, Z., Humpoliček, P., Lehocky, M., Todorović Marković, B.M., Marković, Z.M.: Antibacterial and antibiofouling properties of light triggered fluorescent hydrophobic carbon quantum dots Langmuir–Blodgett thin films. ACS Sustain. Chem. Eng. **6**:4154−4163 (2018)
77. Dong, X., Al Awak, M., Tomlinson, N., Tang, Y., Sun, Y.P., Yang, L.: Antibacterial effects of carbon dots in combination with other antimicrobial reagents. PLoS ONE **12**, e0185324 (2017)
78. Al Awak, M.M., Wang, P., Wang, S., Tang, Y., Sun, Y.P., Yang, L.: Correlation of carbon dots' light-activated antimicrobial activities and fluorescence quantum yield. RSC Adv. **7**, 30177–30184 (2017)
79. Yang, J., Zhang, X., Ma, Y.H., Gao, G., Chen, X., Jia, H.R., Li, Y.H., Chen, Z., Wu, F.G.: Carbon dot-based platform for simultaneous bacterial distinguishment and antibacterial applications. ACS Appl. Mater. Inter. **84**, 732170–732181 (2016)
80. Roy, A.K., Kim, S.M., Paoprasert, P., Park, S.Y., In, I.: Preparation of biocompatible and antibacterial carbon quantum dots derived from resorcinol and formaldehyde spheres. RSC Adv. **5**, 31677–31682 (2015)
81. Otis, G., Bhattacharya, S., Malka, O., Kolusheva, S., Bolel, P., Porgador, A., Jelinek, R.: Selective labeling and growth inhibition of *Pseudomonas aeruginosa* by aminoguanidine carbon dots. ACS Infect. Dis. **8**, 292–302 (2019)
82. Strateva, T., Yordanov, D.: *Pseudomonas aeruginosa*-a phenomenon of bacterial resistance. J. Med. Microbiol. **58**, 1133–1148 (2009)
83. Kuang, W., Zhong, Q., Ye, X., Yan, Y., Yang, Y., Zhang, J., Huang, L., Tan, S., Shi, Q.: Antibacterial nanorods made of carbon quantum dots-ZnO under visible light irradiation. J. Nanosci. Nanotechnol. **19**, 3982–3990 (2019)

84. Thakur, M., Pandey, S., Mewada, A., Patil, V., Khade, M., Goshi, E., Sharon, M.: Antibiotic conjugated fluorescent carbon dots as a theranostic agent for controlled drug release, bioimaging, and enhanced antimicrobial activity. J. Drug Deliv. **2014**, 282193 (2014)
85. Moradlou, O., Rabiei, Z., Delavari, N.: Antibacterial effects of carbon quantum dots@hematite nanostructures deposited on titanium against Gram-positive and Gram-negative bacteria. J. Photoch. Photobio. A **379**, 144–149 (2019)
86. Reichel, V.: Functionalization of cellulose acetate surfaces for removal the of endocrine disruption compounds, Diploma thesis, Department of Colloid Science: University of Graz, Austria (2012)
87. Yao, B., Huang, H., Liu, Y., Kang, Z.: Carbon dots: a small conundrum. Trends Chem. **1**, 235–246 (2019)
88. Kim, S.Y., Park, J.W.: Cellular defense against singlet oxygeninduced oxidative damage by cytosolic NADP+-dependent isocitrate dehydrogenase. Free Radical Res. **37**, 309–316 (2003)
89. Lushchak, V.: Adaptive response to oxidative stress: bacteria, fungi, plants and animals. Comp. Biochem. Physiol. Part C Toxicol. Pharmacol. **153**, 175–190 (2011)
90. Katsikogianni, M., Missirlis, Y.F.: Concise review of mechanisms of bacterial adhesion to biomaterials and of techniques used in estimating bacteria-material interactions. Eur. Cell Mater. **8**, 37–57 (2004)
91. Ammar, Y., Swailes, D.C., Bridgens, B.N., Chen, J.: Influence of surface roughness on the initial formation of biofilm. Surfl Coat Tech. **284**, 410–416 (2015)
92. Prekodravac, J., Vasiljević, B., Markovića, Z., Jovanović, D., Kleut, D., Špitalský Z, Mičušik, M, Danko, M, Bajuk–Bogdanović, D., Todorović–Marković, B.: Green and facile microwave assisted synthesis of (metal-free) N-doped carbon quantum dots for catalytic applications. Ceramic Int. **45**:17006–17013 (2019)
93. Budimir, M., Marković, Z., Jovanović, D., Vujisić, M., Mičušik, M., Danko, M., Kleinova, A., Svajdlenkova, H., Spitalsky, Z., Todorović Marković, B.: Gamma ray assisted modification of carbon quantum dot/polyurethane nanocomposites: structural, mechanical and photocatalytic study. RSC Adv. **9**, 6278–6286 (2019)
94. Zhou, Y., Zahran, E.M., Quiroga, B.A., Perez, J., Mintz, K.J., Peng, Z., Liyanage, P.Y., Pandey, R.R., Chusuei, C.C., Leblanc, R.M.: Size-dependent photocatalytic activity of carbon dots with surface-state determined photoluminescence. Appl. Catal. B Environ **248**, 157–166 (2019)
95. Saud, P.S., Pant, B., Alam, A.M., Ghouri, Z.K., Park, M., Kim, H.Y.:Carbon quantum dots anchored TiO_2 nanofibers: effective photocatalyst for waste water treatment. Ceramic Int. **l41**, 11953–11959 (2015)
96. Zhang, J., Zhang, X., Dong, S., Zhou, X., Dong, S.: N-doped carbon quantum dots/TiO_2 hybrid composites with enhanced visible light driven photocatalytic activity toward dye wastewater degradation and mechanism insight. J. Photoch. Photobio. A **325**, 104–110 (2016)
97. Elkodous, M.A., Hassaan, A., Pal, K., Ghoneim, A.I., Abdeen, Z. (2018) C-dots dispersed macro-mesoporous TiO_2 photocatalyst for effective waste water treatment. Charact. Appl. Nanomater **1**
98. Martins, N.C.T., Ângelo, J., Girão, A.V., Trindade, T., Andrade, L., Mendes, A.: N-doped carbon quantum dots/TiO_2 composite with improved photocatalytic activity. Appl. Catal. B Environ. **193**, 67–74 (2016)
99. Zhang, Y.Q., Ma, D.K., Zhang, Y.G., Chen, W., Huang, S.M.: N-doped carbon quantum dots for TiO_2-based photocatalysts and dye-sensitized solar cells. Nano Energy **2**, 545–552 (2013)
100. Miao, R., Luo, Z., Zhong, W., Chen, S.Y., Jiang, T., Dutta, B., Nasr, Y., Zhang, Y., Sui, S.L.: Mesoporous TiO_2 modified with carbon quantum dots as a high-performance visible light photocatalyst. Appl. Catal. B Environ. **189**, 26–38 (2016)
101. Tian, J., Leng, Y., Zhao, Z., Xia, Y., Sang, Y., Hao, P., Zhan, J., Lid, M., Liu, H.: Carbon quantum dots/hydrogenated TiO_2 nanobelt heterostructures and their broadspectrum photocatalytic properties under UV, visible, and near-infrared irradiation. Nano Energy **11**, 419–427 (2015)

102. Muthulingama, S., Lee, I.H., Uthirakumar, P.: Highly efficient degradation of dyes by carbon quantum dots/N-doped zinc oxide (CQD/N-ZnO) photocatalyst and its compatibility on three different commercial dyes under daylight. J Colloid Interf. Sci. **455**, 101–109 (2015)

103. Bozetine, H., Wang, Q., Barras, A., Li, M., Hadjersi, T., Szunerits, S., Boukherroub, R.: Green chemistry approach for the synthesis of ZnO–carbon dots nanocomposites with good photocatalytic properties under visible light. J. Colloid Interf. Sci. **465**, 286–294 (2016)

104. Zhang, X., Pan, J., Zhu, C., Sheng, Y., Yan, Z., Wang, Y., Feng, B.: The visible light catalytic properties of carbon quantum dots/ZnO nanoflowers composites. J. Mater. Sci. Mater. Electron. **26**, 2861–2866 (2015)

105. Ding, D., Lan, W., Yang, Z., Zhao, X., Chen, Y., Wang, J., Zhang, X., Zhang, Y., Su, Q., Xie, E.: A simple method for preparing ZnO foam/carbon quantum dots nanocomposite and their photocatalytic applications. Mat. Sci. Semicon. Proc. **47**, 25–31 (2016)

106. Li, Y., Zhang, B.P., Zhao, J.X., Ge, Z.H., Zhao, X.K., Zou, L.: ZnO/carbon quantum dots heterostructure with enhanced photocatalytic properties. Appl. Surf. Sci. **279**, 367–373 (2013)

107. Zhang, H., Huang, H., Ming, H., Li, H., Zhang, L., Liu, Y., Kang, Z.: Carbon quantum dots/Ag_3PO_4 complex photocatalysts with enhanced photocatalytic activity and stability under visible light. J. Mater. Chem. **22**, 10501–10506 (2012)

108. Zhang, H., Ming, H., Lian, S., Huang, H., Li, H., Zhang, L., Liu, Y., Kang, Z., Le, S.T.: Fe_2O_3/carbon quantum dots complex photocatalysts and their enhanced photocatalytic activity under visible light. Dalton Trans. **40**, 10822–10825 (2011)

109. Tadesse, A., Devi, D.R., Hagos, M., Battub, G.R., Basavaiah, K.: Synthesis of nitrogen doped carbon quantum dots/magnetite nanocomposites for efficient removal of methyl blue dye pollutant from contaminated water. RSC Adv. **8**, 8528–8536 (2018)

110. Hazarika, D., Saikia, D., Gupta, K., Mandal, M., Karak, N.: Photoluminescence, self cleaning and photocatalytic behavior of waterborne hyperbranched polyester/carbon dot@TiO_2 nanocomposite. Chem. Select **3**, 6126–6135 (2018)

111. Zhang, Z., Zheng, T., Xua, J., Zeng, H., Zhang, N.: Carbon quantum dots/Bi_2MoO_6 composites with photocatalytic H_2 evolution and near infrared activity. J. Photoch. Photobio. A **346**, 24–31 (2017)

112. Wang, J., Tang, L., Zeng, G., Deng, Y., Dong, H., Liu, Y., Wang, L., Peng, B., Zhang, C., Chen, F.: 0D/2D interface engineering of carbon quantum dots modified Bi_2WO_6 ultrathin nanosheets with enhanced photoactivity for full spectrum lightutilization and mechanism insight. Appl. Catal. B Environ. **222**, 115–123 (2018)

113. Kannan, R., Kim, A.R., Eo, S.K., Kang, S.H., Yoo, D.J.: Facile one-step synthesis of cerium oxide-carbon quantum dots/RGO nanohybrid catalyst and its enhanced photocatalytic activity. Ceramic Int. **43**, 3072–3079 (2017)

114. Liu, Y., Yu, Y.X., Zhang, W.D.: Carbon quantum dots-doped CdS microspheres with enhanced photocatalytic performance. J. Alloy Compd. **569**, 102–110 (2013)

115. Wang, X., Cao, L., Lu, F., Meziani, M.J., Li, H., Qi, G., Zhou, B., Harruff, B.A., Kermarrec, F., Sun, Y.P.: Photoinduced electron transfers with carbon dots. Chem. Commun. 3774–3776 (2009)

116. Koilraj, P., Kamura, Y., Sasaki, K.: Carbon-dot-decorated layered double hydroxide nanocomposites as a multifunctional environmental material for Co-immobilization of SeO_4^{2-} and Sr_2^+ from aqueous solutions. ACS Sustainable Chem Eng **5**, 9053–9064 (2017)

117. Yang, P., Zhao, J., Zhang, L., Li, L., Zhu, Z.: Intramolecular hydrogen bonds quench photoluminescence and enhance photocatalytic activity of carbon nanodots. Chemistry **21**, 8561–8568 (2015)

118. Wu, X., Song, Y., Yan, X., Zhu, C., Ma, Y., Du, D., Lin, Y.: Carbon quantum dots as fluorescence resonance energy transfer sensors for organophosphate pesticides determination. Biosens. Bioelectron. **94**, 292–297 (2017)

119. Fung, M.C.M.: The development of carbon nanodots as fluorescent receptor for the detection of organophosphate pesticides. MS thesis, Faculty of Engineering, Computing and Science Swinburne University of Technology Sarawak Campus, Malaysia (2016)

120. Panda, S., Jadav, A., Panda, N., Mohapatra, S.: A novel carbon quantum dot-based fluorescent nanosensor for selective detection of flumioxazin in real samples. New J. Chem. **42**, 2074–2080 (2018)

121. Lin, B., Yan, Y., Guo, M., Cao, Y., Yu, Y., Zhang, T., Huang, Y., Wu, D.: Modification-free carbon dots as turn-on fluorescence probe for detection of organophosphorus pesticides. Food Chem. **245**, 1176–1182 (2018)

122. Korram, J., Dewangan, L., Nagwanshi, R., Karbhal, I., Ghosha, K.K., Satnami, M.L.: A carbon quantum dot–gold nanoparticle system as a probe for the inhibition and reactivation of acetylcholinesterase: detection of pesticides. New J. Chem. **43**, 6874–6882 (2019)

Diamond-Based Nanostructured Materials for Detection of Water Contaminants

A. V. Girão, M. A. Neto, F. J. Oliveira and R. F. Silva

Abstract Sustainability is presently one of the most heard watchwords. Water environment is of great importance to all living beings and climate changes, with cleaning and preservation of our largest natural resource as the number one priority. It can only be attained if environmental conscience is put into practice and, involuntarily or not, the scientific community inherited the additional responsibility to cope with the urgency of this matter. This chapter offers a closer look at diamond-based nanostructured materials for detection of water contaminants. This group of synthetic materials has remarkable electrochemical properties towards sensing and analysis of water pollutants such as heavy metals, pesticides or pharmaceutical compounds. Moreover, doped-diamond thin films provide endless applications as electrodes in environmental monitoring but also as strong tools already being applied in environmental remediation. Diamond-based nanostructured electrodes are suitable for the wide range of electrochemical techniques providing fast, straightforward, sensitive, reproducible and robust means of electroanalysis. The versatility of these electrodes also enables in situ and real time application as sensitive and stable sensors also if being tested for determination of contaminants in complex water matrices.

Keywords Diamond · Thin films · Electrochemical sensors · Environmental monitoring · Electroxidation · Water treatment

1 Environmental Monitoring

Sustainability is probably the most noticeable up-to-date word. Its concept is quite complex, particularly if we are no longer able to maintain a good quality level of ecological balance. The human and biosphere relationship has been severely compromised, we are now starting to pay for the constant human abuse of the ecosystem and

A. V. Girão (✉) · M. A. Neto · F. J. Oliveira · R. F. Silva
CICECO—Department of Materials and Ceramic Engineering, Universidade de Aveiro,
Campus de Santiago, 3810-193 Aveiro, Portugal
e-mail: avgirao@ua.pt

© Springer Nature Switzerland AG 2019   147
G. A. B. Gonçalves and P. Marques (eds.), *Nanostructured Materials*
for Treating Aquatic Pollution, Engineering Materials,
https://doi.org/10.1007/978-3-030-33745-2_6

global warming is here to stay with unknown endurance. Thus, it is now imperative that the economic, environmental and social aspects of sustainability are encouraged in the right direction enabling the future generations the capacity to meet their needs and retain that behaviour repeatedly.

Environmental sustainability consists in the rates of renewable resource return, pollution creation and non-renewable resources exhaustion that can be continued indefinitely, as defined by Daly in the 1990s [1]. Nevertheless, one of the three pillars of sustainability has always been the overlooked one. Economic well-being and social fulfilment grew at the cost of environment degradation. The latter is mainly due to human pollution of the environment, with the lack of vision and/or concern for the short, medium and long-term implications.

Pollution can be defined as the direct (or indirect) contamination of the environment in such matter that it causes potential or real harm or damage to all living beings on Earth. Either in the form of streams, lakes, rivers or oceans, water covers around 70% of our planet, becoming the main resource affected by pollution. Moreover, the hydrological cycle is the perpetual circulation of pollutants between atmosphere, soil water, surface water and ground water. Currently, aquatic pollution has become such a multifaceted problem that it needs to be tackled in its different fronts via diverse tactics. Most of the occurring, or emerging, water pollutants have been acknowledged but it still is necessary to identify, quantify and, if possible, accurately eliminate their presence.

The first approach can be carried out by monitoring our aquatic environment quality in order to control it. The process of studying and data collection result in valuable information on the environment state which, in its turn, provides better understanding of the pollution problem. Monitoring can only be carried out using new and better technology, keeping up with the fast advances in microminiaturization of the analytical instrumentation. Sensors, as any other type of analytical equipment, needs to fulfil several requisites in order to be feasible aiming at its recognition and consequent commercialization. Thus, obtaining real-time and in situ information leads to early warning and precise risk assessment with consequences in environment policies, enabling the making of well-informed political, economic and social decisions. Hence, the previously disregarded pillar becomes the starting point to a balanced sustainability.

The main objective of this chapter is to provide insight on the effective application of diamond-based micro/nanostructures used as sensors for aquatic pollutants. This group of materials used as sensors has already proven to be a valuable alternative to the more elaborated and expensive sensors for pollutants detection. The chapter begins to explore the nanoscience opportunities in environmental monitoring, followed by a brief inspection at diamond electrochemistry, in the nanoscale range. The application of diamond-based sensors is also presented according to the nature of the environmental pollutants. The reader will also be provided with general application of such materials onto technology for environmental remediation. Ultimately, the concluding remarks include the challenges and future perspectives for diamond-based micro/nanostructured materials in aquatic pollution.

1.1 Nanoscience and Sensors

Nanotechnology unlocked a wide range of opportunities for environmental purposes. Nanomaterials typically present size and shape dependent physicochemical properties, high reactivity level and large specific surface area [2]. Thus, chemists and engineers have long exploited and tailored the properties of nanomaterials accordingly to their application, including environmental remediation and monitoring. Nanostructures were primarily used in separation processes as sorbents or membranes for removal of sulphur compounds from fuels or heavy metals from wastewater. Heavy metals nanoadsorbents included clays, metal oxides, carbonaceous or metal-based nanomaterials [3]. New trends in analytical analysis were rapidly developed and the sensors area of application broadened environmental monitoring from gas sensors (metal oxides or graphene) to biosensors (enzymes, aptamers or antibodies) [4–8]. Consequently, new and more effective environmental remediation processes have been applied. Therefore, nanotechnology implementation brought numerous advantages in terms of sensors with higher sensitivity and selectivity and enabling measurements at the nanoscale range as well as reducing sampling needed for analysis.

Sensors are tools or devices that are sensitive and responsive to a certain external interaction that takes place via heat, pressure, strain, magnetic or electric field, radiation or chemical reaction. The response and information are then sent to electronics which processes and converts them to a well-known signal for interpretation. Depending on their application, there are several types of nanomaterials-based sensors for environmental monitoring. The existent groups of sensors can be divided according to the type of physicochemical interaction between the sensor and the analyte besides the different signal transduction mechanisms. There are optical sensors based in the phenomena of fluorescence, colorimetry, plasmonics or surface/enhanced Raman scattering. In addition, there are also optofluidic or microfluidic sensors or electrochemical sensors [4–9]. Sensors performance is assessed according to several requirements such as sensitivity, selectivity or detection limit, fast response, stability and robustness. Thus, reliable sensing technology is highly needed including the design and fabrication of platforms for the sensors, real time and in situ monitoring, multitarget detection, portability and possible remote control of such devices. In a sensor quality assessment, production and maintenance costs are also accountable [4–11].

This chapter focuses on the use of diamond-based micro/nanostructured materials as aquatic environmental monitoring electrochemical-sensors. As in any other environmental monitoring media, sensing in aquatic environments must be followed by reliable quantification measurements of the pollutants in order to suitably control water pollution. The next section introduces general concepts regarding diamond-based micro/nanomaterials and their effective application as electrochemical sensors in environmental monitoring of water media. These sensors offer several advantages over many other sensing techniques and nearly fulfilling most of the requirements for a suitable sensor performance.

2   Diamond Electrochemistry

Diamond has always attracted the attention of scientists besides the public one. It is an extremely hard carbon crystalline form with application in many research areas. Its interest has been further extended by the production of polycrystalline diamond (sp^3 carbon) films presenting mechanical and electronic properties like those of naturally occurring diamond. Currently, synthesis of diamond thin films uses the chemical vapor deposition (CVD) technique in which gaseous mixtures of hydrocarbons and hydrogen are thermally, radiofrequency or electrically activated, resulting in diamond growth onto the surface of a nondiamond substrate, at sub-atmospheric pressure. Thus, depending on the gas activation process the CVD technique for diamond growth is better known by direct current (DC)-plasma [12], radiofrequency (RF)-plasma [13], microwave plasma enhanced (MPE)-CVD [14], electron cyclotron resonance (ECR)-plasma [15] and hot-filament (HFCVD) [16]. Other non-CVD methods for diamond thin film deposition have also been employed such as low- and high-energy carbon ion beams, low pressure or low temperature diamond growth or electron bombardment of the substrate [17–19], as illustrated in Fig. 1.

Synthetic strategies have also been reported for distinct diamond micro and nanostructures such as nanoneedles, rods or forests, nanotextures like nanograss, winkles or cones, and networks (Fig. 2), typically tailored recurring to masks or bottom-up approaches, with effective advanced electrochemistry applications [17–19].

Electroanalytical methods use low cost instrumentation, are highly sensitive and present outstanding low detection limits for the multidetermination of trace amounts of many compounds. Therefore, electrochemical and bioelectrochemical methods have become excellent techniques to make use of diamond-based electrodes. Traditional carbon electrodes like glassy carbon, graphite, or carbon fibers, cloths and

Fig. 1 Diagram illustrating different diamond thin film deposition techniques

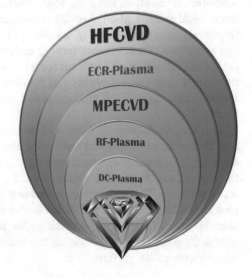

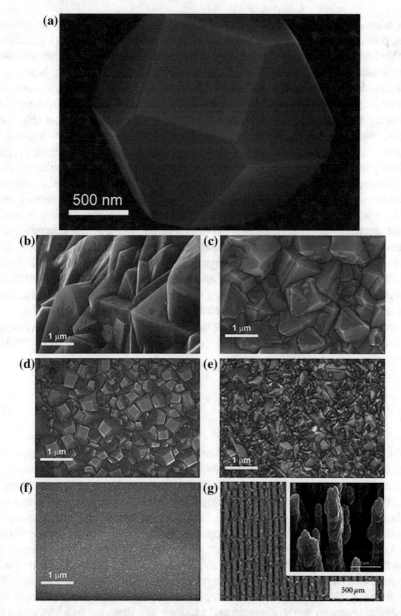

Fig. 2 Scanning electron microscopy images illustrating different diamond micro and nanostructures: **a–d** microcrystalline diamond thin film; **e** sub-microcrystalline diamond thin film; **f** nanocrystalline diamond thin film; **g** nanorod forest. Reproduced with permission from reference [20], Published by Hindawi Publishing Corporation

nanotubes, are low cost and of simple preparation, presenting large surface area and wide potential window of water stability. Although with application in batteries and double-layer capacitors or sensors, these materials still have downsides such as long-term stability or limited detection of compounds oxidizing at relatively high anodic potentials in water. Thus, interest in testing and using diamond thin films as electrodes naturally came into sight bearing in mind the extraordinary properties that nanomaterials have to offer. Moreover, smaller dimension within the micro/nano range benefits the electrodes performance since, theoretically, the measurement rates and mass transport are enhanced and Ohmic resistance is reduced as well as charging currents and the poisonous effects of electrolyte resistance. The interfacial properties and electrochemical behavior of diamond nanostructures have been characterized by different electrochemical techniques and it is generally accepted that the dimensions, morphology and surface termination of the nanostructures play a significant role on diamond electrochemical performance [17–19]. Diamond-based nanostructured electrodes have been explored to some extent as electrochemical and bioelectrochemical sensors for analytes such as nitrite or hydrogen peroxide, and glucose or dopamine [17, 19, 21]. In addition, diamond is chemically and biochemically inert, with little or no cytotoxic response, which has widely opened its use in biomedical instrumentation and devices [22, 23]. Finally, diamond-based nanostructures have also found use in electrochemical energy storage as capacitors/supercapacitors, in electrocatalysis, filtration systems, corrosion inhibition or surface-sensitive adsorption phenomena [19].

2.1 Doped Diamond

In the 1980s, Iwaki et al. [24] was the first one reporting on the use of ion implanted diamond as an electrochemical electrode, and Pleskov et al. [25] studied the photo-electrochemical behavior of an electrode made of semiconducting diamond, followed in 1992 by Fujishima and coworkers [26]. Diamond is a semiconductor with a very wide bandgap (5.47 eV, at 300 K) but like many "insulators", its bandgap can be significantly narrowed down by appropriate doping. The latter is usually carried out using either nitrogen (n-type dopant, activation energy of 1.7 eV) or boron (p-type dopant, activation energy of 0.37 eV), amongst others. Boron is the most used dopant due to its lower activation energy and efficient replacing of carbon within the diamond crystalline structure. Heavy boron doping is then possible, imprinting a metallic character to the diamond and resulting in a diamond thin film with high electric conductivity. Consequently, much of electrochemical research also turned its purposes at exploring boron doped diamond (BDD) electrodes [17, 18, 21, 27]. CVD is the most used synthetic route to prepare BDD thin films since it enables effective control of the doping process and follows the same principle as previously described. Basically, the use of a hot filament (or microwaves) creates a plasma whose resulting high temperature generates carbon radicals and relatively high concentrations of disassociated hydrogen; the carbon/hydrogen gas phase source is fed to the chamber then a boron gas phase is simultaneously fed into the CVD chamber; and, under suitable surface kinetics conditions, BDD growth takes place. Equally important in diamond thin film CVD processes, the deposition conditions (gas composition, temperature, time, pressure, etc.) are vital since they will dictate the final properties of the films such as the surface morphology, growth rate, diamond quality (a sp^2 to sp^3 ratio) [28, 29], grain size and, consequently, orientation of the polycrystalline BDD thin film. A noble gas such as argon can also be added to the gas phase mixture enabling the control of the grain size range of the BDD crystallites. Thus, it is possible to obtain ultra-nanocrystalline (UNC) BDD with grain size <10 nm, nanocrystalline (NC) BDD with grain size 10 nm–1 μm and microcrystalline (MC) BDD with grain size >1 μm. Additionally, one can only assume that the boron content is also affected by the crystallites orientation and that polycrystalline BDD thin films are heterogeneously doped. Ultimately, boron doping plays an important role on the final electrochemical properties and consequent applications [17, 18, 21, 27, 30].

Surface termination (H– or O–) is an important aspect in BDD thin films [31]. It affects the electron transfer kinetics of inner sphere redox processes (surface sensitive), wetting properties and polarity of the electrode surface. In H-terminated surfaces (grown samples in most CVD techniques), both valence and conduction band energy levels are increased, but the conduction band is still above the vacuum level revealing a negative electron affinity at the surface which, in its turn, once the electrode contacts an aqueous solution, a positively charged layer is formed at the surface resulting in a measurable surface conductivity. Hydrogenated surfaces are hydrophobic to a certain extent, though they are quite stable in air for a considerable amount of time, slow oxidation always takes place and with time they start showing a less hydrophobic character [32]. Surface hydrogenation can also be achieved by hydrogen plasma treatment or cathodic polarization. In contrast, O-termination of the surfaces are typically hydrophilic (opposite effects of those of H-termination) and is easily accomplished through oxygen plasma treatment, anodic polarization, acid attack at boiling point or alumina polishing [21, 27, 30]. Figure 3 schematically shows the contact angle in hydrogenated and O-terminated BDD surfaces.

Surface assessment and accurate chemical information can be given by X-ray photoelectron spectroscopy (XPS). Another important aspect that should also be considered is surface reactivity attained by chemical, photochemical, thermal or electrochemical functionalization methods, depending on the desired application [18, 21, 27].

BDD thin films offer exceptional electrochemical properties with many advantages over the classic and expensive metal or other sp^2 electrodes. They have a very

Fig. 3 Scheme illustrating the different contact angles observed in hydrogenated (left) and O-terminated (right) BDD surfaces

wide potential window, corrosion stability in aggressive media, and low background currents in aqueous and non-aqueous electrolyte media. Therefore, BDD thin films are subject of considerable interest for the last few decades as an electrode material with effective application in electrochemical and bioelectrochemical techniques. Different morphologies or synergistic effects with other materials to improve the BDD thin films performance have also been explored [33–35]. Other applications such as microelectrodes arrays tested as new amperometric sensors for high-performance liquid chromatography detectors [36, 37], neuroengineering of neural interfaces and networks [38], in vivo nerve electrical recordings [39], or detection of dissolved oxygen and pH [40], or microfluidics [41] were also reported.

3 Doped Diamond Sensors

The identification and/or determination of certain adverse effects on human health is far from straightforward. Frequently, the effect is identified but its risk factor(s) existing in the environment are not always notorious. Therefore, risk factor assessment is essential and completed in four steps: the "*bad*" substance with adverse health effects (hazard) is identified; the circumstances for exposure to the substance are documented; the health effects are categorized; and the probability of occurrence (or risk) of such health effects is estimated. Then, based on that risk assessment, the limit of acceptable concentration of a certain hazard is decided [42]. Many variables are involved, it may take years or decades to accurately assess a risk factor and every so often such concentration can only be determined, in retrospective, like it happened with the Chernobyl nuclear reactor accident.

Water pollutants may be categorized as *point source* corresponding to pollutants generated in industrial facilities and municipal wastewater treatment plants enters watercourses, and *nonpoint source* like storm drainage or farm overflow, construction locations or other land disturbances [43]. The extensive list of aquatic pollutants includes heavy metals such as lead, cadmium, copper or zinc; pesticides like ziram, picloram or propoxur; phenol and its derivatives; pharmaceutical compounds such as neurotransmitters or antibiotics; pathogens like *Escherichia coli*; and potentially hazardous biomolecules such as harmaline or caffeine.

This section is fully dedicated to the application of doped diamond sensors for the detection and/or quantification of specific aquatic pollutants. So far, BDD thin films have been extensively explored and the nanostructured material presenting the most effective application in environmental monitoring. Figure 4 illustrates the different environmental monitoring application of doped-diamond thin films.

Fig. 4 Diagram illustrating
the different environmental
monitoring applications of
doped-diamond thin films

3.1 Heavy Metals

Metal ions, including heavy metals, existing in water above the recommended critical values turn out to be highly toxic [44]. Thus, control of their concentration in water is essential for human health and safety. In the last years, a few BDD sensors have been developed for fast, sensitive and stable identification and/or quantification of metal ions by means of electrochemical and bioelectrochemical techniques. Table 1 summarizes representative results for BDD-based sensors applied in metals environmental monitoring [45].

Ivandini et al. [46] developed an iridium implanted BDD electrode for arsenic detection in a buffered deionized water solution with a limit of detection (LOD) of 20 nM (1.5 ppb), using cyclic voltammetry (CV). Additionally, it was also found that the electrode was also applicable for As analysis in tap water where a significant number of other ions are present as interferents.

Cadmium detection in buffered ionized water was investigated by ultrasound-assisted square wave anodic stripping voltammetry (SWASV) with a LOD of 1.0 $\mu g\ L^{-1}$ [47]. Copper ion is an interferent being simultaneously detected but established method has been successively applied for real-water samples. A similar study was carried out by Sardinha et al. [48] finding a lower limit of detection of 0.016 $\mu g\ L^{-1}$ for cadmium using a NC-BDD electrode. Fierro et al. [49] created a method for generation of calibration curves for cadmium high accuracy detection using an amalgam of Cu–Cd. The latter is deposited onto the electrode surface and then stripped (SWASV) enabling a LOD of 10 ppb. The calibration curves for cadmium quantification remain accurate in the presence of other interferents like Pb, Se, B or Cr. Banks et al. [50] have shown that sonoelectrochemical treatment increased sensitivity for Cd detection with a LOD of 10^{-9} M using SWASV.

One of the first studies applying BDD electrodes for the electrochemical response evaluation of lead content was carried out by Peilin et al. [51] with a LOD of

Table 1 Representative results for application of doped diamond-based sensors to metals environmental monitoring

Metal	Electrochemical technique	Model/real system	Limit of detection (LOD)	Sensitivity	Ref.
Arsenic	CV[a]	Buffered deionized water	20 nM	93 nA μM^{-1} cm^{-2}	[46]
Cadmium	Ultrasound assisted SWASV[b]	Buffered deionized water	1.0 μg L^{-1}	–	[47]
	SWASV[b]	Buffered deionized water	0.016 μg L^{-1}	–	[48]
	SWASV[b]	Buffered deionized water	10 ppb	–	[49]
	SWASV[b]	Buffered deionized water	10^{-9} M	3.78 μA μM^{-1}	[50]
Lead	CV[a]	Deionized water	1.78 μg L^{-1}	–	[51]
	SWASV[b], LSV[d]	Tap water	2 nM	–	[52]
	SWASV[b]	Tap water	0.3 μg L^{-1}	–	[53]
	SWASV[b]	Buffered deionized water	0.57 μg L^{-1}	1.57 μA μg L^{-1}	[54]
	SWASV[b], LSV[d]	River sediment	187.1 mg kg^{-1}	–	[55]
Lead/Copper	SWASV[b], LSV[d]	Deionized water	2.5 × 10^{-6} M–10^{-4} M[e]	–	[56]
Lead/Cadmium	DPASV[c]	Buffered deionized water	0.25 μM (Pb); 0.31 μM (Cd)	–	[57]

(continued)

Table 1 (continued)

Metal	Electrochemical technique	Model/real system	Limit of detection (LOD)	Sensitivity	Ref.
Pb/Cd/Ag/Cu/Zn	DPASV[c]	Lake, river, well and tap waters	Tap water: 29.1 ppb (Cu) Well water: 28.4 ppb (Cu); 64.1 ppb (Zn)	–	[58]
	DPASV[c]	Buffered deionized water	50 ppb (Zn); 1.0 ppb (Cd); 5.0 ppb (Pb); 10 ppb (Cu); 1.0 ppb (Ag)	9.48 nA ppb^{-1} (Zn); 14.7 nA ppb^{-1} (Cd); 13.0 nA ppb^{-1} (Pb); 7.14 nA ppb^{-1} (Cu); 9.12 nA ppb^{-1} (Ag)	[59]
Pb/Cd/Cu/Zn	DPASV[c]	Buffered deionized water	–	7.7 nA ppb^{-1} (Zn); 9 nA ppb^{-1} (Cd); 21.4 nA ppb^{-1} (Pb); 10 nA ppb^{-1} (Cu)	[60]
	DPASV[c]	Seawater	5.1 μgL^{-1} (Cu); 10.65 μgL^{-1} (Pb); 0.99 μgL^{-1} (Cd); 72.45 μgL^{-1} (Zn)	–	[61]
Silver	SWASV[b], LSV[d]	Deionized water	10^{-9} M	–	[62]
Actinides	Electroprecipitation	Buffered deionized water	0.5 BqL^{-1}	–	[63]

[a]Cyclic voltammetry; [b]square wave anodic stripping voltammetry; [c]differential pulse anodic stripping voltammetry; [d]linear sweep voltammetry; [e]response range

1.78 $\mu g\,L^{-1}$. Lead detection has been correlated to the morphology and boron doping level of MC/NC-BDD electrodes with a LOD of 0.57 $\mu g\,L^{-1}$ for the NC-BDD morphology with less boron doping [54]. In fact, Compton's group reported again on the improvement that using the ultrasound technique enables towards lead detection [55]. Dragoe et al. [52] have performed lead detection in tap water with a LOD of 2 nM and, similarly, Chooto et al. [53] proposed a method with which they obtained a LOD of 0.3 $\mu g\,L^{-1}$.

Prado et al. [56] studied the interaction between Pb and Cu during simultaneous detection by SWASV. An extra peak attributed to hydrogen evolution on copper was observed. The interaction between Pb and Cd during Differential pulse anodic stripping voltammetry (DPASV) analysis was investigated by Manivannan et al. [57] and their findings include LOD values of 0.25 μM (Pb) and 0.31 μM (Cd) but a slight increase of those values was verified when the two metals are present, explained by the proposed model based on 3D calibration curve to avoid cross-interference between these two metals. Sonthalia et al. [58] analyzed much more complex matrices in even more complex media such as lake, well and tap waters concluding that further optimization of the BDD electrode was necessary since the stripping peaks were rather broad and the intermetallic compound formation also needs to be addressed. Similarly, McGaw and Swain [59] compared the performance between a BDD electrode and that of Hg-coated glassy carbon (Hg-GC) in a complex matrix containing Pb/Cd/Ag/Cu/Zn. They concluded that BDD had a as good or superior performance to that for Hg-GC, also demonstrating that BDD lower sensitivity was not due to incomplete metal oxidation or metal phase detachment from its surface. Another complex system with Zn, Cd, Pb and Cu was investigated in buffered water samples and difficulties were encountered since copper caused a lot of interference with other metals becoming the major problem for the analysis [60]. Zhao et al. [61] reported on an in situ experiment for Pb/Cd/Cu/Zn simultaneous detection in seawater. It was possible to effectively obtain repeatable results proving suitability of the BDD electrode for detection of heavy metals in seawater.

Applying the same sonoelectrochemical technique, Compton's group also reported on the improvement of silver detection [62]. Detection of trace amounts of radionuclides has been explored by de Sanoit et al. [63] in the design of an electrochemically assisted radiation BDD-based sensor. Direct alpha-particles counting, and alpha spectrometry were used to determine the electrodeposition of the actinides onto the BDD thin films with a LOD of 0.5 $Bq\,L^{-1}$ a work in the process of being patented. Finally, a review on the major contribution of the French scientific academic community in the field of electrochemical sensors and electroanalytical methods within the last 20 years should also be mentioned [44].

As previously observed for several other analytical techniques commonly used in the determination of metals in complex matrices, the competition effect between the ions is a major drawback for selectivity. Nonetheless, BDD electrodes appear to mostly overcome specially because samples derivatization prior to analysis is not essential.

3.2 Pesticides

Pesticides are being extensively used in most sectors of agriculture, livestock or households and their toxicities have been assessed via different assays or using models such as in vitro, in vivo or in situ strategies. Mostly, pesticides are employed to incapacitate, prevent or even kill pest damage to plants. The major problem associated to such compounds is their accumulation environmental persistence. Moreover, there are still not enough devices or approaches to detect pesticides in aquatic media or a large-scale remedy to eliminate them [64]. This type of synthetic compounds can be classified according to their chemical nature and/or their functional application. Thus, pesticides can be subdivided in three groups: insecticides, herbicides and fungicides. For practical reasons, sensors based in BDD for detection of pesticides in water will be divided accordingly to those groups and presented in this sub-section.

As indicated by the name, fungicides are a group of pesticides designed to prevent or destroy parasitic fungi. To the best of our knowledge, we were only able to retrieve one reported work on BDD application for detection of a fungicide. In this work, Stankovich and Kalcher [65] reported on sensitive and selective quantification of ziram using an amperometric procedure with a limit of detection (LOD) estimated to be 2.7 nM. The method was then tested in two river water samples spiked with the contaminant and it showed good potential as an electroanalytical approach for ziram quantification.

Herbicides are used to control or kill unwanted plants in agricultural crops. It should be noted that in most of the literature here presented, the detection methods include testing in specific but real water samples, or spiked ones if the contaminant is not present in the real system. Table 2 summarizes representative results for BDD-based sensors applied in pesticides environmental monitoring.

Spătaru et al. [66] studied the detection of aniline by cathodic stripping voltammetry with satisfactory long-term stability of the response and LOD of 5 μM. A microfluidic lab-on-a-chip was developed by Medina-Sánchez et al. [67] for electrochemical detection and degradation of herbicide atrazine. Chronoamperometry revealed a limit of detection of 3.5 pM for atrazine and remediation was carried out using wide potential window with higher oxygen overvoltage. Another work illustrating the versatility of BDD as a simple, fast, sensitive and accurate technique for environmental monitoring is that for bentazone detection [68]. The sensor showed a LOD of 0.5 μM when using DPV and 1.2 μM if SWASV is the selected technique. Paraquat is a well-known herbicide and it was investigated by Tyszczuk-Rotko et al. [69] by square-wave voltammetry with a low detection limit of 1.5×10^{-9} mol L^{-1}. Another herbicide, picloram, was studied using a BDD electrode in cyclic and differential pulse voltammetry [70]. The obtained LOD was 70 nmol L^{-1} and was successfully applied in analysis of environmental tap and natural water samples as well as in human urine (spiked samples). Triclopyr was investigated by DPV and SWASV electrochemical techniques, giving a lower limit of detection of 0.82 μM (DPV) [71]. The proposed method was then verified by analysis of spiked water and urine with excellent results.

Table 2 Representative results for application of doped diamond-based sensors to pesticides environmental monitoring

	Electrochemical technique	Model/real system	Limit of detection (LOD)	Sensitivity	Ref.
Fungicide					
Ziram	Flow injection amperometry	Buffered deionized water	2 nM	–	[65]
Herbicide					
Aniline	LSV[a]	Buffered deionized water	5 μM	0.08 μA μM^{-1}	[66]
Atrazine	Chronoamperometry	Buffered deionized water	3.5 pM	–	[67]
Bentazone	DPV[b], SWASV[c]	Buffered deionized water	1.2 μM (DPV), 0.5 μM (SWASV)	–	[68]
Paraquat	SWASV[c]	Buffered deionized water	1.5×10^{-9} mol L^{-1}	–	[69]
Picloram	DPV[b]	Deionized water	70 nmol L^{-1}	–	[70]
Triclopyr	DPV[b], SWASV[c]	Tap and natural waters	0.82 μM (DPV); 1.85 μM (SWASV)	–	[71]
Insecticide					
Azamethiphos	SWASV[c]	Ground and river waters	0.45 μM	–	[72]
Carbaryl	SWASV[c]	Deionized water	2.08 μM	–	[73]
Imidacloprid	SWASV[c]	Deionized water	8.60 μM	–	[74]
Mancozeb	PA[d]	Buffered deionized water	5.14×10^{-7} mol L^{-1}	–	[75]
Methomyl	DPV[b], SWASV[c]	Buffered deionized water	1.2×10^{-6} mol L^{-1} (DPV); 1.9×10^{-5} mol L^{-1} (SWASV)	–	[76]

(continued)

Table 2 (continued)

	Electrochemical technique	Model/real system	Limit of detection (LOD)	Sensitivity	Ref.
Pirimicarb	DPV[b], SWASV[c]	Buffered deionized water	1.24 μM (DPV)	–	[77]
Propoxur	DPV[b]	Buffered deionized water	0.50 μM	–	[78]
Imidacloprid/paraoxon	Chronoamperometry	Tap water	33×10^{-2} and 19×10^{-2} μM[e]	–	[79]
Methyl parathion/chlorpyrifos	DPV[b]	Buffered deionized water	1.29×10^{-13} and 4.9×10^{-13} M	–	[80]

[a]Linear sweep voltammetry; [b]differential pulse anodic stripping voltammetry; [c]square wave anodic stripping voltammetry; [d]pulsed amperometry; [e]detection threshold

BDD electrodes have also found wide application in the detection of insecticides used to exterminate insects and corresponding eggs and/or larvae. Azamethiphos is an insecticide and its quantification using boron-doped diamond electrode in SWASV has been reported [72]. The limit of detection was estimated to be 0.45 μM in ground and river water (spiked) samples and negligible effect of the possible interfering compound was observed. Garbellini et al. [73] investigated the use of ultrasound irradiation in association with BDD electrode to determine carbaryl. The SWASV studies revealed that sonication decreased the LOD from 2.96 to 2.08 μM which is explained by the enhancement of amount of analyte reaching the electrode and the latter being simultaneously cleaned reaching a faster equilibrium between the two rates. Brahim et al. [74] reported on the determination of imidacloprid using SWASV and observed that detection is mainly controlled by adsorption. The obtained detection and quantification limits were 8.60 and 28.67 mmol L^{-1}. Mancozeb was studied through batch injection for electrochemical detection by pulsed amperometry [75]. The selected method of analysis revealed fast response, high sensitivity with good repeatability and reproducibility with a LOD of 5.14 $\times$ 10^{-7} mol L^{-1}. Costa et al. [76] explored determination of methomyl, a carbamate insecticide, through DPV and SWASV choosing the first of three oxidation processes. The detection limits and concentration ranges of the SWASV and DPV methods were 1.9 $\times$ 10^{-5} and 1.2 $\times$ 10^{-6} mol L^{-1}, and 6.6–42.0 $\times$ 10^{-5} to 5.0–410.0 $\times$ 10^{-6} mol L^{-1}, respectively. The developed method was tested in river and tap waters with success. Another carbamate-based insecticide pirimicarb was determined with a LOD of 1.24 μM [77]. It was found that this insecticide shows three irreversible oxidation processes and the two first processes are pH dependent. Selva et al. [78] proposed a method for an electrochemical study using cathodically pre-treated BDD electrode in the detection of propoxur by DPV. The found LOD value was around 0.50 μmol L^{-1}. A BDD electrode and Pt and Ir nanoparticles composite has been tested for detection and selectivity in a mixture of paraoxon and imidacloprid in tap water [79]. The authors found that chronoamperometry revealed that partial selectivity was verified, and the detection threshold was 33 $\times$ 10^{-2} μM (imidacloprid) and 19 $\times$ 10^{-2} μM (paraoxon). Lastly, Wei et al. [80] developed an acetylcholinesterase-based biosensor based on a boron-doped diamond electrode modified with gold nanoparticles and carbon spheres for the detection of organophostate pesticides such as chlorpyrifos and methyl parathion. The LOD value found for methyl parathion was 1.29 $\times$ 10^{-13} and 4.9 $\times$ 10^{-13} M for chlorpyrifos.

It appears that, when comparing DPV and SWASV electrochemical techniques, the former gives lower limits of detection for this pollutants group. Nevertheless, one should bear in mind that the BDD electrode physicochemical properties play a major role in the detection and quantification process.

3.3 Phenol and Derivatives

Phenol and its derivatives, particularly phenolic compounds, are an organic compounds family with a characteristic hydroxyl group attached to an aromatic ring. These compounds are common by-products of the industrial processes of petroleum, paper, plastics or drugs, but also used as food antioxidants. However, these are often discharged along with effluents from the industry and considered as water pollutants [81].

Lv et al. [82] used an as-grown and nanograss array of BDD electrodes to electrochemically detect catechol in buffered deionized water. A LOD of about 2.8 and 1.3 μM were found for the as-grown BDD electrode and the nanograss array BDD electrode, respectively, and demonstrating that the nanograss array BDD was more efficient towards catechol detection. A fast and direct determination of a phenolic antioxidant used in biodiesel, butylated hydroxyanisole (BHA), by batch injection analysis with amperometric detection was proposed by Tormin et al. [83]. It was one of the first works indicating the application of BDD electrodes for detection of compounds in non-aqueous electrolytes. The LOD obtained for the analyte was 100 ng of BHA per g of biodiesel. Janegitz et al. [84] have reported on a tyrosinase-based biosensor for phenol determination using a BDD electrode with gold nanoparticles electrodeposited onto the surface. The obtained results using SWASV led to a LOD of 0.07 μM with the sensor exhibiting good sensitivity, stability and reproducibility. A BDD anodically pretreated electrode was tested for 4-chloro-3-methylphenol detection, in the absence and presence of a cationic surfactant, mimicking disinfectant or preservative liquid media [85]. Different electrochemical techniques like CV, direct current voltammetry (DCV), differential pulse voltammetry (DPV) or SWASV were used in this study with LOD values of around 0.85, 0.46 and 0.34 μM for DCV, DPV and SWASV, respectively. Moreover, DPV studies showed that the relationship between the boron content and LOD was not linear. Another investigation of electrochemical behavior of 4-nitrophenyl triazole labeled nucleosides at BDD electrode by means of voltammetric techniques [86]. There was a highly stable response of the –NHOH/–NO redox couple and strong adhesion of the tested powerfully adsorbing nucleoside conjugates to the BDD surface was verified.

3.4 Pharmaceutical Compounds

The environmental impact of both human and veterinary pharmaceuticals has dramatically increased in the last decade. Unfortunately, only now the risks are slightly better understood but it still is vital to manage them. As with any other type of pollutant, awareness is not enough, and immediate actions are required. Uncontrolled entrance of pharmaceutical compounds in the environment takes place during their production, consumption and disposal. Again, the fate of most pollutants ends in the aquatic environment: antibiotics and analgesic/anti-inflammatories mainly present in surface water and wastewater; high levels of active pharmaceutical ingredients, excreted by human patients or improper disposal; and increasing municipal discharges of wastewater onto the oceans with consequent impact in marine systems and human health. Thus, individual identification and determination of pharmaceuticals in aquatic environment is crucial as well as those of drug mixtures since there are already evidence that the aggregate toxicity is higher [87, 88].

In this sub-section, representative reported works on BDD thin films applied for the detection and quantification for pharmaceutical compounds are presented and separated according to their physiological system of action. Thus, the active substances will be mainly divided into antimicrobial drugs, drugs affecting the cardiovascular, central nervous, digestive and reproductive/endocrine systems as well as drugs with impact on the skeletal and soft muscles [89].

Antimicrobial drugs are used in the prevention or treatment of disease caused by bacteria, viruses or fungi amongst others. Antibiotics such as Cephalexin [90], Enrofloxacin [91], Levofloxacin [92], Metronidazole [93] and Penicillin V [94] have been identified and determined by several electrochemical techniques like SWASV and DPV, using BDD electrodes. The latter showed good limits of detection in buffered water samples and the presence of interferents was also evacuated. After validation of the method experimental parameters, electroanalysis of real water samples was carried out. If case of absence of the analyte, the real samples were spiked. Table 3 resumes some of the findings in reported works for BDD electrodes used in identification and quantification of pharmaceutical compounds.

Simultaneous detection of Penicillin V and Paracetamol was reported by Švorc et al. [95] finding good limits of detection for both analytes and obtaining successful selectivity with no need to chemical or electrochemical modification of the BDD electrode. Pereira et al. [96] developed a fast batch injection analysis method for simultaneous determination of Phenazopyridine, Sulfamethoxazole, and Trimethoprim. The analysis method was efficient and selective towards the three antibiotics with the major advantage of reduced analysis time and probability of the batch injection system.

Cardiovascular medication is used in cardiovascular diseases including hypertension or atherosclerosis and drugs affecting blood may also be comprised such as anticoagulants. Amlodipine [97], Furosemide [98] and Pindolol [99] are examples of anti-hypertensive drugs which have been detected by DPV and SWASV using BDD electrodes, showing very good limits of detection in buffered water samples and preserving their selectivity in the presence of interferents. Rosuvastatin calcium reduces low-density lipoprotein (LDL) cholesterol in blood and it was determined by DVP and SWASV with low LOD of 1.04 mg L^{-1}. Binary systems have also been investigated [101], particularly with Hydrochlorothiazide [102–105]. These studies have showed that the applied methods of analysis were successful towards quantification and selective identification of the analytes, with low limits of detection and effective application in real systems like urine or human serum. A quaternary system containing amlodipine besylate, amiloride hydrochloride, hydrochlorothiazide, and atenolol [106]. An anodically pre-treated BDD electrode was employed in SWASV with very well-resolved and reproducible oxidative peaks demonstrating that coupling a BDD

Table 3 Representative results for application of doped diamond-based sensors to pharmaceutical compounds environmental monitoring

System	Electrochemical technique	Real system	Limit of detection (LOD) (model system)	Ref.
Antimicrobial				
Cephalexin	SWASV[a], DPV[b]	River water	34.74 ng L^{-1}	[90]
Enrofloxacin	SWASV[a]	Urine	0.0057 μg L^{-1}	[91]
Levofloxacin	SWASV[a], CV[c]	Urine/Serum	10.01 μg L^{-1} (CV), 2.88 μg L^{-1} (SWASV)	[92]
Metronidazole	SWASV[a], CV[c]	Urine	0.065 μmol L^{-1}	[93]
Penicillin V	DPV[b]	Deionized water	0.25 μM	[94]
Penicillin V/Paracetamol	SWASV[l]	Urine	0.21 and 0.32 μM	[95]
Sulfamethoxazole Trimethoprim Phenazopyridine	MPA[d]	Deionized water	0.20 mg L^{-1} 0.15 mg L^{-1} 0.05 mg L^{-1}	[96]
Cardiovascular				
Amlodipine	DPV[b]	Urine	0.07 μM	[97]
Furosemide	SWASV[a]	Urine/Serum	3.0×10^{-7} mol L^{-1}	[98]
Pindolol	DPV[b]	Urine/Serum	26 nmol L^{-1}	[99]
Rosuvastatin calcium	SWASV[l]	Urine/Serum	1.04 mg L^{-1}	[100]
Amlodipine/valsartan	SWASV[a]	Urine	0.0764 and 0.193 μmol L^{-1}	[101]
Hydrochlorothiazide/valsartan	SWASV[a]	Deionized water	0.639 and 0.935 μmol L^{-1}	[102]
Hydrochlorothiazide/Enalapril	MPA[d]	Deionized water	0.20 and 0.01 μmol L^{-1}	[103]
Hydrochlorothiazide/captopril	MPA[d]	–	0.27 and 0.14 μmol L^{-1}	[104]
Hydrochlorothiazide/metoprolol	SWASV[a], DPV[b]	Urine	0.376 μmol L^{-1} and 0.077 μmol L^{-1} (DPV)	[105]
Amlodipine besylate (AML) Amiloride hydrochloride (AMI) Hydrochlorothiazide (HCTZ) Atenolol (ATN)	SWASV[a]	Tap water/Urine	0.30 μmol L^{-1} 0.09 μmol L^{-1} 0.08 μmol L^{-1} 0.06 μmol L^{-1}	[106]

(continued)

Table 3 (continued)

System	Electrochemical technique	Real system	Limit of detection (LOD) (model system)	Ref.
Central nervous system				
Ibuprofen	SWASV[a], DPV[b]	Urine	4.1×10^{-7} mol L^{-1} (DPV), 9.3×10^{-7} mol L^{-1} (SWASV)	[107]
Imipramine	SWASV[a]	Deionized water	4.35×10^{-7} mol L^{-1}	[108]
Naphazoline	SWASV[a]	–	0.04 μmol L^{-1}	[109]
Nicotine	DPV	Deionized water	0.04 μM	[110]
Sibutramine	SWASV[a]	Deionized water	0.08–1.94 mg L^{-1}	[111]
Alprazolam/Bromazepam	DPV[b]	Deionized water	6.4×10^{-7} and 3.1×10^{-7} mol L^{-1}	[112]
Digestive				
Omeprazole	CV[c]	Deionized water	9 nmol L^{-1}	[113]
Diphenhydramine 8-Chlorotheophylline Pyridoxine	Amperometry	Deionized water	0.60 μmol L^{-1} 0.19 μmol L^{-1} 0.54 μmol L^{-1}	[114]
Reproductive/endocrine				
Tadalafil	SWASV[a]	Deionized water	19.5 nmol L^{-1}	[115]
Skeletal muscles				
Rapamycin	DPV[b]	Urine	0.22 μM	[116]
Leucovorin	DPV[b]	Urine	6.7×10^{-8} mol L^{-1}	[117]
Soft muscles				
Tramadol	SWASV[a]	Urine	0.004 μM	[118]
Tramadol/Acetaminophen	MPA[d]	Urine	0.04 and 0.03 μM	[119]

[a]Square wave anodic stripping voltammetry; [b]Differential pulse anodic stripping voltammetry; [c]Cyclic voltammetry; [d]Multi pulse amperometry; [e]detection threshold

electrode and the SWASV technique result in an advanced electroanalytical platform with effective application in water media systems.

Central nervous system drugs include anesthetics and psychiatric drugs that influence neurotransmitters or behavior like antidepressants or antianxiety agents. Ibuprofen [107], Imipramine [108], Naphazoline [109], Nicotine [110] and Sibutramine [111] have been determined using BDD electrodes and DPV or SWASV electrochemical techniques. The proposed methods of analysis provided good LOD values and had effective application in real aquatic systems. A mixture between Alprazolam/Bromazepam [112] was also investigated and quite good limits of detection were found for each of the analytes, 6.4×10^{-7} mol L^{-1} and 3.1×10^{-7} mol L^{-1}, respectively.

BDD thin films have also been successfully used in the detection and quantification of drugs affecting the digestive system altering its movements or the gastrointestinal fluids such as antidiarrheal drugs, laxatives or antacids, including proton pump inhibitors like Omeprazole [113], amongst others [114]; reproductive/endocrine system drugs include to treat erectile disfunction Tadalafil [115]; skeletal muscles drugs used for treatment of diseases related to the spinal cord and like the immunosuppressor Rapamycin [116] or a cancer treatment drug like Leucovorin [117]; and soft muscles drugs affecting the internal body organs muscles like Tramadol [118] and Acetaminophen [119].

3.5 Pathogens

As previously described, diamond is inert and innocuous towards bacillus, bacteria and other pathogenic microorganisms. Marciano et al. [120] has reported on diamond-like carbon, a metastable form of amorphous carbon, films produced with high deposition rates showing antimicrobial activity towards *Escherichia coli*; and synergism between nanocrystalline diamond and graphene treated with fluorine plasma showed antibacterial activity [121]. A biosensor based on BDD electrode poised at a positive potential has been exploited for the detection of *E. coli* contamination in artificially spiked samples of meat, milk and tap water [122]. Through a specific enzyme, β-galactosidase, after a total analysis time of less than 1.5 h, the detection limit of 4×10^4 cells/mL was obtained. Deep-ultraviolet (DUV) light sources are being investigated on sterilization of *E. coli* since direct irradiation results in a significant inhibition of growth of such microorganisms [123]. Thus, DUV diamond light emitting diodes (LEDs) may become a valuable germicidal light source.

BDD electrodes (DiaCell® technology) have been successfully tested against *Legionella pneumophila* infection in several tap waters [124]. To the best of our knowledge, we were not able to find reported work using BDD electrodes for detection of any *Bacillus* type. As it will be seen in Sect. 4, BDD electrodes demonstrate effective application in environmental remediation via water electrolysis and/or electroxidation.

3.6 Biomolecules

Environmental consciousness has also to do with balance and equilibrium in the ecosystem. Pollutants with anthropogenic origin are the major concerns because nature always had self-management skills. In this small section, we report on a few biomolecules with increasing toxicological significance. This is mainly due to their use in a wide range of sources including medicinal purposes acting as hallucinogenic, such as β-carboline alkaloids, also produced in cigarette smoke [125]. Sensitive determination of harmaline was carried out using a BDD electrode in DPV and SWASV techniques with LOD of 0.08 and 0.2 μM, respectively. Two other biomolecules of increasing interest like vanillin (VAN) and caffeine (CAF) were simultaneously determined using an anodically pretreated BDD electrode [126]. SWASV studies provided simultaneous determination of VAN and CAF, with detection limits of 0.234 and 0.071 μM, respectively.

4 Environmental Remediation

The previous section summarized some of recent work developed around the application of boron-doped diamond (BDD) thin films for environmental monitoring, particularly in aquatic media. The real efficiency of BDD as an electrode providing fast, straightforward, sensitive, reproducible and robust electrochemical analysis of numerous pollutants has been demonstrated. BDD thin films are rather versatile materials due to their unique properties and their electrochemical excellency, they have been further explored in electroxidation of contaminants. This confers BDD thin films the ability to work simultaneously as environmental sensors and cleaners. This important aspect and functionality of BDD thin films is briefly explored in this section.

4.1 Advanced Electrochemical Oxidation Processes

The development of efficient, simple and cost-effective processes to remove and/or destroy pollutants is a priority. Most conventional wastewater treatment plants are not prepared to efficiently eliminate all types of contaminants becoming another source of environmental pollution into natural waters, particularly the ocean. Subsequently, the best option is to eliminate or preferably extinguish the pollutants before being discharged. Advanced oxidation processes are an available technology to effectively perform that purpose. It is based on in situ production of strong oxidants like hydroxyl radicals which can oxidize any organic contaminant until its final mineralization. These hydroxyl radicals can be produced via chemical means like the Fenton process, photochemical methods (heterogeneous catalysis, UV photolysis and photo-Fenton) or electrochemically. The latter are the most recent advanced oxidation technology which include electroxidation and electro-Fenton processes.

Electroxidation is based on anodic oxidation and BDD thin films have been extensively investigated for this purpose already with effective application in this type of green technology. Additionally, BDD thin films are robust and stable enough to support electrical potentials above that of water electrolysis. Thus, hydroxyls are produced from water splitting and BDD also performs mineralization of such organic pollutants. In fact, BDD thin films have been tested in real systems such as wastewater effluents or winery wastewater, amongst others [127–132].

The electro-Fenton process is based on Fenton's reaction combined with electrochemistry which, ultimately, originate further hydroxyl species enough to also mineralize the contaminants. Once more, BDD thin films are being investigated as cathode or anode electrodes in the electro-Fenton process [133–135].

BDD thin films are an all-purpose nanostructured material with wide and effective application in environmental science.

5 Concluding Remarks

The main objective of this chapter is to provide the reader the current state of scientific research around diamond-based nanostructured materials used as electrochemical sensors for aquatic contaminants. CVD diamond can be produced with variable morphologies and with different surface chemical terminations which, in their turn, will determine the electrochemical performance. This versatility naturally enables tailoring of their properties towards all sorts of purposes. Application of these valuable materials in electrochemical sensing and analysis has been clearly demonstrated due to their remarkable electrochemical capabilities as verified by the immense amount of work that has been carried out on the subject. Diamond displays most of the requirements to be considered the elected material to be employed for a certain sensing application. As an electrode for electrochemical purposes, its performance is as good as that of noble and expensive metals and is commercially available. Moreover, as an electrochemical sensor it provides superior results than those delivered by conventional analytical techniques.

Nevertheless, not all is perfect in the diamond world. There also some drawbacks and challenges to overcome. Its effective application in environmental remediation will always be contested by emerging contaminants or merely by its cost or availability. Simultaneous sensing of numerous contaminants with similar redox behavior with in situ and at real time suitable signal deconvolution is not straightforward. The same applies to miniaturization and portability of the sensor devices or remote access to them.

Still, all points out to a positive future for synthetic diamond-based nanostructured materials and their real application as sensors in environmental monitoring and as massive cleaning tools in environmental remediation.

References

1. Daily, H.E.: Ecol. Econ. **2**, 1–6 (1990)
2. Arregui, F.J.: Sensors Based on Nanostructured Materials. Springer, Berlin (2009)
3. Hussain, C.M., Kharisov, B. (eds.): Advanced Environmental Analysis. Royal Society of Chemistry, Cambridge (2016)
4. Li, M., Gou, H., Al-Ogaidi, I., Wu, N., Sustain, A.C.S.: Chem. Eng. **1**, 713–723 (2013)
5. Singh, R.C., Singh, M.P., Virk, H.S.: Solid State Phenom. **201**, 131–158 (2013)
6. Justino, C.I.L., Duarte, A.C., Rocha-Santos, T.A.P.: Sensors (Switzerland) 17 (2017)
7. Hernandez-Vargas, G., Sosa-Hernández, J.E., Saldarriaga-Hernandez, S., Villalba-Rodríguez, A.M., Parra-Saldivar, R., Iqbal, H.M.N.: Biosensors **8**, 1–21 (2018)
8. Deng, Y.: Semiconducting Metal Oxides for Gas Sensing (2019)
9. Hussain, C.M., Kharisov, B, (eds.): Advanced Environmental Analysis, Royal Society of Chemistry, vol. 2. Cambridge (2016)
10. Andrade, F.J., Blondeau, P., Macho, S., Riu, J., Rius, F.X.: Encycl. Anal. Chem. 1–17 (2014)
11. Luo, X., Yang, J.: J. Sensors (2019)
12. Rujuan, Z., Kelin, G., Zuping, Z., Yanxia, W., Jiezhou, L., Zhilin, X., Hongtu, L., Zhiqiang, W., Jian, Y., Guien, Z., Changsui, W.: Chin. Phys. Lett. **7**, 445–448 (1990)
13. Swartzlander, A.B., Nelson, A.J., Woollam, J.A.: J. Mater. Res. **3**, 1397–1403 (1988)
14. Mortet, V., Hubicka, Z., Vorlicek, V., Jurek, K., Rosa, J., Vanecek, M.: Phys. Status Solidi Appl. Res. **201**, 2425–2431 (2004)
15. Tsai, W., Reynolds, G.J., Hikido, S., Cooper, C.B.: Appl. Phys. Lett. **60**, 1444–1446 (1992)
16. Haubner, R., Lux, B.: Diam. Relat. Mater. **2**, 1277–1294 (1993)
17. Brillas, E., Martínez-Huitle, C.A.: Synthetic Diamond Films: Preparation, Electrochemistry, Characterization, and Applications (2011)
18. Mildren, R.P., Rabeau, J.R.: Optical Engineering of Diamond (2013)
19. Yang, N., Foord, J.S., Jiang, X.: Carbon N. Y. **99**, 90–110 (2016)
20. Yu, Y., Zhou, Y., Wu, L., Zhi, J.: Int. J. Electrochem. **2012**, 1–10 (2012)
21. Yang, N., Yu, S., MacPherson, J.V., Einaga, Y., Zhao, H., Zhao, G., Swain, G.M., Jiang, X.: Chem. Soc. Rev. **48**, 157–204 (2019)
22. Garrett, D.J., Tong, W., Simpson, D.A., Meffin, H.: Carbon N. Y. **102**, 437–454 (2016)
23. Nistor, P.A., May, P.W.: J. R. Soc. Interface **14** (2017)
24. Iwaki, M., Sato, S., Takahashi, K., Sakairi, H.: Nucl. Instruments Methods Phys. Res. **209–210**, 1129–1133 (1983)
25. Pelskov, Y.V., Sakharova, A.Y., Krotova, M.D., Bouilov, L.L., Spitsyn, B.V.: J. Electroanal. Chem. **228**, 19–27 (1987)
26. Patel, K., Hashimoto, K., Fujishima, A.: J. Photochem. Photobiol. A Chem. **65**, 419–429 (1992)
27. Macpherson, J.V.: Phys. Chem. Chem. Phys. **17**, 2935–2949 (2015)
28. Garcia-Segura, S., Vieira Dos Santos, E., Martínez-Huitle, C.A.: Electrochem. Commun. **59**, 52–55 (2015)
29. Ayres, Z.J., Newland, J.C., Newton, M.E., Mandal, S., Williams, O.A., Macpherson, J.V.: Carbon N. Y. **121**, 434–442 (2017)
30. Schwarzová-Pecková, K., Vosáhlová, J., Barek, J., Šloufová, I., Pavlova, E., Petrák, V., Zavázalová, J.: Electrochim. Acta **243**, 170–182 (2017)
31. Kavan, L., Vlckova Zivcova, Z., Petrak, V., Frank, O., Janda, P., Tarabkova, H., Nesladek, M., Mortet, V.: Electrochim. Acta **179**, 626–636 (2015)
32. Safaie, P., Eshaghi, A., Bakhshi, S.R.: J. Non. Cryst. Solids **471**, 410–414 (2017)
33. Hébert, C., Mazellier, J.P., Scorsone, E., Mermoux, M., Bergonzo, P.: Carbon N. Y. **71**, 27–33 (2014)
34. Lee, S.K., Song, M.J., Lim, D.S.: J. Electroanal. Chem. **820**, 140–145 (2018)
35. Vlčková Živcová, Z., Mortet, V., Taylor, A., Zukal, A., Frank, O., Kavan, L.: Diam. Relat. Mater. **87**, 61–69 (2018)

36. Mahé, E., Devilliers, D., Dardoize, F.: Talanta **132**, 641–647 (2015)
37. Rusinek, C.A., Becker, M.F., Rechenberg, R., Schuelke, T.: Electrochem. Commun. **73**, 10–14 (2016)
38. Piret, G., Hébert, C., Mazellier, J.P., Rousseau, L., Scorsone, E., Cottance, M., Lissorgues, G., Heuschkel, M.O., Picaud, S., Bergonzo, P., Yvert, B.: Biomaterials **53**, 173–183 (2015)
39. Halpern, J.M., Cullins, M.J., Chiel, H.J., Martin, H.B.: Diam. Relat. Mater. **19**, 178–181 (2010)
40. Read, T.L., Cobb, S.J., Macpherson, J.V.: ACS Sens. **4**, 756–763 (2019)
41. Forsberg, P., Jorge, E.O., Nyholm, L., Nikolajeff, F., Karlsson, M.: Diam. Relat. Mater. **20**, 1121–1124 (2011)
42. Chartres, N., Bero, L.A., Norris, S.L.: Environ. Int. **123**, 231–239 (2019)
43. Ginige, T.: T.A. Convention, Advanced Environmental (2011)
44. Pujol, L., Evrard, D., Groenen-Serrano, K., Freyssinier, M., Ruffien-Cizsak, A., Gros, P.: Front. Chem. **2**, 1–24 (2014)
45. Gumpu, M.B., Sethuraman, S., Krishnan, U.M., Rayappan, J.B.B.: Sens. Actuators B Chem. **213**, 515–533 (2015)
46. Ivandini, T.A., Sato, R., Makide, Y., Fujishima, A., Einaga, Y.: Anal. Chem. **78**, 6291–6298 (2006)
47. Zhang, T., Li, C., Mao, B., An, Y.: Ionics (Kiel) **21**, 1761–1769 (2015)
48. Sardinha, A.F., Arantes, T.M., Cristovan, F.H., Ferreira, N.G.: Thin Solid Films **625**, 70–80 (2017)
49. Fierro, S., Watanabe, T., Akai, K., Yamanuki, M., Einaga, Y.: J. Electrochem. Soc. **158**, F173 (2011)
50. Banks, C.E., Hyde, M.E., Tomčík, P., Jacobs, R., Compton, R.G.: Talanta **62**, 279–286 (2004)
51. Peilin, Z., Jianzhong, Z., Shenzhong, Y., Xikang, Z., Guoxiong, Z.: Anal. Bioanal. Chem. **353**, 171–173 (1995)
52. Dragoe, D., Spătaru, N., Kawasaki, R., Manivannan, A., Spătaru, T., Tryk, D.A., Fujishima, A.: Electrochim. Acta **51**, 2437–2441 (2006)
53. Innuphat, C., Chooto, P.: ScienceAsia **43**, 33–41 (2017)
54. Arantes, T.M., Sardinha, A., Baldan, M.R., Cristovan, F.H., Ferreira, N.G.: Talanta **128**, 132–140 (2014)
55. Saterlay, A.J., Agra-Gutiérrez, C., Taylor, M.P., Marken, F., Compton, R.G.: Electroanalysis **11**, 1083–1088 (1999)
56. Prado, C., Wilkins, S.J., Marken, F., Compton, R.G.: Electroanalysis **14**, 262–272 (2002)
57. Manivannan, A., Kawasaki, R., Tryk, D.A., Fujishima, A.: Electrochim. Acta **49**, 3313–3318 (2004)
58. Sonthalia, P., McGaw, E., Show, Y., Swain, G.M.: Anal. Chim. Acta **522**, 35–44 (2004)
59. McGaw, E.A., Swain, G.M.: Anal. Chim. Acta **575**, 180–189 (2006)
60. El Tall, O., Jaffrezic-Renault, N., Sigaud, M., Vittori, O.: Electroanalysis **19**, 1152–1159 (2007)
61. Zhao, B., Ren, Y., Li, J., Yu, X., Zhang, J.: Surf. Rev. Lett. **1850179**, 1–9 (2018)
62. Saterlay, A.J., Marken, F., Foord, J.S., Compton, R.G.: Talanta **53**, 403–415 (2000)
63. de Sanoit, J., Quang Tran, T., Pomorski, M., Pierre, S., Mer-Calfati, C., Bergonzo, P.: Appl. Radiat. Isot. **80**, 32–41 (2013)
64. Prieto Garcia, F., Cortés Ascencio, S.Y., Oyarzun, J.C.G., Hernandez, A.C., Alavarado, P.V.: J. Res. Environ. Sci. Toxicol. **1**, 2315–5698 (2012)
65. Stanković, D.M., Kalcher, K.: Sens. Actuators B Chem. **233**, 144–147 (2016)
66. Spătaru, T., Spătaru, N., Fujishima, A.: Talanta **73**, 404–406 (2007)
67. Medina-Sánchez, M., Mayorga-Martinez, C.C., Watanabe, T., Ivandini, T.A., Honda, Y., Pino, F., Nakata, A., Fujishima, A., Einaga, Y., Merkoçi, A.: Biosens. Bioelectron. **75**, 365–374 (2016)
68. Jevtić, S., Stefanović, A., Stanković, D.M., Pergal, M.V., Ivanović, A.T., Jokić, A., Petković, B.B.: Diam. Relat. Mater. **81**, 133–137 (2018)

69. Tyszczuk-Rotko, K., Bęczkowska, I., Nosal-Wiercińska, A.: Diam. Relat. Mater. **50**, 86–90 (2014)
70. Bandžuchova, L., Švorc, L., Sochr, J., Svítkova, J., Chýlkova, J.: Electrochim. Acta **111**, 242–249 (2013)
71. Janikova-Bandzuchova, L., Šelešovská, R., Schwarzová-Pecková, K., Chýlková, J.: Electrochim. Acta **154**, 421–429 (2015)
72. Vukojević, V., Djurdjić, S., Jevtić, S., Pergal, M.V., Marković, A., Mutić, J., Petković, B.B., Stanković, D.M.: Int. J. Environ. Anal. Chem. **98**, 1175–1185 (2018)
73. Garbellini, G.S., Avaca, L.A., Salazar-Banda, G.R.: Quim. Nov. **33**, 2261–2265 (2010)
74. Ben Brahim, M., Belhadj Ammar, H., Abdelhédi, R., Samet, Y.: Chinese Chem. Lett. **27**, 666–672 (2016)
75. Silva, R.A.G., Silva, L.A.J., Munoz, R.A.A., Richter, E.M., Oliveira, A.C.: J. Electroanal. Chem. **733**, 85–90 (2014)
76. Costa, D.J.E., Santos, J.C.S., Sanches-Brandão, F.A.C., Ribeiro, W.F., Salazar-Banda, G.R., Araujo, M.C.U.: J. Electroanal. Chem. **789**, 100–107 (2017)
77. Selva, T.M.G., de Araujo, W.R., Bacil, R.P., Paixão, T.R.L.C.: Electrochim. Acta **246**, 588–596 (2017)
78. Selva, T.M.G., Paixão, T.R.L.C.: Diam. Relat. Mater. **66**, 113–118 (2016)
79. Belghiti, D.K., Zadeh-Habchi, M., Scorsone, E., Bergonzo, P.: Procedia Eng. **168**, 428–431 (2016)
80. Wei, M., Zeng, G., Lu, Q.: Microchim. Acta **181**, 121–127 (2014)
81. Soto-Hernandez, M., Palma-Tenango, M., del Garcia-Mateos, M.R.: Phenolic Compounds—Natural Sources, Importance and Applications. InTech (2017)
82. Lv, M., Wei, M., Rong, F., Terashima, C., Fujishima, A., Gu, Z.Z.: Electroanalysis **22**, 199–203 (2010)
83. Tormin, T.F., Gimenes, D.T., Richter, E.M., Munoz, R.A.A.: Talanta **85**, 1274–1278 (2011)
84. Janegitz, B.C., Medeiros, R.A., Rocha-Filho, R.C., Fatibello-Filho, O.: Diam. Relat. Mater. **25**, 128–133 (2012)
85. Brycht, M., Lochyński, P., Barek, J., Skrzypek, S., Kuczewski, K., Schwarzova-Peckova, K.: J. Electroanal. Chem. **771**, 1–9 (2016)
86. Vosáhlová, J., Koláčná, L., Daňhel, A., Fischer, J., Balintová, J., Hocek, M., Schwarzová-Pecková, K., Fojta, M.: J. Electroanal. Chem. **821**, 111–120 (2018)
87. Kümmerer, K.: Pharmaceuticals in the Environment, (2008)
88. Aga, D.S.: Fate of Pharmaceuticals in the Environment and in Water Treatment Systems. CRC Press, Boca Raton (2008)
89. Wells, B.G., DiPiro, J.T., Schwinghammer, T.L., DiPiro, C. V.: Pharmacotherapy Handbook, n.d
90. Feier, B., Gui, A., Cristea, C., Săndulescu, R.: Anal. Chim. Acta **976**, 25–34 (2017)
91. Dönmez, F., Yardım, Y., Şentürk, Z.: Diam. Relat. Mater. **84**, 95–102 (2018)
92. Rkik, M., Ben Brahim, M., Samet, Y.: J. Electroanal. Chem. **794**, 175–181 (2017)
93. Ammar, H.B., Ben Brahim, M., Abdelhédi, R., Samet, Y.: Mater. Sci. Eng. C **59**, 604–610 (2016)
94. Švorc, Ľ., Sochr, J., Rievaj, M., Tomčík, P., Bustin, D.: Bioelectrochemistry **88**, 36–41 (2012)
95. Švorc, Ľ., Sochr, J., Tomčík, P., Rievaj, M., Bustin, D.: Electrochim. Acta **68**, 227–234 (2012)
96. Pereira, P.F., Da Silva, W.P., Muñoz, R.A.A., Richter, E.M.: J. Electroanal. Chem. **766**, 87–93 (2016)
97. Švorc, L., Cinková, K., Sochr, J., Vojs, M., Michniak, P., Marton, M.: J. Electroanal. Chem. **728**, 86–93 (2014)
98. Medeiros, R.A., Baccarin, M., Fatibello-Filho, O., Rocha-Filho, R.C., Deslouis, C., Debiemme-Chouvy, C.: Electrochim. Acta **197**, 179–185 (2016)
99. Pereira, G.F., Deroco, P.B., Silva, T.A., Ferreira, H.S., Fatibello-Filho, O., Eguiluz, K.I.B., Salazar-Banda, G.R.: Diam. Relat. Mater. **82**, 109–114 (2018)
100. Silva, T.A., Pereira, G.F., Fatibello-Filho, O., Eguiluz, K.I.B., Salazar-Banda, G.R.: Diam. Relat. Mater. **58**, 103–109 (2015)

101. Mansano, G.R., Eisele, A.P.P., Dall'Antonia, L.H., Afonso, S., Sartori, E.R.: J. Electroanal. Chem. **738**, 188–194 (2015)
102. Eisele, A.P.P., Mansano, G.R., De Oliveira, F.M., Casarin, J., Tarley, C.R.T., Sartori, E.R.: J. Electroanal. Chem. **732**, 46–52 (2014)
103. Lourencao, B.C., Medeiros, R.A., Fatibello-Filho, O.: J. Electroanal. Chem. **754**, 154–159 (2015)
104. Gimenes, D.T., Marra, M.C., De Freitas, J.M., Abarza Muñoz, R.A., Richter, E.M.: Sens. Actuators B Chem. **212**, 411–418 (2015)
105. Salamanca-Neto, C.A.R., Eisele, A.P.P., Resta, V.G., Scremin, J., Sartori, E.R.: Sens. Actuators B Chem. **230**, 630–638 (2016)
106. Moraes, J.T., Salamanca-Neto, C.A.R., Švorc, Ľ., Sartori, E.R.: Microchem. J. **134**, 173–180 (2017)
107. Švorc, Ľ., Strežová, I., Kianičková, K., Stanković, D.M., Otřísal, P., Samphao, A.: J. Electroanal. Chem. **822**, 144–152 (2018)
108. Oliveira, S.N., Ribeiro, F.W.P., Sousa, C.P., Soares, J.E.S., Suffredini, H.B., Becker, H., de Lima-Neto, P., Correia, A.N.: J. Electroanal. Chem. **788**, 118–124 (2017)
109. Oliveira, T.D.C., Freitas, J.M., Abarza Munoz, R.A., Richter, E.M.: Talanta **152**, 308–313 (2016)
110. Švorc, Ľ., Stanković, D.M., Kalcher, K.: Diam. Relat. Mater. **42**, 1–7 (2014)
111. Freitas, J.M., Oliveira, T.C., Santana, M.H.P., Banks, C.E., Munoz, R.A.A., Richter, E.M.: Sens. Actuators B Chem. **282**, 449–456 (2019)
112. Samiec, P., Švorc, Ľ., Stanković, D.M., Vojs, M., Marton, M., Navrátilová, Z.: Sens. Actuators B Chem. **245**, 963–971 (2017)
113. Stefano, J.S., Tormin, T.F., da Silva, J.P., Richter, E.M., Munoz, R.A.A.: Microchem. J. **133**, 398–403 (2017)
114. Freitas, J.M., Oliveira, T.D.C., Gimenes, D.T., Munoz, R.A.A., Richter, E.M.: Talanta **146**, 670–675 (2016)
115. Sartori, E.R., Clausen, D.N., Pires, I.M.R., Salamanca-Neto, C.A.R.: Diam. Relat. Mater. **77**, 153–158 (2017)
116. Stanković, D.M., Kalcher, K.: Electrochim. Acta **168**, 76–81 (2015)
117. Šelešovská, R., Kränková, B., Štěpánková, M., Martinková, P., Janíková, L., Chýlková, J., Vojs, M.: J. Electroanal. Chem. **821**, 2–9 (2018)
118. Afkhami, A., Ghaedi, H., Madrakian, T., Ahmadi, M., Mahmood-Kashani, H.: Biosens. Bioelectron. **44**, 34–40 (2013)
119. Santos, A.M., Vicentini, F.C., Figueiredo-Filho, L.C.S., Deroco, P.B., Fatibello-Filho, O.: Diam. Relat. Mater. **60**, 1–8 (2015)
120. Marciano, F.R., Bonetti, L.F., Da-Silva, N.S., Corat, E.J., Trava-Airoldi, V.J.: Synth. Met. **159**, 2167–2169 (2009)
121. Oh, H.G., Lee, J.Y., Son, H.G., Kim, D.H., Park, S.H., Kim, C.M., Jhee, K.H., Song, K.S.: Results Phys. **12**, 2129–2135 (2019)
122. Majid, E., Male, K.B., Luong, J.H.T.: J. Agric. Food Chem. **56**, 7691–7695 (2008)
123. Matsumoto, T., Tatsuno, I., Hasegawa, T.: Water (Switzerland) 11 (2019)
124. Furuta, T., Tanaka, H., Nishiki, Y., Pupunat, L., Haenni, W., Rychen, P.: Diam. Relat. Mater. **13**, 2016–2019 (2004)
125. Švorc, L., Cinková, K., Samphao, A., Stanković, D.M., Mehmeti, E., Kalcher, K.: J. Electroanal. Chem. **744**, 37–44 (2015)
126. Ali, H.S., Abdullah, A.A., Pınar, P.T., Yardım, Y., Şentürk, Z.: Talanta **170**, 384–391 (2017)
127. Garcia-Segura, S., Keller, J., Brillas, E., Radjenovic, J.: J. Hazard. Mater. **283**, 551–557 (2015)
128. Garcia-Segura, S., Ocon, J.D., Chong, M.N.: Process Saf. Environ. Prot. **113**, 48–67 (2018)
129. Loos, G., Scheers, T., Van Eyck, K., Van Schepdael, A., Adams, E., Van der Bruggen, B., Cabooter, D., Dewil, R.: Sep. Purif. Technol. **195**, 184–191 (2018)
130. Lan, Y., Coetsier, C., Causserand, C., Groenen Serrano, K.: Chem. Eng. J. **333**, 486–494 (2018)

131. Candia-Onfray, C., Espinoza, N., Sabino da Silva, E.B., Toledo-Neira, C., Espinoza, L.C., Santander, R., García, V., Salazar, R.: Chemosphere **206**, 709–717 (2018)
132. He, Y., Lin, H., Guo, Z., Zhang, W., Li, H., Huang, W.: Sep. Purif. Technol. 802–821 (2019)
133. Olvera-Vargas, H., Oturan, N., Oturan, M.A., Brillas, E.: Sep. Purif. Technol. **146**, 127–135 (2015)
134. Moreira, F.C., Boaventura, R.A.R., Brillas, E., Vilar, V.J.P.: Appl. Catal. B Environ. **202**, 217–261 (2017)
135. Kanakaraju, D., Glass, B.D., Oelgemöller, M.: J. Environ. Manage. **219**, 189–207 (2018)

Toxic Effects of Metal Nanoparticles in Marine Invertebrates

Joana C. Almeida, Celso E. D. Cardoso, Eduarda Pereira and Rosa Freitas

Abstract The extensive use of nanomaterials, namely metal and metal oxide nanoparticles (NPs), in a variety of application areas—such as electronics, medicine, energy, environment, industry, information, security, among others—leads to the end-up of these materials into the aquatic environments. Once there, NPs accumulate in organisms and may amplify along the food chain, inducing effects on these organisms and humans. Due to the relevance of this issue, works concerning NPs potential effects to the aquatic organisms have been published in the literature. This chapter starts to explore the main applications and the synthesis methods of NPs, as well as their impact in the environment. Then, common parameters used to evaluate ecotoxicological impacts are described. Lastly, research undertaken on the biological toxic impacts of titanium dioxide, zinc oxide and silver NPs in marine invertebrates is reviewed, based on the most recent literature. The selection of these NPs was based on the evaluation of nanomaterials most used in consumer products.

Keywords Toxic effects · Invertebrates · Metal nanoparticles · Organisms · Toxicity

1 Contextualization

Nanotechnology has grown exponentially in recent years, playing a crucial role in our society and the way people live by improving many technology and industry sectors

J. C. Almeida · C. E. D. Cardoso
Chemistry Department, CICECO-Aveiro Institute of Materials, University of Aveiro, Campus de Santiago, 3810-193 Aveiro, Portugal

J. C. Almeida · C. E. D. Cardoso (✉) · E. Pereira
Chemistry Department, CESAM & LAQV-REQUIMTE, University of Aveiro, Campus de Santiago, 3810-193 Aveiro, Portugal
e-mail: cedc@ua.pt

R. Freitas
Biology Department, CESAM, University of Aveiro, Campus de Santiago, 3810-193 Aveiro, Portugal

© Springer Nature Switzerland AG 2019
G. A. B. Gonçalves and P. Marques (eds.), *Nanostructured Materials for Treating Aquatic Pollution*, Engineering Materials,
https://doi.org/10.1007/978-3-030-33745-2_7

as computing, medicine, cosmetics, sunscreens, textiles, agriculture, sensing, energy production, and environmental protection, thus providing products constituted by nanoparticles (NPs), with novel and unique functions [1]. The exponential development of nanotechnology and its corresponding increase in the use of commercial products have led to major concerns regarding the potential risks on human health as well as on the environment and inhabiting organisms. The wide use of NPs, either quantitatively or in a diversity of products, lead to an increase of its production and, consequently, a diversification in emission sources into the environment [2]. Furthermore, the lack of legislation for nanotechnologies leads to a lack of concern of manufacturers and to careless and ineffective approaches that can prevent NPs from reaching the environment. Thus, as the production of NPs increases as well as its use in consumer products, the greater will be the release into the environment as a result of spills, use of products, or post-consumer degradation of material. Therefore, NPs can enter the aquatic environment throughout its life cycle in three different periods: during the production of raw material and consumer products, during its use and after disposal of NP-containing products (waste handling) [2]. Thus, NPs can reach rivers, lakes and oceans, harming aquatic life. In this way, toxic impacts of NPs into organisms are becoming an urgent issue as reviewed by Canesi and Corsi [3]. When exposed to NPs, the aquatic organisms can suffer from oxidative stress, stimulated by reactive oxygen species (ROS), which is known to produce a range of harmful effects upon cells. ROS can damage a wide variety of cellular components (membranes, protein and DNA), which results in peroxidation of lipids, distortions of the conformation of proteins, disruption of DNA, interference with signal-transduction pathways, and modulation of gene transcription [4], with consequences on cells functioning and, therefore, organisms health and behavior.

Invertebrates represent over 90% of animal species, widespread in very different ecosystems, which make them suitable target organisms and excellent models for evaluating the environmental impact of NPs. Among the invertebrate marine organisms, bivalves are frequently chosen to perform ecotoxicological studies since they are sessile filter feeders and have a complex reproductive cycle. Also, their capacity of readily accumulate contaminants, but metabolize and eliminate them more slowly than vertebrates and also sharing many of the physiological functions found in higher vertebrates lead them to be vastly used in these types of studies.

To ensure the survival and maintenance of the quality of life of aquatic organism, it is crucial to understand the toxicity of NPs in aquatic environment and inhabiting aquatic organisms. In particular, a thorough knowledge about environmental safety guidelines, biochemical responses at environmental NPs dose exposures, and interaction between NPs and biological systems is required. This chapter aims to highlight and discuss the issues concerning the importance of NPs nowadays as well as the toxicological effects of NPs with focus on the effects of metal and metal oxide NPs in marine invertebrates.

2 Nanoparticles

In 1959, the physicist Richard Feynman presented a lecture—entitled There's plenty of room at the bottom—about the control and manipulation of matter at the atomic scale, which anticipated the development in nanotechnology, constituting an historical and scientific milestone [5, 6]. Since then, the term "nanotechnology" began to be used, and the research into Nanotechnology has been developing increasingly in the last decades—mostly due to new nanomaterials research, including studies involving particles with nanoscale dimensions.

Nanotechnology is the name given to the development and application of nanomaterials, i.e., materials and structures with at least one dimension of 1-100 nm. When this rule is found for the three, two or one dimension, the material is designated as a nanoparticle, nanofiber or nanoplate, respectively [7]. These nanomaterials tend to exhibit unique physical and chemical properties. For example, its mechanical, optical, magnetic and chemical properties—which depend heavily on shape, size, surface characteristics and inner structure of nanomaterials—are much different from those of particles and macroscopic surfaces of similar composition, due to size effects and surface (that become evident at the nanoscale) [6, 8]. Most chemical reactions happen at the surface, so higher surface area to volume ratio of nanomaterials leads to much higher chemical activity. Moreover, it is possible the nanomaterials coating with several types of specific ligands [5, 6] to increase its potential for strategic applications.

Nanomaterials have been widely used and its incorporation into consumer products has been notorious in the last years. In this context, Nanotechnology Consumer Product Inventory (CPI) was created in 2005 by the Woodrow Wilson International Center for Scholars and the Project on Emerging Nanotechnology, in order to evaluate how is nanotechnology entering in the market [9]. This inventory is based on the composition (in nanomaterials) of consumer products, and it was concluded that the largest group of nanomaterials present in these products are metal and metal oxide NPs (Fig. 1).

Since titanium dioxide, zinc oxide and silver NPs are the most used metal and metal oxide NPs, a greater focus will be given to these NPs, in particular their toxicity in marine organisms.

2.1 Applications

The unique properties of NPs make them potentially important in numerous applications and these are the reason of its constant increase of interest [5]. Therefore, nanotechnology operates in strategic areas such as cosmetics and sunscreens, electronics, catalysis, medicine, food, construction, renewable energy and environmental remediation, as represented in Fig. 2.

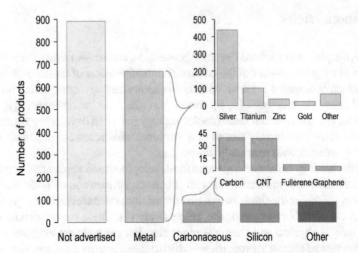

Fig. 1 Composition of nanomaterials listed in the Nanotechnology Consumer Product Inventory, grouped into five major categories: not advertised, metal and metal oxides, carbonaceous nanomaterials, silicon-based nanomaterials, and other (such as organics, polymers and ceramics) [9]

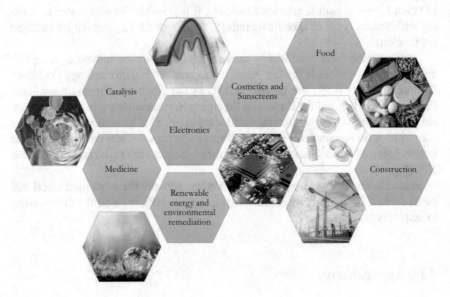

Fig. 2 Main applications of NPs

Examples of the most significant NPs used in each application mentioned above are the following:

- **Cosmetics and sunscreens**: titanium dioxide and zinc oxide are used in some sunscreens since they act as UV filters (absorbing and reflecting UV light). Also, its transparency to visible light results in a transparent formulation, pleasing to the consumer [10]. Silver NPs, due to their bactericidal action, are used in cosmetics such as deodorants and toothpaste [11]. Iron oxide NPs are used in lipsticks as a pigment [12].

- **Electronics**: The increasing development of portable electronic equipment (mobile phones, laptop computers, among others) requires lightweight, compact and high capacity batteries. The use of NPs in separator plates of batteries originates batteries able to store more energy due to their foam like structure.'In the case of the display technology, the light emitting diodes (LED) of modern displays for TV screens and computer monitors are encouraging the use of NPs (such as lead telluride, cadmium sulphide, zinc selenide and sulphide) [12].

- **Catalysis**: NPs, such as gold, nickel oxide and cobalt oxide NPs [13, 14], are used as efficient catalyst in the production of chemicals since their high surface area offers higher catalytic activity. Platinum NPs are used in the automotive catalytic converters to reduce significantly the cost and the amount of Pt and also to improve the converter performance [12].

- **Medicine**: drug delivery and tissue engineering are some of the medical applications which have been improved with nanotechnology. Biopolymeric NPs, liposomes, polymeric micelles, dendrimers, inorganic NPs (silver, gold, iron oxide and silica) and quantum dots have been used in the drug delivery systems since allows a site-specific, and target-oriented delivery of drugs, reducing its side effects [15]. In the field of tissue engineering, silver and gold NPs have been incorporated into the hydrogel matrices to form scaffolds for soft, bone, and cardiac tissues regeneration [16].

- **Food**: the food industry has been using NPs for the production, processing, protection and packaging of food [17]. For example, the use of nanosensors in the packaging warns when the food spoils, changing the colour. The use of silver NPs confers antimicrobial properties to the package. Also, silver, magnesium oxide and zinc oxide NPs are usually incorporated to increase the barrier properties of packaging materials. Besides that, NPs are used in supplements for enhance absorption and bioavailability.

- **Construction**: nanotechnology have been improving the construction processes by making them quicker, inexpensive and safer. In the case of concrete, nanosilica (SiO_2) can improve its mechanical properties and its durability when mixed with the normal concrete. The addition of haematite (Fe_2O_3) NPs can also increase the strength of the concrete. Another important construction material to which NPs can add valour is glass, since it can provide a better blocking of light and heat penetrating through the windows. In addition, the NPs can also provide abilities of self-healing, corrosion resistance and insulation when added to the paints. It is also be possible to use these paints—which have hydrophobic properties and then repel water—to apply, for example, in coating of metal pipes to offer protection from salt water attack [12].

- **Environmental remediation**: NPs have, in general, excellent properties to sorb metals and organic substances from waters due to their large surface area [18]. NPs are also used for soil and air remediation. In the case of soil contamination,

NPs—like iron, titanium dioxide and carbon NPs—are injected into specific target locations in soil to remove the contaminants [19]. Regarding air remediation, NPs have been used in catalysts or membranes. For example, a catalyst of gold NPs embedded in a porous manganese oxide can be used to breakdown volatile organic compounds in air [20].

- **Renewable energy**: Renewable energy sector can also benefit with the use of NPs since they allow to increase considerable the efficiency of solar cells through higher light and UV absorption with a very low reflection coating. Also, the hydrophobic property of some NPs has led to self-cleaning solar cells [12].

2.2 Methods of Synthesis and Structural Characterization

NPs can be prepared with high control of size and/or morphology by the concerted interaction of atoms or molecules during the synthesis process. By adjusting the experimental variables—such as reagent concentration, temperature, pH, presence of additives, properties of the solvent, addition of nucleating seeds—the characteristics of colloidal particles formed from homogeneous solutions in a liquid phase can be easily manipulated, which is particularly interesting. The influence of these factors during the nucleation and growth processes of the particles will affect the size and/or morphology of the final particles [6].

In this way, the selection of the synthetic methodology is important concerning, mainly, morphologies, particle sizes and its properties. Additionally, some methods are faster, cheaper or greener, and consequently, more attractive for industry. In general, the synthesis methods can be classified as top-down or bottom-up. Top-down method is "destructive", which means that a bulk material is reduced to a powder and then to NPs. On the contrary, the bottom-up method is "constructive", and therefore, the material is formed from atoms to clusters until finally to the nanoparticle which is formed due to the physical forces operating at the nanoscale during self-assembly [2, 12]. There are several top-down methods, such as mechanical milling/grinding, nanolithography, laser ablation, sputtering and thermal decomposition. From these approaches, mechanical milling is one of the most used since it can be used for the synthesis of metal, oxide and polymer NPs. However, these methods cannot be applied in large scale production of NPs since they are costly and slow processes [12]. Regarding bottom-up methods, NPs can be synthesized through sol-gel, spinning, chemical vapour deposition, pyrolysis, microemulsion, precipitation, hydro- and solvothermal, electrochemical, biosynthesis, among other methods. Sol-gel is one the most used bottom-up method, since it allows to synthesize most of the NPs (carbon, metal and metal oxide-based NPs) and is easy to use [12].

Nanoparticles focused in this chapter belongs to two different classes of inorganic NPs: metal NPs (silver NPs) and metal oxide NPs (titanium and zinc oxide NPs). As mentioned above there are several methods to obtain metallic NPs: physical, biological and chemical approaches. The method most used to synthetize metal NPs (like silver, gold and platinum NPs) is the chemical reduction method because of its simple equipment, low-cost and high yield. In general, the chemical reduction process of silver NPs in the solution includes metal precursors, reducing agents, and stabilizing/capping agents. Also, in this method, coprecipitation and the following substeps as reduction, nucleation and growth occur. Note that NPs size is controlled by interrupting the precipitation method [21, 22].

In the case of metal oxide NPs synthesis (as titanium and zinc oxide NPs), the most used method is sol-gel. Briefly, this method consists on the sol—a colloidal solution of solids suspended in a liquid phase—and the gel—a solid macromolecule submerged in a solvent. In this type of process, it is used a chemical solution to act as precursor (metal oxides and chlorides), which is further disperse in the "sol". As a result, it is obtained a system containing a liquid and a solid phase. Lastly, it is performed a phase separation to recover the NPs by different methods—sedimentation, filtration or centrifugation—and then, the moisture is removed by drying [12].

The importance of NPs in aforementioned applications is determined by its characteristics, namely its size distribution, shape, crystallinity, composition, porosity, surface area, surface functionality, surface charge, surface speciation, agglomeration state and concentration. All of these characteristics can be measured using a plethora of characterisation techniques, as follows:

- Transmission Electron Microscopy (TEM)—Size distribution and shape of nanoparticle;
- Scanning Electron Microscopy (SEM)—Size distribution and morphology of nanoparticle;
- Dynamic Light Scattering (DLS)—Average particle size; overall charge on the nanoparticle;
- X-ray Diffraction (XRD)—Crystalline structure; NP size;
- Fourier Transform Infrared Spectroscopy (FTIR)—Nature and strength of the bonds;
- X-ray Photoelectron Spectroscopy (XPS)—Elemental composition at NP surface;
- Brunauer-Emmett-Teller (BET) method—Surface area, porosity and mean diameter;
- Magnetic Studies (for magnetic NPs)—Magnetic properties: saturation magnetization (that represents magnetic intensity of magnetic materials), coercivity and squareness.

2.3 Impact in the Environment and Toxicity

Nanotechnology is an area which makes intensive use of new materials and chemical substances, therefore carries some risk either to the environment or to human health [8]. In addition, to our knowledge, information on NPs toxicity is still limited, there are still few specific regulations for their use. However, the evaluation of NPs toxicity—which depends on their aggregation, agglomeration, dispensability, size, solubility, surface area, surface charge and surface chemistry [23]—is crucial to assess the potential risks of these materials on the environment and health [24].

NPs reach the environment through natural sources (forest fires, volcanic activities, weathering, formation from clay minerals, soil erosion by wind and water, dust storms from desert) or anthropogenic activities. Regarding anthropogenic activities, NPs can enter the environment in three different stages of their life cycle: during the production of raw material and consumer products, during its use and after disposal of NPs-containing products (waste handling) [25]. Furthermore, the diversification of NPs emission sources into the environment is expected to increase with the increase of NPs production and application (either in terms of quantity or diversity of products) [25]. Currently, NPs can reach the environment directly or indirectly through wastewater treatment plants or landfills. The release pattern and masses depend on the NPs type and its application. When present in the environment, NPs behave differently in organisms, soil and water, since the exposure to different medium associated with a large number of physicochemical processes leads to a series of possible interactions that will change the NPs. Also, they accumulate in various organisms (plants, fish and other aquatic organisms) and environmental matrices (air, water, soils and sediments), and may enter into the food chain. Thus, humans can be exposed to NPs through feeding or other pathways of exposure such as inhalation, dermal contact and injection [23].

Based on NPs life cycle, material flow models have been used to predict NPs emission and environmental concentration levels [26]. Results obtained for concentrations of titanium dioxide, zinc oxide and silver NPs in waste streams and environmental compartments in the EU in 2014 and 2020 are presented in Table 1. Despite that, the production volume of NPs may give a good indication of the emission of specific NPs. Comparing the available data on production volumes, it is noted that the main difference is on the way of data collection. Titanium dioxide NPs is one of the most relevant materials regarding worldwide production volumes with more than 10,000 tons, in 2010. The worldwide production volume of zinc oxide and silver NPs was estimated to be between 100 and 1000 tons, and approximately 55 tons, respectively, in 2010 [25]. Therefore, mathematical models are based not only on NPs life cycle, but also on its production volumes.

The dominating emission pathway of titanium dioxide NPs occurs via wastewater, corresponding to 85% of total titanium dioxide NPs emissions [25, 27]. This fraction is divided into three different ways of emission: 36% of titanium dioxide NPs emissions occurs through its accumulation in sewage sludge during wastewater treatment; in addiction, 30% is deposited onto landfills directly or after the incineration of the sewage sludge; and lastly, 33% occurs via wastewater effluent [26]. Like titanium dioxide NPs, the dominating emission pathway of zinc oxide NPs occurs via wastewater since both are used in cosmetics. On contrary, silver NPs are emitted to both landfills and wastewater [25].

Using computational models, it was possible to estimate that the NPs concentrations in surface waters were in the ng/L or µg/L range depending on the type of NPs. Regarding the NPs focused in this chapter, it was estimated that NPs mean concentration in surface water for the EU in 2014 were approximately 2.2 µg/L for titanium

Table 1 Predicted (accumulated) concentrations of titanium dioxide (TiO_2), zinc oxide (ZnO) and silver (Ag) NPs in waste streams and environmental compartments in the EU in 2014 and 2020. Values shown are mean values and are rounded off to three significant digits [26]

	EU (2014)			EU (2020)		
	TiO_2	ZnO	Ag	TiO_2	ZnO	Ag
STP effluent (μg/L)	44.4	0	2.65×10^{-3}	106	0	6.70×10^{-3}
STP sludge (g/kg)	1.60	0	61.3×10^{-6}	4.37	0	175×10^{-6}
Solid waste to landfill (mg/kg)	12.9	1.69	79.0×10^{-3}	37.4	4.58	223×10^{-3}
Solid waste to WIP (mg/kg)	10.3	1.27	19.7×10^{-3}	29.5	3.43	55.3×10^{-3}
WIP bottom ash (g/kg)	395×10^{-3}	6.43×10^{-3}	170×10^{-6}	1.12	17.5×10^{-3}	496×10^{-6}
WIP fly ash (g/kg)	543×10^{-3}	12.2×10^{-3}	340×10^{-6}	1.53	32.0×10^{-3}	1.01×10^{-3}
Surface water (μg/L)	2.17	0.38	1.51×10^{-3}	5.76	0.94	4.15×10^{-3}
Sediment (mg/kg)	43.1	6.97	30.1×10^{-3}	123	20.2	88.2×10^{-3}
Natural and urban soil (μg/kg)	2.94	1.82	0.02	8.57	5.28	0.06
Sludge treated soil (g/kg)	61.1×10^{-3}	1.82×10^{-6}	2.31×10^{-6}	0.18	5.28×10^{-6}	6.94×10^{-6}
Air (ng/m^3)	2.05	0.94	0.01	5.48	2.33	0.03

dioxide, 0.38 μg/L for zinc oxide NPs and about 1.5 ng/L for silver NPs (Table 1). Note that, in 2018, the predicted concentration for the EU are two to three times higher than in 2014. Since these values were obtained through mathematical models, the results accuracy is limited because either by life cycle information or production volumes are not available in enough detail. Because of that, the values obtained by computational models were then investigated with further studies assessing the actual presence of NPs in the aquatic environment [25], confirming the modelling results.

3 Ecotoxicological Impacts

Nanoparticles can be toxic for organisms since they interact with them through, for example, intracellular uptake. After the entrance of NPs into a cell (Fig. 3)—through diffusion or endocytosis—they can promote the generation of reactive oxygen species (ROS). The internalization of NPs through endocytosis can occurs in three different main ways: clathrin-mediated endocytosis (vesicle size ~120 nm), clathrin independent endocytosis (~60 nm), and phagocytosis and macro-pinocytosis (<1 mm) [28]. Note that clathrin is the protein involved in the formation of coated vesicles. Then, the free metal ions and the ROS can cross the membrane of the cell nucleus, inducing DNA damage. As a consequence, either the DNA damage is repaired, or it occurs an irreversible damage on the chromosomes or the cell death (apoptosis) [29].

To evaluate the effects of NPs into organisms, several biochemical parameters are used as markers. Most relevant and used parameters to evaluate the biochemical alterations are described below.

3.1 Energy Reserves and Metabolic Capacity

Organisms have reserves of energy, that corresponds to the energy released by any reduced carbon stored in compounds such as fatty acids in triglycerides of adipose tissue, glucose in glycogen of liver/muscle, and amino acids in protein of muscles. But how they obtain these reserves? After digestion and absorption of triglycerides, glycogen and protein—obtain from the ingestion of food—, these are converted into fatty acids, glucose and amino acids, respectively. Through catabolic reaction, Acetyl Coenzyme A (Acetyl-CoA)—a key intermediate compound in cellular metabolism—is formed and reacts with oxaloacetate, forming citrate. Then, two carbons of citrate

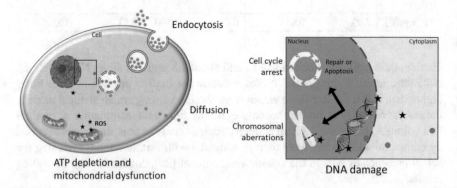

Fig. 3 Mechanism of interaction between NPs and organisms through intracellular uptake (left) and DNA damage induced by free metal ions and ROS (right). ●= represents metal/metal oxide NP; ●= free metal ions; ◯= adenosine triphosphate (ATP); ★= ROS

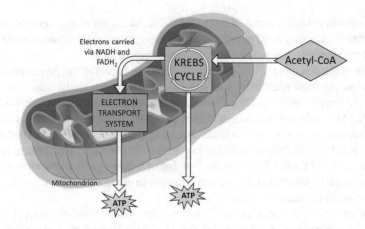

Fig. 4 Mechanism of production of ATP inside mitochondrion

are oxidized to CO_2 with the formation of succinyl-CoA, which is further converted into oxaloacetate (through oxidation). Lastly, this cycle (Krebs cycle) repeats. Part of the energy released from oxidation reactions that occur during the cycle are stored in succinyl-CoA and used to form adenosine triphosphate (ATP). The other part is released as electrons, which are collected by NADH and $FADH_2$ and introduced in the electron transport system (ETS), inside mitochondria. ETS consists on a cluster of proteins that transfer the electrons carried via NADH and $FADH_2$, and with the flux of electrons, a large amount of energy is released and used to produce ATP (Fig. 4). The ATP formed is used by the cell in metabolic processes for cellular functioning [30].

Therefore, energy reserves of organisms can be evaluated through the determination of protein (PROT), glycogen (GLY) and lipid (LIP) contents. On the other side, ETS activity gives indications concerning any metabolic changes that could occur on organisms. For instance, in bivalves exposed to a mild stress, ETS may decrease and energy reserves increase. They can use this strategy for a short time as it involves filtering and breathing less [31–33]. But, if exposed to a high stress, organisms increase metabolism and use reserve energies to fuel up defense mechanisms [34].

3.2 Antioxidant Defenses and Biotransformation Mechanisms

The exposure of an organism to toxic substances leads to the production of ROS by the mitochondrial electron-transport chain, as represented in Figs. 3 and 4. Superoxide anion ($O_2{}^-$), hydroxyl radical ($\cdot OH$), and hydrogen peroxide (H_2O_2) are the main ROS formed [4]. The imbalance between oxidants (ROS) and cell's antioxidants in favour of oxidants is designated oxidative stress. To block harmful effects of ROS, in order to avoid oxidative stress of cells, organisms have antioxidant defense systems, which means that they have enzymatic and nonenzymatic antioxidants that maintain the balance between the production and elimination of ROS.

Superoxide dismutase (SOD), catalase (CAT) and glutathione peroxidase (GPx) are commonly used enzymatic antioxidants, which are naturally produced by our body. The enzyme SOD catalyses the conversion of the radical superoxide (O_2^-) to hydrogen peroxide (H_2O_2), which is then reduced to water using CAT as catalyst. The reduction of hydrogen peroxide is also catalysed by other enzymes, such as GPx (Fig. 5) [4]. Thus, SOD is an immediate defense (with the removal of superoxide) that together with CAT and GPx are in the first line of defense (with the neutralization of H_2O_2 or organic peroxides). In this way, changes in the activity of these antioxidants allow to evaluate oxidative stress of cells.

Non-enzymatic defenses include low-molecular-weight compounds like vitamins, β-carotene, uric acid and reduced glutathione (GSH). Besides its high abundance in all cell compartments, GSH is the principal soluble antioxidant. GSH is related to GPx, since to neutralize H_2O_2 into H_2O, GPx takes hydrogens from two GSH molecules resulting in two H_2O and one oxidized glutathione (GSSG) molecules. Once oxidized, the GSSG is converted in GSH using glutathione reductase (GR) [4].

Biotransformation consists on the enzyme-catalysed conversion of a substance from a chemical form to another, within an organism body. This process is an important defense mechanism as it converts toxic substances into less harmful substances or substances that can be excreted from the body. The glutathione S-transferases (GSTs) is a group of enzymes commonly used in this process.

In general, after exposure of organisms to a certain level of stress, the antioxidant and biotransformation enzymes can increase [31–33]; but if stress is too high enzymes are inhibited [35].

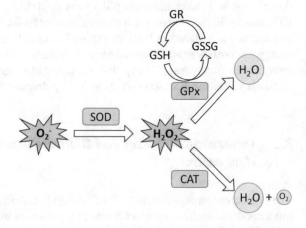

Fig. 5 Role of SOD, CAT and GPx in the neutralization of ROS, and its relationship with ration between GSH and GSSG

3.3 Cellular Damage

When the production of ROS exceeds the capacity of antioxidant defense mechanisms, oxidative stress can damage a wide variety of cellular components (membranes, protein and DNA) leading to lipid peroxidation, protein oxidation, and genetic damage [4]. Lipid peroxidation (LPO) is an useful indicator to evaluate the cellular damage, corresponding to the oxidation of the lipids of cellular membrane. As a consequence of LPO, there are byproducts that could be formed, such as Thiobarbituric acid reactive substances (TBARS), which can be used as markers of LPO (Fig. 5). Furthermore, protein carbonylation (PC) is a biomarker commonly used to measure the oxidation of proteins caused by ROS. Thus, aldehyde or ketone groups are formed through direct oxidation of side chains or its reaction with lipid and sugar oxidation products. Consequently, carbonyl groups can affect, for example, the cell signal transduction [36]. Besides, to evaluate DNA damages, the activity of DNA repair enzymes, such as tumour protein p53 and phosphate dikinase regulatory protein (PDRP)—which are involved in minimizing DNA lesions, mutation accumulation, and chromosomal aberrations—, is monitored [37]. Some studies have demonstrated that the levels of cellular damage decreased with the increase of antioxidant defenses aforementioned [31, 34]. However, others showed that these indicators increased even with an increase of antioxidant enzymes activity [32–34, 38], which means that the activation of antioxidants defences is not enough to eliminate ROS.

3.4 Redox Balance

According to Fig. 5, the ratio between GSH and GSSG depends on the reaction of hydrogen peroxide reduction using GPx, which takes hydrogens from GSH molecules to form H_2O and GSSG molecules. In this context, the ratio GSH/GSSG decreases with the increase of ROS [31–33], with lower values indicating loss of redox homeostasis.

3.5 Neurotoxicity

Neurotoxicity is a change in the normal activity of nervous system that causes damage to nervous tissue, leading to neurons disturbance or death. This can be triggered by the exposure to contaminants like NPs.

Neurons communicate among each other in two ways: they can send an electrical signal directly to their neighbours—very fast—or release small neurotransmitter molecules into the synaptic cleft (recognized by receptors on their neighbours). In the last one (which is the most common), after neurotransmitters delivered the message, they are broken down by enzymes and transported out of synaptic cleft through neurotransmitter transporters. Acetylcholinesterase (AChE) is a key enzyme in the nervous system [39] used to breaks down the neurotransmitter acetylcholine (ACh)— into acetic acid and choline—in the synaptic cleft (Fig. 6). ACh is a neurotransmitter used at neuromuscular junction to cause muscle contraction; it is also used in the autonomic nervous system; and plays important roles in memory, arousal and attention. So, AChE stops the signal of ACh and allows the recycling and rebuilding of choline into new neurotransmitters for the next message. However, AChE is a very sensitive enzyme [39], so the exposure to NPs will inhibit its activity. In this way, the monitorization of AChE activity gives indications about neurotoxicity.

Besides AChE, there are other major neurotransmitters, whose activity is monitored to evaluate effects of contaminants into organisms, such as dopamine (DA), norepinephrine (NE), serotonin, gamma-aminobutyric acid (GABA), and glutamate.

Typically AChE activity decreases or is inhibit when organisms are exposed to contaminants [31, 38, 40, 41] but may enhance their activity if the stress is too high [42].

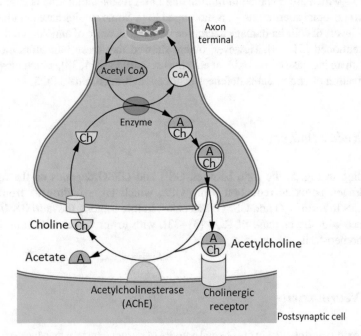

Fig. 6 Role of AChE on the formation of acetylcholine, an essential neurotransmitter to the nervous system

3.6 Immunotoxicity

The exposure to NPs can also induce adverse on immune function, which is defined as immunotoxicity. NPs are known to be taken up by hemocytes, which are the cells of hemolymph responsible for immune defense and the first line of defense against contaminants, namely in marine invertebrates. In this context, hemocytes can be used to performed functional and toxicological tests to evaluate NPs effects [43, 44]. Among the different hemocytes parameters, total hemocyte count (THC), differential hemocyte count (DHC), hemocyte viability, phagocytic activity and lysosomal membrane stability have been used to evaluate immunomodulatory effects of NPs [43]. Typically, when organisms are exposed to NPs, while low concentrations can stimulate the hemocyte parameters, high levels of NPs can inhibit them [45].

3.7 Embryotoxicity

Embryotoxicity consists on toxic effects on embryos caused by NPs that cross the placental barrier. The embryo development is the most sensitive stage of the organism to environmental perturbations [46]. Because of this, exposure of NPs to the organisms during this life stage allows to identify the substances that pose a greater threat to marine species. The embryotoxicity tests are performed through the exposition of NPs to eggs [47] or sperm [48], to fertilized eggs [44] or even to embryos (Fig. 7). After fertilization, the embryo formed is tested for any effect such as malformations, growth retardation, mortality, among others. The endpoints commonly used in bivalves are the percentage of normal D-larvae and the fertilization success, both through microscopical observation. The first endpoint consists on counting normally and abnormally developed larvae; a normal larva is the one in which the shell is D-shaped (straight hinge) and the mantle does not protrude out of the shell. The fertilization success is the percentage of fertilized eggs in respect to the total [49].

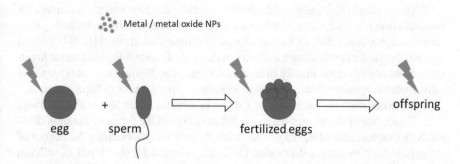

Fig. 7 Embryotoxicity tests: exposure of NPs during different life stages of organisms

All these conventional biomarkers are commonly used to assess the effects of NPs in aquatic organisms, namely to evaluate ROS-mediated NPs injury. Besides them, there are other biomarkers sometimes used to evaluate effects into organisms such as ethoxyresorufin-O-deethylase (EROD), glutathione reductase (GR), metallothionein (MT), heat stress proteins (HSPs), caspase-3, acid phosphatase (ACP), alkaline phosphatase (ALP), lysosomal neutral red retention time (NRRT), multixenobiotic-resistance (MXR), among others.

4   Assessment of Nanoparticles Toxicity in Marine Organisms

As already mentioned, titanium dioxide, zinc oxide and silver NPs are the most used metal and metal oxide NPs. In this way, high amounts of these NPs end up into the aquatic ecosystems, being extremely important to know its impact on aquatic life, namely their effects on invertebrate marine organisms. The choice of this group of organisms was based on their recognition as significant biological targets for NPs. Moreover, among the different marine invertebrate groups—bivalves, polychaetes, copepods, sea urchin, amphipods, artemia, rotifer and cnidarians—the group of bivalves is the most widely studied for evaluation of nanomaterials effects [3]. This observation is due to the fact they are filter feeder animals and consequently can accumulate toxic substances present in the waters.

There are only a few dozens of works about toxicity of titanium dioxide, zinc oxide and silver NPs into marine invertebrates published in the literature. Despite that, there is an increase awareness from the scientific community for the impact and effects of NPs in the environment and further studies have to be developed. Toxicity data retrieved from the most recent literature is summarized in Table 2, according to the type of NPs. The effects of NPs depends on exposure time [37, 50, 51], mode of exposure [52, 53], particle size and concentration [50, 51, 54], uptake pathway [36] and season of the year [55].

Titanium dioxide NPs effects into marine invertebrates have become a concern for researchers due to its increase use in a wide diversity of applications, including sunscreens, cosmetics, catalysis and environmental remediation. Besides two artemia species, the studies evaluating the effects of titanium dioxide NPs into marine invertebrates were performed mainly in bivalve species. The *Tegillarca granosa* was the most studied organism in recent works, followed by the *Mytilus galloprovincialis*.

The studies were performed mostly using titanium dioxide NPs with size lower than 50 nm; one of the studies used two different crystalline forms of titanium dioxide NPs (anatase and rutile) [56]; and another study further evaluate the impact of using gold for NPs surface decoration [57], being observed higher levels of MDA in gills and digestive gland of *Ruditapes decussatus* after exposure to $AuTiO_2$ NPs than TiO_2 NPs. Concerning exposure, only one study tested both waterborne and dietary exposures [53], concluding that the major route of NP uptake and accumulation was through waterborne exposure. The studies were performed between 4 h and 30 days, and the concentration of NPs varied between 50 μg/L and 100 mg/L, being 100 μg/L titanium dioxide NPs the most commonly used dose. The choice of 100 μg/L titanium dioxide NPs is based on the most extensive data previously obtained about effects of these NPs. However, in order to extrapolate the results obtained to realistic environment, the studies on NPs effects must be performed at the predicted environmental relevant concentrations of NPs. So, the main problems regarding the exposure conditions are the use of either an unrealistic high dose of NPs (usually above 1 mg/L) or short duration exposures (usually less than 7 days). Regarding the evaluation of effects, Xia et al. [42] evaluate oxidative stress and neurotoxicity using SOD, CAT, MDA and AChE and then summarized these conventional biomarkers in a multivariate data to apply the integrated biomarker response (IBR) index. High values of IBR can indicate the enhanced biological responses and poor health condition of the organisms [42] and, therefore, this approach is useful to evaluate the toxic effects of NPs on the marine organisms.

Titanium dioxide NPs induced effects on immune function and at cellular level, but no studies on the effects on embryo development of marine invertebrates were performed. In terms of immunomodulatory effects, these were confirmed by alterations on the microbiota composition of mussels hemolymph [58]. Also, toxicological tests in hemocytes of *Tegillarca granosa* showed a significant reduction in immune response, as well as a reduction of the expressions of immune-related genes [59, 60]. Regarding the effects at cellular level—oxidative stress and cell injury in proteins, membrane and DNA damage—, these increase with the use of higher NPs concentration and with exposure time [51, 53], as expected. Also, titanium dioxide NPs were responsible for oxidative stress, as confirmed by an increase of H_2O_2 and NO levels, and an increase of SOD and CAT activities [42, 57]. Moreover, evident neurotoxic impact of titanium dioxide NPs was observed through monitorization of DA, ACh and GABA neurotransmitters concentrations; AChE activity was also monitored, being that it decreases in one work and increased in another work [42, 61]. Although a few works reflected no or minimum effects of titanium dioxide NPs [56, 62], this can be due to the short-term exposures of organisms to new stressors without given them time to adapt to that conditions. For instance, the exposure of *Crassostrea virginica* hemocytes to TiO_2 NPs for up to 4 h under dark and light conditions produce minimal apparent effects upon cell mortality, phagocytosis, and ROS production. Furthermore, effects of variations on physicochemical parameters of water were studied; namely, Shi et al. [63] evaluated the effects of titanium dioxide into bivalves under predicted acidification of seawater conditions. It was concluded that for lower pH, titanium dioxide contents accumulated in bivalves increased.

Lastly, few references are made to the toxicity mechanism of titanium dioxide NPs, which remains unclear and therefore further investigation is strongly needed. Nevertheless, ROS production has been pointed as one of the main toxicity mechanisms [42, 53, 57].

Regarding the studies assessing the zinc oxide NPs toxicity, a diversity of organisms—four different species of bivalves, two species of sea urchins and one species of artemia and cnidarians—were used as test models to evaluate the impacts of this type of NPs.

The studies were performed using zinc oxide NPs from 14 nm to 100 nm. Like for titanium dioxide NPs, the impact of using gold in NPs surface was evaluated, resulting in a significant increase in H_2O_2 level, SOD and CAT activities and MDA content [54]. In terms of exposure conditions, the invertebrates (including their sperm or eggs) were exposed mostly through water contaminated with zinc oxide NPs. The longest essay was performed for 4 weeks, while the shortest lasted 10 min. The concentration of NPs varied between 6.5 µg/L and 100 mg/L. The range of concentrations 1–10 mg/L was the commonly used in laboratory research since it triggers physiological responses in aquatic organisms [64]; other researchers focused on works with ecological relevance, by choosing environmentally relevant concentrations (a few µg/L zinc oxide NPs) [48, 65].

The studies carried out in the different species of bivalves revealed that zinc oxide NPs induce effects on immune function and at cellular level. An increase of mortality along time was observed in *Mytilus galloprovincialis* exposed to 10 and 100 mg/L NPs; but note that these concentration values are extremely high and don't reflect the environmental concentrations. In terms of immunotoxicity studies, the exposure of *Mytilus coruscus* to high concentrations of zinc oxide NPs resulted in a decrease of hemocytes parameters (total hemocyte count, phagocytosis and lysosomal membrane stability) and an increase of hemocyte mortality and ROS [66]. The increase of ROS is corroborated by the increase of SOD, CAT, GPx, ACP and ALP in another study after the exposure of *Mytilus coruscus* to the same conditions [64]. Regarding pH, the mussel *Mytilus coruscus* was exposed to different concentrations of zinc oxide NPs at two pH levels: 7.3 (corresponding to the extreme low pH predicted for hypoxic zones by the year 2100) and 8.1 (which is the present pH average). It was concluded that low pH (7.3) causes a decrease of immune parameters of the mussels, but its effect is not so strong as high concentration of NPs [66]. It was also observed that cellular stress responses are influenced by the origin of mussels population [67]. The study of zinc oxide NPs effects on the artemia *Artemia salina* also revealed an expected dependence of toxicity on the exposure time to NPs, as well as an inhibition of vitality and body length [68]. Zinc oxide NPs may also impact cellular fitness by causing alteration in the membrane lipid, as observed in the coral *Seriatopora caliendrum*.

The studies performed with sea urchin organisms were mainly focused on embryos development, which is the most sensitive stage of the organism to environmental perturbations. Also, the use of embryos of marine invertebrates complements the use of adult specimens on studies evaluating the toxicity of NPs [3]. In this sense, *Paracentrotus lividus* were exposed to different concentrations of 14 and 100 nm zinc oxide NPs, including through the ingestion of contaminated food [48, 65, 69]. The results shown a size-dependent toxicity, i.e. an increase of toxicity with NPs size reduction, which is general observed in this kind of works [48, 69]. Since smaller NPs have a higher surface area, they can probably induce a higher production of ROS (that is the main contributor for the NPs cytotoxicity). So, zinc oxide NPs with 14 nm, in comparison to NPs with 100 nm, caused more DNA damages in spermatozoa and, consequently, more transmissible effects to offspring (malformed larvae). Also, with increase of NPs concentration, miotic activity decreased, more chromosomal aberrations were observed, and larval development were more affected [65]. Concerning the effect of temperature, marine larvae may be more resilient to negative effects of high zinc oxide NPs concentration at optimal temperatures (27 °C) compared to lower (25 °C) or higher (29 °C) temperatures [47]. When high temperature is combined with high concentration of zinc oxide NPs, larval growth is affected.

Regarding the toxicity mechanism, toxicity of zinc oxide NPs is frequently associated to the release of zinc ions. This assumption was confirmed by measurement of the dissolution and aggregation of zinc oxide NPs and the direct effect of Zn^{2+} ions on sea urchin larvae *Tripneustes gratilla* [47]. The results obtained suggested that Zn^{2+} ions were the major contributor to the negative effects of zinc oxide NPs on larval development and growth. The information about toxic mechanisms is still unclear, but the particular characteristics of the NPs (such as size, surface, and shape) have an important role in the study of NPs toxicity.

Similar to the studies evaluating the effects of titanium dioxide NPs into marine invertebrates, the effects of silver NPs are mainly evaluated in bivalves, especially *Mytilus galloprovincialis*.

These studies were performed either using silver NPs or polymer-coated silver NPs, with sizes between 5 and 85 nm. The effect of poly-vinyl-pyrrolidone (PVP)-coated silver NPs was evaluated in different studies due to its widely use in biomedical applications as biosensor or virucidal agent, for example. It was also studied a silver NP-enabled product [70]; the study of this kind of products instead of pristine silver NPs is crucial to address any risks associated to its widespread use, disposal and uncontrolled release into the aquatic environment. In terms of exposure conditions, the invertebrates were exposure either through food or water contaminated with silver NPs. In the first case, the assays were of 21 or 28 days, while on exposure through contaminated water, the essays lasted between 30 min and 1 week. The concentration of NPs varied between 0.0001 µg/L and 100 mg/L, being 10 µg/L and 100 µg/L Ag NPs the most commonly used doses. The choice of 100 µg/L Ag NPs is based on the ability of this dose of NPs in producing adverse effects and, being usually used in ecotoxicity tests on aquatic species, it enables the correlation of the results with those obtained in previous works [43]. Regarding the evaluation of effects, besides the conventional biomarkers, some researches have been using proteomics-based methods tools as a complement of information given by the conventional methods as well as to better understand the NPs toxicology mechanism [36, 55].

Effects of silver NPs on immune function, at cellular level and on development of bivalves—mainly *Mytilus galloprovincialis*—were observed. Similar to what happen with titanium dioxide NPs, the application of (functional or toxicological) tests in hemocytes of bivalves has been proved to be effective on the evaluation of immunomodulatory/immunotoxic effects [43, 44, 52] of silver NPs. For instance, a small lysosomal membrane destabilization in hemocytes of *Mytilus galloprovincialis* can be induced by silver NPs after only 30 min of exposure. Also, silver NPs are responsible for oxidative stress, as confirmed by an increase of GST, LPO and PC [36, 50]. The bioaccumulation of silver increases with the increase of exposure time; on contrary, higher concentration of silver NPs induced less bioaccumulation in gills and digestive glands of mussels, because they have more tendency to aggregate and, consequently, undergo less dissolution [50]. Moreover, the effects of NPs in mussels were evaluated at two different seasons (in autumn and spring) since different seasons means different developmental stages and, consequently present different protein expression profiles [55]. It was also concluded that uptake routes have a great impact in toxicity: the blockade of clathrin-mediated endocytosis allows to prevent ROS-related disturbance [36].

Lastly, malformation of embryos was observed after parental exposure to 1 μg/L and 10 μg/L Ag NPs through fed [52]. Also, exposure of fertilized eggs to 0.001–1000 μg/L Ag NPs caused malformations and developmental delay, but no mortality of embryos [44].

In terms of mechanism, despite the toxicity mechanism be still unclear, toxicity of silver NPs is associated to the release of silver ions [3]. The study of silver NPs effects on the sea urchin *Strongylocentrotus droebachiensis* revealed that the toxicity mechanism of silver NPs could be a Trojan-horse mechanism, in the way that silver NPs react with ROS to form silver ions, which are toxic [71]. However, it is not well understood if the effects of NPs can only be explained by the released of silver ions. The lack of information about toxicity mechanism requires its exploration in further studies. To elucidate this, it is necessary to know actual internal concentrations of silver in the organism after exposure. In this context, the analysis of hemolymph indicates the amount of silver absorbed and distributed within the animal. Amphipod *Parhyale hawaiensis* showed an increase of silver concentration in hemolymph with the increase of exposure time to NPs [72] (Table 2).

So far, toxic effects of different NPs have been tested on different species of marine invertebrates. Nevertheless, there are a variety of exposure conditions, and different studies test different NPs concentrations, times of exposure, pH, salinity of water, among others. In this context, there are some questions that need to be addressed in further studies. The concentration of NPs is one of the most important aspects in this kind of studies. Many studies are performed at high levels of exposure concentration, which are unrealistic and therefore do not reflect the actual scenarios of environmental ecosystems. Also, with the use of short-term exposures, the organisms can not be able to adapt to the new conditions or allowed to respond to stress.

Moreover, despite all the assumptions and theories about toxicity mechanism, the interaction between NPs and biological systems is still unclear. Nevertheless, to have a complete picture of NPs impact, different pattern of exposure, as well as NPs transformations (e.g. its dissolution) should be considered [69]. There are different route of internalization for the same type of NPs and NPs properties (size, shape, surface area) affect both uptake and interactions. More attention should be given to this subject.

Lastly, to fully understand the effects of a determined nanomaterial into different organisms, a standardization of exposure condition is required, in order to be possible to compare the results. This will allow to define environmental safety guidelines concerning real exposure scenarios. Also, a research with environmental relevance

Table 2 Summary of *in vitro* and *in vivo* studies of the metal/metal oxide NPs most present into aquatic ecosystems: titanium dioxide, zinc oxide and silver NPs. Toxicity data are present according to the type of metal/metal oxide NP, specie of marine invertebrate organism and exposure conditions

NPs	Species	Exposure conditions	Effects	Ref.
TiO$_2$ NPs (30 nm)	*Tegillarca granosa, Meretrix meretrix, Cyclina sinensis* (bivalves)	Exposure to 100 µg/L for 21 days; salinity 21, 25 °C, pH 7.4, 7.8 and 8.1, 12/12 h light/dark cycle, with aeration	All species tested accumulated significantly greater amount of TiO$_2$ in pCO$_2$-acidified seawater. TiO$_2$ contents accumulated in the gills, foot and mantles of bivalves after 21 days at pH 7.4 and pH 7.8 were, respectively, about 1.34 and 1.16 times greater than those raised in the ambient pH of 8.1	[63]
TiO$_2$ NPs, AuTiO$_2$ NPs (10–15 nm)	*Ruditapes decussatus* (bivalve)	Exposure to 50 and 100 µg/L for 14 days; salinity 32, 18 °C, 12/12 h light/dark cycle, with aeration	Titanium was significantly accumulated in both gills and digestive gland after 2 weeks of exposure The use of gold as a noble metal to decorate the surface of TiO$_2$ NPs contributes to titanium accumulation in both organs TiO$_2$ NPs alter the behaviour of the clams by reducing filtration and respiration rates. The highest concentration of TiO$_2$ NPs induces an overproduction of H$_2$O$_2$ in gills and digestive gland and NO production only in gills AuTiO$_2$ NPs presented higher MDA levels in gills and digestive gland than the TiO$_2$ NPs	[57]

(continued)

Table 2 (continued)

NPs	Species	Exposure conditions	Effects	Ref.
TiO$_2$ NPs (10–65 nm)	*Mytilus galloprovincialis* (bivalve)	Exposure to 100 µg/L for 4 days; salinity 36, 16 °C, pH 7.9–8.1	The exposure to TiO$_2$ NPs affected the microbiota composition of mussels hemolymph: the abundance of some genera decreased (e.g. *Shewanella, Kistimonas, Vibrio*) while others increased (e.g. *Stenotrophomonas*). The increase in the bactericidal activity of whole hemolymph confirms the immunomodulatory effects of TiO$_2$ NPs	[58]

(continued)

Table 2 (continued)

NPs	Species	Exposure conditions	Effects	Ref.
TiO$_2$ NPs (40 nm)	*Tegillarca granosa* (bivalve)	Exposure to 0.05 and 0.1 mg/L for 10 days; salinity 21, 28 °C, pH 8.1	The phagocytosis of hemocytes was significantly hampered in all treatment groups, either using 0.05 or 0.1 mg/L of TiO$_2$ NPs, indicating a significant reduction in immune response. Regarding the total counts of hemocytes, cell type composition and ALP levels, no significant changes were observed when clams were exposed to TiO$_2$ NPs alone. However, when combined with endocrine disrupting chemicals (EDCs), it promotes significant reduction in the total counts and phagocytosis of hemocytes The results of the impacts of TiO$_2$ NPs exposure on the expression of five immune related genes (TRAF2, TRAF3, TRAF6, and NFκβ1 from the NFκβ and Toll-like receptor signalling pathway) demonstrated a significantly reduce of the expressions of all the tested immune-related genes	[59]

(continued)

Table 2 (continued)

NPs	Species	Exposure conditions	Effects	Ref.
TiO_2 NPs (35 nm)	*Tegillarca granosa* (bivalve)	Exposure to 0.1, 1 and 10 mg/L for 4 days; salinity 18, 32 °C, pH 8.09	The results showed that the in vivo concentrations of the neurotransmitters were significantly increased when exposed to high doses of TiO_2 NPs (1 mg/L for DA and 10 mg/L for ACh and GABA). The clams showed significantly low AChE activity The expression of genes encoding modulatory enzymes (AChE, GABAT, and MAO) and receptors (mAChR3, GABAD, and DRD3) for the neurotransmitters tested were all significantly down-regulated after TiO_2 NPs exposure	[61]

(continued)

Table 2 (continued)

NPs	Species	Exposure conditions	Effects	Ref.
TiO$_2$ NPs (30 nm)	*Mytilus galloprovincialis* (bivalve)	Exposure to 1, 10 and 100 mg/L for 8 days; salinity 32, 18–19 °C, 12/12 h light/dark cycle, with aeration	No considerable effect was found in the digestive glands of any of the groups treated with TiO$_2$ NPs with concentration gradients ranging from 1 to 100 mg/L. The level of O$_2^-$ anion, the activity of SOD and the GSH/GSSG ratio showed no significantly differences in digestive glands of all treated groups compared to the control. However, slight modifications were observed in the gills at high concentrations	[62]

(continued)

Table 2 (continued)

NPs	Species	Exposure conditions	Effects	Ref.
Anatase NPs (7.4 nm) n-Titan (86 nm)	*Crassostrea virginica* (bivalve)	Exposure of hemolymph to 0.1, 0.5 and 1.0 mg/L for 2 and 4 h; salinity 35.7, 20 °C, full dark or full light	Exposure of hemocytes to TiO_2 NPs for up to 4 h under dark and light conditions produce minimal apparent effects upon cell mortality, phagocytosis, and ROS production. Even at 1.0 mg/L, few effects were observed after 4 h of exposure. Photo-activation of the TiO_2 NPs did not increase mortality or sub-lethal effects upon the hemocytes. Hemocyte phagocytosis was significantly changed under some conditions: exposure to anatase NPs for 4 h under dark conditions, and to UV–Titan for 2 h under light conditions	[56]

(continued)

Table 2 (continued)

NPs	Species	Exposure conditions	Effects	Ref.
TiO_2 NPs (21 nm)	*Artemia salina* (artemia)	Exposure to 0.1, 1 and 10 mg/L for 48 h; fed with algae cells previously exposed to 0.1, 1 and 10 mg/L for 48 h; 23 °C, 16/8 h light/dark cycle, with aeration	Waterborne exposure was the major route of NP uptake and accumulation than that of dietary exposure. Higher availability of NPs in waterborne exposure enhanced the uptake and accumulation (LC_{50} 48 h 4.21 mg/L) Waterborne exposure seemed to cause higher ROS production and antioxidant enzyme (SOD and CAT) activity as compared to dietary exposure of TiO_2 NPs. The mortality of *Artemia* by dietary exposure was found to be dependent on the exposure concentration	[53]
TiO_2 NPs (25 nm)	*Chlamys farreri* (bivalve)	Exposure to 1 mg/L for 14 days; 18 °C, pH 7.95, with aeration	TiO_2 NPs caused oxidative damage on the scallops by the significantly elevated SOD and CAT activities, and MDA contents. The increased AChE activities reflected neurotoxicity of TiO_2 NPs. The histopathological analysis revealed alterations in the gill and digestive gland, such as dysplastic and necrosis	[42]

(continued)

Table 2 (continued)

NPs	Species	Exposure conditions	Effects	Ref.
TiO$_2$ NPs (35 nm)	*Tegillarca granosa* (bivalve)	Exposure to 10 and 100 mg/L for 30 days; salinity 21.5, 25 °C, pH 8.10	The total number, the phagocytic activity, and red granulocytes ratio of the hemocytes were significantly reduced after 30 days of exposure. The expressions of genes encoding Pattern Recognition Receptors and downstream immune-related molecules were significantly down-regulated by TiO$_2$ NPs exposures	[60]

(continued)

Table 2 (continued)

NPs	Species	Exposure conditions	Effects	Ref.
TiO_2 NPs (21 nm)	*Artemia franciscana* (artemia)	Exposure to 0.5–64 mg/L for 24, 48, 72 and 96 h; salinity 32 and 34, 25 °C, pH 8.27–8.31, darkness or starvation	TiO_2 NPs evidenced concentration-response relationship with effects generally increasing with contact time except for concentration values >50 mg/L. After 24 h, no toxic effects were observed up to 16 mg/L, while between 16 and 50 mg/L mortality was very high (97% at 50 mg/L); no effect was present at 64 mg/L. After 48 h, the toxicity trend was similar to 24 h exposure, but significant lethality effects were anticipated at approximately 20 mg/L. The LC_{50} of TiO_2 NPs after 24, 48, 72 and 96 h was 28, 18, 18, and 15 mg/L, respectively The absence of light generally lowered TiO_2 NPs effects while starvation increased its toxicity	[51]

(continued)

Table 2 (continued)

NPs	Species	Exposure conditions	Effects	Ref.
ZnO NPs (14 and 100 nm)	*Paracentrotus lividus* (sea urchin)	Exposure of dry sperm to 6.5, 33, 65, 196, 654 and 1961 µg/L for 30 min; salinity 38, 18 °C, pH 8	ZnO NPs induced DNA damages in spermatozoa after 30 min of exposure. While the sperm fertilization capability was not affected, morphological alterations (skeletal alterations) in offspring were observed and a positive correlation between sperm DNA damage and offspring quality was reported Smaller ZnO NPs (14 nm) induced greater effects than bigger ZnO NPs (100 nm)	[48]

(continued)

Table 2 (continued)

NPs	Species	Exposure conditions	Effects	Ref.
ZnO NPs (polydisperse NPs: 20–50 and >100 nm)	*Mytilus coruscus* (bivalve)	Exposure to 0, 2.5 and 10 mg/L for 1, 3, 7 and 14 days, followed by a recovery period of more 7 days; salinity 25, 25 °C, pH 7.3 and 8.1, with aeration	At high ZnO NPs concentrations, total hemocyte counting, phagocytosis, esterase, and lysosomal content were significantly decreased whereas hemocyte mortality and ROS were increased. Although low pH also significantly influenced all of the immune parameters of the mussels, its effect was not as strong as that of ZnO NPs. Interactive effects observed between pH and ZnO NPs in most hemocyte parameters during the exposure period shown a slight recovery from the stress of ZnO NPs and pH, as well as significant carry-over effects of low pH and ZnO NPs	[66]
ZnO NPs (<100 nm)	*Artemia salina* (artemia)	Exposure to 10, 50 and 100 mg/L for 24 and 96 h; 25 °C, darkness Exposure to 0, 0.03, 0.06, 0.12, 0.25 and 0.5 mg/L for 3, 5, 7, 9, 12 and 14 days; 25 °C, 14/10 h light/dark cycle, with aeration	ZnO NPs toxicity was strictly dependent on the exposure time and ZnO NPs were relatively less toxic towards crustaceans (EC$_{50}$ 96 h 58 mg Zn/L) than towards algae, for example During the 14-days chronic exposure, ZnO NPs had a significant inhibition of vitality and body length (EC$_{50}$14d 0.02 mg Zn/L)	[68]

(continued)

Table 2 (continued)

NPs	Species	Exposure conditions	Effects	Ref.
ZnO NPs (100 nm)	*Mytilus galloprovincialis* (bivalve)	Exposure to 0.01, 0.1, 1, 10 and 100 mg/L for 4 weeks; salinity 25, 23 °C, pH 7.9, with aeration	Starting from 72 h, increasing mortality values along the exposure time were observed. The 100% mortality for the NPs occurred at 14 days in the 100 mg/L ZnO NPs and at 21 days in 10 mg/L ZnO NPs. After 28 days, NPs showed (LC_{50}) = 0.78 mg/L ZnO NPs. The relative expression of investigated genes evidenced that distinct actions of apoptosis and antioxidation occurred in *M. galloprovincialis* exposed to ZnO NPs with a peculiar pattern dependent on exposure time and concentration	[37]

(continued)

Table 2 (continued)

NPs	Species	Exposure conditions	Effects	Ref.
ZnO NPs (15–25 nm)	*Mytilus coruscus* (bivalve)	Exposure to 0, 2.5 and 10 mg/L for 1, 3, 7 and 14 days, followed by a recovery period of more 7 days; salinity 25, 25 °C, pH 7.3 and 8.1	Most biochemical indexes (SOD, CAT, GPx, ACP and ALP) measured in gills and hemocytes were increased when the mussels were subject to low pH or high concentration of nano-ZnO, suggesting oxidative stress responses. No significant interactions between the two stressors were observed for most measured parameters. After a 1-week recovery period, low pH and nano-ZnO had less marked impact for SOD, GPx, ACP and ALP in hemocytes as compared to the end of the 14 days exposure. However, no recovery was observed in gills	[64]

(continued)

Table 2 (continued)

NPs	Species	Exposure conditions	Effects	Ref.
ZnO NPs	*Unio tumidus* (bivalve)	Exposure to 201 µg/L for 14 days; 18 and 25 °C, with aeration	Zn-containing exposures resulted in the elevated concentrations of total and Zn-bound metallothionein (MT and Zn-MT) in the digestive gland, an increase in the levels of non-metallated MT (up to 5 times) and ALP and lysosomal membrane destabilization in hemocytes. A common signature of ZnO NPs exposures was modulation of the MXR protein activity (a decrease in the digestive gland and increase in the gills) The origin of population strongly affected the cellular stress responses of mussels. Mussels from one of the reservoirs showed depletion of caspase-3 in the digestive gland and up-regulation of HSP70, HSP72 and HSP60 levels in the gill during most exposures, whereas in the mussels from the other reservoir caspase-3 was up-regulated and HSPs either down regulated or maintained stable. Also, mussels from the second reservoir were less adapted to heating shown by a glutathione depletion at elevated temperature (25 °C)	[67]

(continued)

Table 2 (continued)

NPs	Species	Exposure conditions	Effects	Ref.
ZnO NPs (50–70 nm)	*Seriatopora caliendrum* (cnidarian)	Exposure to 50, 100 and 200 µg/L for 24 h; salinity 36.6, 27 °C, pH 8.09, 12/12 h light/dark cycle, with aeration	Exposure of coral to ZnO NPs leads to an increasing of lyso-GPCs, docosapentaenoic acid-possessing GPCs and docosahexaenoic acid-possessing GPCs and decreasing of arachidonic acid-possessing GPCs (GPC means glycerophosphocholine). A backfilling of polyunsaturated plasmanylcholines was observed in the coral exposed to ZnO NPs levels over a threshold	[73]

(continued)

Table 2 (continued)

NPs	Species	Exposure conditions	Effects	Ref.
Au-ZnO NPs (30 nm)	*Ruditapes decussatus* (bivalve)	Exposure to 50 and 100 µg/L for 14 days; salinity 32, 19 °C, 12/12 h light/dark cycle, with aeration	Au-ZnO NPs induce biochemical and histological alterations within either the digestive gland or gill tissues at high concentration. This was deduced from the significant increase in H_2O_2 level, SOD and CAT activities and MDA content. Furthermore, the toxicity of Au-ZnO NPs was linked with the increase of intracellular iron and calcium levels in both tissues. Histological alterations in gill and digestive gland were more pronounced with 100 µg/L Au-ZnO NP and this is likely related to oxidative mechanisms. Gill and digestive gland are differentially sensitive to Au-ZnO NPs if the exposure concentration is higher than 50 µg/L	[54]

(continued)

Table 2 (continued)

NPs	Species	Exposure conditions	Effects	Ref.
ZnO NPs (14 and 100 nm)	*Paracentrotus lividus* (sea urchin)	Exposure to 6.5, 33, 65, 196, 654 and 1961 µg/L for 24 h; salinity 38, 18 °C, pH 8	ZnO NPs interfere with cell cycle inducing a dose dependent decrease of mitotic activity and chromosomal aberrations at higher concentrations (1961 µg/L). Moreover, the larval development was affected by ZnO NPs 100 nm in a dose-dependent way. ZnO particles at lower concentrations exerted morphological alterations in larvae. Considering embryotoxicity data, ZnO NPs 14 nm resulted the less toxic	[65]
ZnO NPs (30 nm)	*Tripneustes gratilla* (sea urchin)	Exposure of eggs to 0.001, 0.01, 0.1, 1 and 10 mg/L for 10 min; salinity 35, 25, 27 and 29 °C, pH 8.1	High concentrations of ZnO NPs had a negative effect, but this impact was less pronounced for sea urchins reared at their preferred temperature of 27 °C compared to 25 or 29 °C. Larval growth was also impacted by combined stress of elevated temperature and ZnO NPs	[47]

(continued)

Table 2 (continued)

NPs	Species	Exposure conditions	Effects	Ref.
ZnO NPs (14 and 100 nm)	*Paracentrotus lividus* (sea urchin)	Fed with macroalgae and agar-agar mixed with 1 and 10 mg NPs/kg food for 3 weeks	The assumption of food containing ZnO NPs 100 nm provoked in adult echinoids damages to immune cells (33% of damaged nucleus) and transmissible effects to offspring (75.5% of malformed larvae). Instead food with ZnO NPs 14 nm provoked 64% of damaged nucleus in immune cells and 84.7% of malformed larvae	[69]
Ag NPs (58.6 nm)	*Parhyale hawaiensis* (amphipod)	Fed with 155 mg NPs/kg food during 1 h for 0, 7, 14 and 28 days; salinity 30, 24 °C, 12/12 h light/dark cycle, without aeration	Ag concentrations in the hemolymph were 3.3, 5.1 and 8.4 ng/mg for exposure during 7, 14 and 28 days, respectively	[72]

(continued)

Table 2 (continued)

NPs	Species	Exposure conditions	Effects	Ref.
Ag NPs (20 nm)	*Saccostrea glomerata* (bivalve)	Exposure to 12.5 and 125 μg/L for 0, 3, 5 and 7 days; salinity 32–35, 20 °C, pH 7.8, with aeration	The higher concentration of Ag NPs induced less bioaccumulation in gills and digestive glands after 7 days of exposure, while the lower concentration showed higher bioaccumulation. Significant differences compared to the initial day of exposure (day 0) were reported in DNA strand breaks after 5 and 7 days of exposure, GST, from the third day of exposure, in all the Ag samples, and in some samples for LPO and GR biomarkers, while no significant induction of EROD was observed. A combined effect for each type of treatment and time of exposure was also reported for DNA strand breaks and GST biomarkers measured at the digestive glands. In general, the significant inductions measured showed the following trend: 125 μg/L Ag NPs > 12.5 μg/L Ag NPs even though bioaccumulation followed the opposite trend	[50]

(continued)

Table 2 (continued)

NPs	Species	Exposure conditions	Effects	Ref.
nanArgen™ (Ag NP-enabled consumer product) (30 nm)	*Mytilus galloprovincialis* (bivalve)	Exposure to 1 and 10 μg/L for 96 h; salinity 40, 18 °C, pH 8, 18/6 h light/dark cycle	A significant concentration-dependent accumulation of Ag was found in mussels' whole soft tissue in agreement with a concentration-dependent decrease in NRRT and an increase of micronucleus frequency in hemocytes and GST activities in digestive glands. A significant increase in MDA levels and MT via both molecular and biochemical tests, were also observed but only at the highest nanArgen™ concentration (10 μg/L). No changes were observed in CAT activities. Efflux activities of ATP-binding cassette transport proteins in gill biopsies showed a significant decrease only at the lowest concentration (1 μg/L)	[70]

(continued)

Table 2 (continued)

NPs	Species	Exposure conditions	Effects	Ref.
Poly-vinyl-pyrrolidone/polyethyleneimine (PVP/PEI)-coated Ag NPs (5 nm)	*Mytilus galloprovincialis* (bivalve)	Fed with microalgae previously exposed to 10 μg/L during 24 h for 21 days; salinity 28.7, 15.6 °C, pH 7.7, 12/12 h light/dark cycle	Mussels significantly accumulated Ag in both seasons (autumn and spring) and Ag NPs were found within digestive gland cells and gills. 104 differentially expressed protein spots were distinguished in autumn and 142 in spring. Among them, chitinase like protein-3, partial and glyceraldehyde-3-phosphate dehydrogenase were overexpressed in autumn but underexpressed in spring. In autumn, pyruvate metabolism, citrate cycle, cysteine and methionine metabolism and glyoxylate and dicarboxylate metabolism were altered, while in spring, proteins related to the formation of phagosomes and H_2O_2 metabolism were differentially expressed	[55]

(continued)

Table 2 (continued)

NPs	Species	Exposure conditions	Effects	Ref.
Poly-vinyl-pyrrolidone/polyethyleneimine (PVP/PEI)-coated Ag NPs (5 nm)	*Mytilus galloprovincialis* (bivalve)	Exposure of hemocytes to 0.00001, 0.0001, 0.001, 0.01, 1, 10 and 100 mg/L for 24 h; fed with microalgae previously exposed to 1 µg/L and 10 µg/L during 24 h for 21 days; salinity 27.5, 15.5 °C, pH 7.8, 12/12 h light/dark cycle	After 24 h of in vitro exposure, Ag NPs were cytotoxic to mussel hemocytes starting at 1 mg/L Ag NPs (LC_{50}: 2.05 mg Ag/L). Microalgae significantly accumulated Ag after the exposure to both doses and mussels fed for 21 days with microalgae exposed to 10 µg/L Ag NPs significantly accumulated Ag in the digestive gland and gills. Sperm motility and fertilization success were not affected but exposed females released less eggs than non-exposed ones. The percentage of abnormal embryos was significantly higher than in control individuals after parental exposure to both doses	[52]
Poly-vinyl-pyrrolidone (PVP)-coated Ag NPs (41.6 and 85.0 nm)	*Mytilus galloprovincialis* (bivalve)	Exposure to 100 µg/L for 0, 3, 6 and 12 h; salinity 35‰, 16 °C, pH 8.0, 12/12 h light/dark cycle, with aeration	Both sizes of Ag NPs can cause protein thiol oxidation and/or protein carbonylation. However, blockade of endocytotic uptake routes mitigated Ag NPs toxicity	[36]

(continued)

Table 2 (continued)

NPs	Species	Exposure conditions	Effects	Ref.
Ag NPs (47 nm)	*Mytilus galloprovincialis* (bivalve)	Exposure of hemocytes to 0.1, 0.5, 1, 5, 10, 50, 100, 500 and 1000 µg/mL for 30 min; exposure of fertilized eggs to 0.001–1000 µg/L for 48 h; salinity 36, 16 °C, pH 7.9–8.1, with aeration	In vitro short-term exposure (30 min) of hemocytes to Ag NPs induced small lysosomal membrane destabilization (EC$_{50}$ = 273.1 µg/mL) and did not affect other immune parameters (phagocytosis and ROS production). Responses were little affected by hemolymph serum as exposure medium in comparison to artificial saline water. However, Ag NPs significantly affected mitochondrial membrane potential and actin cytoskeleton at lower concentrations. It also affected *Mytilus* embryo development, with EC$_{50}$ = 23.7, causing malformations and developmental delay, but no mortality	[44]

(continued)

Table 2 (continued)

NPs	Species	Exposure conditions	Effects	Ref.
Poly(allylamine)-coated Ag NPs (14 nm)	*Strongylocentrotus droebachiensis* (sea urchin)	Exposure to 100 µg/L for 48 and 96 h; 8 °C, 12/12 h light/dark cycle	Sea urchins showed an increasing resilience to Ag over time. Silver contamination and nutritional state influenced the production of reactive oxygen species. After passing through coelomic sinuses and gut, Ag NPs were found in coelomocytes. Inside blood vessels, apoptosis-like processes appeared in coelomocytes highly contaminated by poly(allylamine)-coated Ag NPs. Increasing levels of Ag accumulated by urchins once exposed to Ag NPs pointed to a Trojan-horse mechanism operating over 12-days exposure. However, under short-term treatments, physical interactions of poly(allylamine)-coated Ag NPs with cell structures might be, at some point, predominant and responsible for the highest levels of stress-related proteins detected	[71]

(continued)

Table 2 (continued)

NPs	Species	Exposure conditions	Effects	Ref.
Poly-vinyl-pyrrolidone (PVP)-coated Ag NPs (41.6 and 85.0 nm)	*Mytilus galloprovincialis* (bivalve)	Exposure to 100 μg/L for 0, 3, 6 and 12 h; salinity 35, 16 °C, pH 8.0, 12/12 h light/dark cycle, with aeration	Differential hemocyte counts revealed significant variations in frequency of different immune cells in mussels exposed for 3 h to either Ag NP size. However, as exposure duration progressed cell levels were subsequently differentially altered depending on particle size (i.e. no significant effects after 3 h with larger Ag NP). Ag NP effects were also delayed/varied after blockade of either clathrin- or caveolae-mediated endocytosis. The results also noted significant negative correlations between changes in levels hyalinocytes and acidophils or in levels basophils and acidophils as a result of Ag NP exposure	[43]
Poly-vinyl-pyrrolidone (PVP)-coated Ag NPs (41.6 and 85.0 nm)	*Mytilus galloprovincialis* (bivalve)	Exposure to 100 μg/L for 0, 3, 6 and 12 h; salinity 35, 16 °C, pH 8.0, 12/12 h light/dark cycle, with aeration	Ag NPs induced histopathological alterations in digestive gland (maximum inflammation 2.75) with Ag NPs 85.0 nm; significant histopathological index with Ag NPs 41.6 and 85.0 nm at different time-points. Significant histopathological indexes were recorded after uptake routes were blockade: Ag NPs 41.6 nm + nystatin and Ag NPs 85.0 nm + amantadine; all time-points	[74]

should include the mimics of environment conditions and test different physicochemical conditions (pH, temperature) in order to predict future ocean warming and/or acidification.

5 Conclusions and Perspectives

NPs are present in our everyday lives and with its intensive use, they will end up in the aquatic environment. As the knowledge about its fate and toxicology in the aquatic environment is still limited, it is needed to access the environmental risk of the pristine NPs and NPs-enabled consumer products. It is also important to know accurate safety levels in order to not endangering biodiversity conservation as consequence of NPs release to environment.

This review summarizes the information published in the most recent literature on the toxicity of the metal and metal oxide NPs most present on aquatic environments. Articles evidence significant biological responses, from immunomodulation and effects at cellular level to developmental defects, in organism exposed to a wide concentration range of different NPs. However, most of the studies were performed in species of organisms living in the water column and, as showed above, NPs concentration is higher in sediments. Then, future studies must include the evaluation of NPs into those species living in close contact with sediments.

Furthermore, the relation between NPs exposure, uptake and depuration, as well as how this relates to toxicity is still unclear. Consequently, future studies must also include the study of NPs toxicology mechanism towards the aquatic organisms.

Lastly, a lot of important endpoints have been studied to evaluate NPs toxicity into marine organisms. However, there are other important endpoints which are not so explored. For instance, there are a lack of studies evaluating the metabolism and energy reserves. But, if organisms do not have energy, they are not able to reproduce and the impact in the population is enormous.

Acknowledgements This work was supported by the National Funding for Science and Technology (FCT) through doctoral grant to Joana C. Almeida [SFRH/BD/139471/2018], and the University of Aveiro, FCT/MEC for the financial support to CESAM and CICECO [UID/AMB/50017/2013; UID/CTM/50011/2013], through national funds and, where applicable, co-financed by the FEDER, within the PT2020 Partnership Agreement. This work was also carried out under the Project inpactus—innovative products and technologies from eucalyptus, Project N.° 21874 funded by Portugal 2020 through European Regional Development Fund (ERDF) in the frame of COMPETE 2020 n°246/AXIS II/2017. This work was also financially supported by the project BISPECIAl: BIvalveS under Polluted Environment and CIImate chAnge PTDC/CTA-AMB/28425/2017 (POCI-01-0145-FEDER-028425) funded by FEDER, through COMPETE2020—Programa Operacional Competitividade e Internacionalização (POCI), and by national funds (OE), through FCT/MCTES.

References

1. Mobasser, S., Firoozi, A.: Review of nanotechnology applications in science and engineering. J. Civ. Eng. Urban. **6**, 84–93 (2016)
2. Blackman, J.: Metallic Nanoparticles. Elsevier Science (2008)
3. Canesi, L., Corsi, I.: Effects of nanomaterials on marine invertebrates. Sci. Total Environ. **565**, 933–940 (2015)
4. Birben, E., Sahiner, U.M., Sackesen, C., Erzurum, S., Kalayci, O.: Oxidative stress and antioxidant defense. World Allergy Organ. J. **5**, 9–19 (2012)
5. Francisquini, E., Schoenmaker, J., Souza, J.A.: Nanopartículas Magnéticas e suas Aplicações. In: Química Supramol. e Nanotecnologia, 1st edn, pp. 269–288 (2014)
6. Martins, M.A., Trindade, T.: Os nanomateriais e a descoberta de novos mundos na bancada do químico. Quim. Nova **35**, 1434–1446 (2012)
7. ISO, International Organization for Standardization (2012). https://www.iso.org/obp/ui/#iso: std:iso:ts:80004:-2:ed-1:v1:en. Accessed 16 Jun 2016
8. Quina, F.H.: Nanotecnologia e o Meio Ambiente: Perspectivas e Riscos. Quim. Nova **27**, 1028–1029 (2004)
9. Vance, M.E., Kuiken, T., Vejerano, E.P., McGinnis, S.P., Hochella, M.F., Rejeski, D., Hull, M.S.: Nanotechnology in the real world: redeveloping the nanomaterial consumer products inventory. Beilstein J. Nanotechnol. **6**, 1769–1780 (2015)
10. Lu, P.J., Huang, S.C., Chen, Y.P., Chiueh, L.C., Shih, D.Y.C.: Analysis of titanium dioxide and zinc oxide nanoparticles in cosmetics. J. Food Drug Anal. **23**, 587–594 (2015)
11. Katz, L.M., Dewan, K., Bronaugh, R.L.: Nanotechnology in cosmetics. Food Chem. Toxicol. **85**, 127–137 (2015)
12. Anu Mary Ealia, S., Saravanakumar, M.P.: A review on the classification, characterisation, synthesis of nanoparticles and their application. IOP Conf. Ser. Mater. Sci. Eng. **263**, 1–15 (2017). https://doi.org/10.1088/1757-899x/263/3/032019
13. Corma, A., Garcia, H.: Supported gold nanoparticles as catalysts for organic reactions. Chem. Soc. Rev. **37**, 2096–2126 (2008)
14. Mishra, P., Singh, L., Islam, M.A., Nasrullah, M., Sakinah, A.M.M., Wahid, Z.A.: NiO and CoO nanoparticles mediated biological hydrogen production: effect of Ni/Co oxide NPs-ratio. Bioresour. Technol. Rep. **5**, 364–368 (2018)
15. Patra, J.K., Das, G., Fraceto, L.F., Campos, E.V.R., Rodriguez-Torres, M.D.P., Acosta-Torres, L.S., Diaz-Torres, L.A., Grillo, R., Swamy, M.K., Sharma, S., Habtemariam, S., Shin, H.S.: Nano based drug delivery systems: recent developments and future prospects. J. Nanobiotechnol. **16**, 1–33 (2018)
16. Tan, H.-L., Teow, S.-Y., Pushpamalar, J.: Application of metal nanoparticle-hydrogel composites in tissue regeneration. Bioengineering **6**, 17 (2019)
17. Bouwmeester, H., Dekkers, S., Noordam, M.Y., Hagens, W.I., Bulder, A.S., de Heer, C., ten Voorde, S.E.C.G., Wijnhoven, S.W.P., Marvin, H.J.P., Sips, A.J.A.M.: Review of health safety aspects of nanotechnologies in food production. Regul. Toxicol. Pharmacol. **53**, 52–62 (2009)
18. Hlongwane, G.N., Sekoai, P.T., Meyyappan, M., Moothi, K.: Simultaneous removal of pollutants from water using nanoparticles: a shift from single pollutant control to multiple pollutant control. Sci. Total Environ. **656**, 808–833 (2019)
19. Li, Q., Chen, X., Zhuang, J., Chen, X.: Decontaminating soil organic pollutants with manufactured nanoparticles. Environ. Sci. Pollut. Res. **23**, 11533–11548 (2016)
20. Sinha, A.K., Suzuki, K., Takahara, M., Azuma, H., Nonaka, T., Fukumoto, K.: Mesostructured manganese oxide/gold nanoparticle composites for extensive air purification. Angew. Chem. Int. Ed. **119**, 2949–2952 (2007)
21. Harish, K.K., Nagasamy, V., Himangshu, B., Anuttam, K.: Metallic nanoparticle: a review. Biomed. J. Sci. Tech. Res. **4**, 3765–3775 (2018)
22. Natsuki, J., Natsuki, T., Hashimoto, Y.: A review of silver nanoparticles: synthesis methods, properties and applications. Int. J. Mater. Sci. Appl. **4**, 325–332 (2016)

23. Smita, S., Gupta, S.K., Bartonova, A., Dusinska, M., Gutleb, A.C., Rahman, Q.: Nanoparticles in the environment: assessment using the causal diagram approach. Environ. Heal. **11**, 11 (2012)
24. Khan, H.A., Shanker, R.: Toxicity of nanomaterials. Biomed. Res. Int. **2015**, 2 (2015)
25. Bundschuh, M., Filser, J., Lüderwald, S., McKee, M.S., Metreveli, G., Schaumann, G.E., Schulz, R., Wagner, S.: Nanoparticles in the environment: where do we come from, where do we go to? Environ. Sci. Eur. **30**, 6–22 (2018)
26. Sun, T.Y., Bornhöft, N.A., Hungerbühler, K., Nowack, B.: Dynamic probabilistic modeling of environmental emissions of engineered nanomaterials. Environ. Sci. Technol. **50**, 4701–4711 (2016)
27. Keller, A.A., McFerran, S., Lazareva, A., Suh, S.: Global life cycle releases of engineered nanomaterials. J. Nanoparticle Res. **15**, 1692–1708 (2013)
28. Châtel, A., Mouneyrac, C.: Signaling pathways involved in metal-based nanomaterial toxicity towards aquatic organisms. Comp. Biochem. Physiol. Part C. **196**, 61–70 (2017)
29. AshaRani, P.V., Mun, G.L.K., Hande, M.P., Valiyaveettil, S.: Cytotoxicity and genotoxicity of silver nanoparticles in human cells. ACS Nano **3**, 279–290 (2009)
30. Nelson, D.L., Cox, M.M.: Lehninger Principles of Biochemistry. Springer, Berlin, Heidelberg (2001)
31. Pinto, J., Costa, M., Leite, C., Borges, C., Coppola, F., Henriques, B., Monteiro, R., Russo, T., Di Cosmo, A., Soares, A.M.V.M., Polese, G., Pereira, E., Freitas, R.: Ecotoxicological effects of lanthanum in *Mytilus galloprovincialis*: biochemical and histopathological impacts. Aquat. Toxicol. **211**, 181–192 (2019)
32. Coppola, F., Almeida, Â., Henriques, B., Soares, A.M.V.M., Figueira, E., Pereira, E., Freitas, R.: Biochemical impacts of Hg in *Mytilus galloprovincialis* under present and predicted warming scenarios. Sci. Total Environ. **601–602**, 1129–1138 (2017)
33. Monteiro, R., Costa, S., Coppola, F., Freitas, R., Vale, C., Pereira, E.: Evidences of metabolic alterations and cellular damage in mussels after short pulses of Ti contamination. Sci. Total Environ. **650**, 987–995 (2019)
34. Freitas, R., de Marchi, L., Moreira, A., Pestana, J.L.T., Wrona, F.J., Figueira, E., Soares, A.M.V.M.: Physiological and biochemical impacts induced by mercury pollution and seawater acidification in *Hediste diversicolor*. Sci. Total Environ. **595**, 691–701 (2017)
35. Matozzo, V., Ballarin, L., Pampanin, D.M., Marin, M.G.: Effects of copper and cadmium exposure on functional responses of hemocytes in the clam, *Tapes philippinarum*. Arch. Environ. Contam. Toxicol. **41**, 163–170 (2001)
36. Bouallegui, Y., Ben Younes, R., Oueslati, R., Sheehan, D.: Role of endocytotic uptake routes in impacting the ROS-related toxicity of silver nanoparticles to *Mytilus galloprovincialis*: a redox proteomic investigation. Aquat. Toxicol. **200**, 21–27 (2018)
37. Li, J., Schiavo, S., Xiangli, D., Rametta, G., Miglietta, M.L., Oliviero, M., Changwen, W., Manzo, S.: Early ecotoxic effects of ZnO nanoparticle chronic exposure in *Mytilus galloprovincialis* revealed by transcription of apoptosis and antioxidant-related genes. Ecotoxicology **27**, 369–384 (2018)
38. De Marchi, L., Neto, V., Pretti, C., Figueira, E., Chiellini, F., Morelli, A., Soares, A.M.V.M., Freitas, R.: Toxic effects of multi-walled carbon nanotubes on bivalves: comparison between functionalized and nonfunctionalized nanoparticles. Sci. Total Environ. **622–623**, 1532–1542 (2018)
39. Lionetto, M.G., Caricato, R., Calisi, A., Giordano, M.E., Schettino, T.: Acetylcholinesterase as a biomarker in environmental and occupational medicine: new insights and future perspectives. Biomed. Res. Int. **2013**, 1–8 (2013)
40. De Marchi, L., Neto, V., Pretti, C., Figueira, E., Chiellini, F., Soares, A.M.V.M., Freitas, R.: The impacts of emergent pollutants on *Ruditapes philippinarum*: biochemical responses to carbon nanoparticles exposure. Aquat. Toxicol. **187**, 38–47 (2017)
41. Nunes, B., Nunes, J., Soares, A.M.V.M., Figueira, E., Freitas, R.: Toxicological effects of paracetamol on the clam *Ruditapes philippinarum*: exposure vs recovery. Aquat. Toxicol. **192**, 198–206 (2017)

42. Xia, B., Zhu, L., Han, Q., Sun, X., Chen, B., Qu, K.: Effects of TiO$_2$ nanoparticles at predicted environmental relevant concentration on the marine scallop *Chlamys farreri*: an integrated biomarker approach. Environ. Toxicol. Pharmacol. **50**, 128–135 (2017)
43. Bouallegui, Y., Ben Younes, R., Turki, F., Oueslati, R.: Impact of exposure time, particle size and uptake pathway on silver nanoparticle effects on circulating immune cells in *Mytilus galloprovincialis*. J. Immunotoxicol. **14**, 116–124 (2017)
44. Auguste, M., Ciacci, C., Balbi, T., Brunelli, A., Caratto, V., Marcomini, A., Cuppini, R., Canesi, L.: Effects of nanosilver on *Mytilus galloprovincialis* hemocytes and early embryo development. Aquat. Toxicol. **203**, 107–116 (2018)
45. Matozzo, V., Gagné, F., Immunotoxicology approaches in ecotoxicology: lessons from mollusks. In: Lessons Immunity. From Single-Cell Organisms to Mammals, pp. 29–51. Elsevier Inc. (2016)
46. Moreira, A., Figueira, E., Libralato, G., Soares, A.M.V.M., Guida, M., Freitas, R.: Comparative sensitivity of *Crassostrea angulata* and *Crassostrea gigas* embryo-larval development to As under varying salinity and temperature. Mar. Environ. Res. **140**, 135–144 (2018)
47. Mos, B., Kaposi, K.L., Rose, A.L., Kelaher, B., Dworjanyn, S.A.: Moderate ocean warming mitigates, but more extreme warming exacerbates the impacts of zinc from engineered nanoparticles on a marine larva. Environ. Pollut. **228**, 190–200 (2017)
48. Oliviero, M., Schiavo, S., Dumontet, S., Manzo, S.: DNA damages and offspring quality in sea urchin *Paracentrotus lividus* sperms exposed to ZnO nanoparticles. Sci. Total Environ. **651**, 756–765 (2019)
49. Fabbri, R., Montagna, M., Balbi, T., Raffo, E., Palumbo, F., Canesi, L.: Adaptation of the bivalve embryotoxicity assay for the high throughput screening of emerging contaminants in *Mytilus galloprovincialis*. Mar. Environ. Res. **99**, 1–8 (2014). https://doi.org/10.1016/j.marenvres.2014. 05.007
50. Carrazco-Quevedo, A., Römer, I., Salamanca, M.J., Poynter, A., Lynch, I., Valsami-Jones, E.: Bioaccumulation and toxic effects of nanoparticulate and ionic silver in *Saccostrea glomerata* (rock oyster). Ecotoxicol. Environ. Saf. **179**, 127–134 (2019)
51. Minetto, D., Libralato, G., Marcomini, A., Volpi Ghirardini, A.: Potential effects of TiO$_2$ nanoparticles and TiCl4 in saltwater to *Phaeodactylum tricornutum* and *Artemia franciscana*. Sci. Total Environ. **579**, 1379–1386 (2017)
52. Duroudier, N., Katsumiti, A., Mikolaczyk, M., Schäfer, J., Bilbao, E., Cajaraville, M.P.: Dietary exposure of mussels to PVP/PEI coated Ag nanoparticles causes Ag accumulation in adults and abnormal embryo development in their offspring. Sci. Total Environ. **655**, 48–60 (2019)
53. Bhuvaneshwari, M., Thiagarajan, V., Nemade, P., Chandrasekaran, N., Mukherjee, A.: Toxicity and trophic transfer of P25 TiO$_2$ NPs from *Dunaliella salina* to *Artemia salina*: effect of dietary and waterborne exposure. Environ. Res. **160**, 39–46 (2018)
54. Sellami, B., Mezni, A., Khazri, A., Bouzidi, I., Saidani, W., Sheehan, D., Beyrem, H.: Toxicity assessment of ZnO-decorated Au nanoparticles in the Mediterranean clam *Ruditapes decussatus*. Aquat. Toxicol. **188**, 10–19 (2017)
55. Duroudier, N., Cardoso, C., Mehennaoui, K., Mikolaczyk, M., Schäfer, J., Gutleb, A.C., Giamberini, L., Bebianno, M.J., Bilbao, E., Cajaraville, M.P.: Changes in protein expression in mussels Mytilus galloprovincialis dietarily exposed to PVP/PEI coated silver nanoparticles at different seasons. Aquat. Toxicol. **210**, 56–68 (2019)
56. Doyle, J.J., Ward, J.E., Wikfors, G.H.: Acute exposure to TiO$_2$ nanoparticles produces minimal apparent effects on oyster, *Crassostrea virginica* (Gmelin), hemocytes. Mar. Pollut. Bull. **127**, 512–523 (2018)
57. Saidani, W., Sellami, B., Khazri, A., Mezni, A., Dellali, M., Joubert, O., Sheehan, D., Beyrem, H.: Metal accumulation, biochemical and behavioral responses on the Mediterranean clams *Ruditapes decussatus* exposed to two photocatalyst nanocomposites (TiO$_2$ NPs and AuTiO$_2$ NPs). Aquat. Toxicol. **208**, 71–79 (2019)
58. Auguste, M., Lasa, A., Pallavicini, A., Gualdi, S., Vezzulli, L., Canesi, L.: Exposure to TiO$_2$ nanoparticles induces shifts in the microbiota composition of *Mytilus galloprovincialis* hemolymph. Sci. Total Environ. **670**, 129–137 (2019)

59. Shi, W., Guan, X., Han, Y., Zha, S., Fang, J., Xiao, G., Yan, M., Liu, G.: The synergic impacts of TiO_2 nanoparticles and 17β-estradiol (E2) on the immune responses, E2 accumulation, and expression of immune-related genes of the blood clam, *Tegillarca granosa*. Fish Shellfish Immunol. **81**, 29–36 (2018)

60. Shi, W., Han, Y., Guo, C., Zhao, X., Liu, S., Su, W., Zha, S., Wang, Y., Liu, G.: Immunotoxicity of nanoparticle $nTiO_2$ to a commercial marine bivalve species, *Tegillarca granosa*. Fish Shellfish Immunol. **66**, 300–306 (2017)

61. Guan, X., Shi, W., Zha, S., Rong, J., Su, W., Liu, G.: Neurotoxic impact of acute TiO_2 nanoparticle exposure on a benthic marine bivalve mollusk, *Tegillarca granosa*. Aquat. Toxicol. **200**, 241–246 (2018)

62. Mezni, A., Alghool, S., Sellami, B., Ben Saber, N., Altalhi, T.: Titanium dioxide nanoparticles: synthesis, characterisations and aquatic ecotoxicity effects. Chem. Ecol. **34**, 288–299 (2018)

63. Shi, W., Han, Y., Guo, C., Su, W., Zhao, X., Zha, S., Wang, Y., Liu, G.: Ocean acidification increases the accumulation of titanium dioxide nanoparticles ($nTiO_2$) in edible bivalve mollusks and poses a potential threat to seafood safety. Sci. Rep. **9**, 1–10 (2019)

64. Huang, X., Liu, Y., Liu, Z., Zhao, Z., Dupont, S., Wu, F., Huang, W., Chen, J., Hu, M., Lu, W., Wang, Y.: Impact of zinc oxide nanoparticles and ocean acidification on antioxidant responses of *Mytilus coruscus*. Chemosphere **196**, 182–195 (2018)

65. Oliviero, M., Schiavo, S., Rametta, G., Miglietta, M.L., Manzo, S.: Different sizes of ZnO diversely affected the cytogenesis of the sea urchin *Paracentrotus lividus*. Sci. Total Environ. **607–608**, 176–183 (2017)

66. Wu, F., Cui, S., Sun, M., Xie, Z., Huang, W., Huang, X., Liu, L., Hu, M., Lu, W., Wang, Y.: Combined effects of ZnO NPs and seawater acidification on the haemocyte parameters of thick shell mussel *Mytilus coruscus*. Sci. Total Environ. **624**, 820–830 (2018)

67. Falfushynska, H.I., Gnatyshyna, L.L., Ivanina, A.V., Sokolova, I.M., Stoliar, O.B.: Detoxification and cellular stress responses of unionid mussels *Unio tumidus* from two cooling ponds to combined nano-ZnO and temperature stress. Chemosphere **193**, 1127–1142 (2018)

68. Schiavo, S., Oliviero, M., Li, J., Manzo, S.: Testing ZnO nanoparticle ecotoxicity: linking time variable exposure to effects on different marine model organisms. Environ. Sci. Pollut. Res. **25**, 4871–4880 (2018)

69. Manzo, S., Schiavo, S., Oliviero, M., Toscano, A., Ciaravolo, M., Cirino, P.: Immune and reproductive system impairment in adult sea urchin exposed to nanosized ZnO via food. Sci. Total Environ. **599–600**, 9–13 (2017)

70. Ale, A., Liberatori, G., Vannuccini, M.L., Bergami, E., Ancora, S., Mariotti, G., Bianchi, N., Galdopórpora, J.M., Desimone, M.F., Cazenave, J., Corsi, I.: Exposure to a nanosilver-enabled consumer product results in similar accumulation and toxicity of silver nanoparticles in the marine mussel *Mytilus galloprovincialis*. Aquat. Toxicol. **211**, 46–56 (2019)

71. Magesky, A., Ribeiro, C.A. de O.M, Beaulieu, L., Pelletier, É.: Silver nanoparticles and dissolved silver activate contrasting immune responses and stress-induced heat shock protein expression in sea urchin. Environ. Toxicol. Chem. **36**, 1872–1886 (2017)

72. Vannuci-Silva, M., Cadore, S., Henry, T.B., Umbuzeiro, G.: Higher silver bioavailability after nanoparticle dietary exposure in marine amphipods. Environ. Toxicol. Chem. **38**, 806–810 (2019)

73. Tang, C.H., Lin, C.Y., Lee, S.H., Wang, W.H.: Membrane lipid profiles of coral responded to zinc oxide nanoparticle-induced perturbations on the cellular membrane. Aquat. Toxicol. **187**, 72–81 (2017)

74. Bouallegui, Y., Ben Younes, R., Bellamine, H., Oueslati, R.: Histopathological indices and inflammatory response in the digestive gland of the mussel *Mytilus galloprovincialis* as biomarker of immunotoxicity to silver nanoparticles. Biomarkers **23**, 277–287 (2017)

Nanotechnology: Environmentally Sustainable Solutions for Water Treatment

Mahesh Kumar Gupta, Praveen Kumar Tandon and Neelam Shukla

Abstract Nanotechnology, which refers to the techniques that suggest the ways to synthesize, design and control properties in the nano region, is the processes and/or technologies which are environmentally benign, improved and can be used in such a way to reduce waste generation, cut pollution and conserve natural resources. Nanotechnology has emerged as a most revolutionary technology in the world that provides efficient, cost-effective and environmentally acceptable solutions for the decontamination of water. Although nanotechnology has been used in almost all the fields like medicine, biotechnology, electronics etc., its major application to drinking water treatment has begun only recently. Although many physical and chemical methods have been developed till now to synthesize a variety of nanomaterials but majority of them have negative effect on the environment and social life. Majority of them require high energy consumption, need toxic and costly chemicals and produce hazardous by-products. Thus, it is desirable to find alternative sources of eco-friendly and renewable materials and to develop sustainable procedures for the treatment of polluted water. As it is not possible to cover all the methods which are being used traditionally or currently to decontaminate water, in the present chapter we have picked up only the recent environmentally sustainable methodologies for the removal of contaminants like dyes, heavy metals, anions and microbial contamination in water in which nanotechnology plays the vital role. In this chapter we have mainly discussed the role of nanotechnology as an effective solution for the decontamination of wastewater in a sustainable way.

Keywords Nanotechnology · Environmentally sustainable · Heavy metals · Adsorption · Photocatalysis · Bioremediation

M. K. Gupta · P. K. Tandon (✉) · N. Shukla
Department of Chemistry, University of Allahabad, Allahabad 211002, Uttar Pradesh, India
e-mail: pktandon1@gmail.com

© Springer Nature Switzerland AG 2019
G. A. B. Gonçalves and P. Marques (eds.), *Nanostructured Materials for Treating Aquatic Pollution*, Engineering Materials,
https://doi.org/10.1007/978-3-030-33745-2_8

1 Introduction

Water is an extremely essential resource for the survival of all living beings on earth but now worldwide water pollution is the major concern for environmentalists. With increasing population of world water pollution is becoming more and more complex and difficult to remove. Increasing demand of fabrics and food has increased the use of dyes to colour the products in industries like textile, leather, paper, cotton, wool, pulp etc. Industrial effluents containing the unspent dyes when discharged into the water bodies damage the environment and are health hazards also as the coloured wastes are carcinogenic and toxic to humans and aquatic life [1]. The wastewater discharged from dying processes has high BOD and COD and contains high amounts of dissolved solids [2]. Coloured water interferes in normal photosynthetic process, inhibits the growth of aquatic life by blocking sunlight and utilizes dissolved oxygen. Water pollution caused by the presence of heavy metals form a very dangerous category due to their toxic and carcinogenic nature. Heavy metals are non-biodegradable, have the property to accumulate biologically in the food chain [3] and hence, tend to accumulate in the environment for long time. Mining, electroplating, metallurgy, chemical plants, agriculture and household wastewater etc. are among the major sources of heavy metal contamination in water. Some of the heavy matal ions like Pb, As, Cr, Hg, Cu, Cd, Zn, Ni etc. can cause severe threat to human's health even at low concentration when taken over a long time period. While a minimum amount of some heavy metal ions like copper, zinc, iron, cobalt, manganese, selenium etc. are essential for proper functioning of human body [4]. Excess of even these essential metal ions produces toxic effects when their intake crosses the prescribed minimum dose [5] (Table 1). Increased agricultural and industrial activities have contributed to the generation of a variety of inorganic aquatic pollutants which even at small concentrations can cause serious environmental problems and may also adversely affect the human health. Although bromate, fluoride, phosphate, nitrate/nitrite, perchlorate etc. are common substances and their presence is sometimes necessary but it must be within certain limits to avoid environmental impact and the development

Table 1 Permissible limit of selected heavy metals in drinking water as reported by U.S. Environmental Agency (U.S. EPA) and World Health Organization (WHO)

Contaminants	EPA limitations		WHO provisional guideline value (mg/L)
	Permissible maximum limit (mg/L)	Goal to be achieved for maximum limit (mg/L)	
Lead	0.015	0	0.01
Chromium	0.1	0.1	0.05
Cadmium	0.005	0.005	0.003
Arsenic	0.01	0	0.01
Mercury	0.002	0.002	0.006
Copper	1.30	1.30	2
Zinc	5	–	3
Nickel	–	–	0.07

of various human diseases [6]. Water contamination by pathogens resulting in water-borne diseases is a worldwide problem [7]. Diseases like diarrhea, typhoid, cholera, gastroenteritis etc. are due to water borne pathogens like *Escherichia coli*, *Shigella spp.*, *Salmonella spp.*, *Vibrio spp.*, protozoa (*Cryptosporidium*), *Giardia lamblia* etc. Some nonpathogenic bacteria (viz. sulfur, crenothrix iron bacteria), although are not harmful but may cause taste and odor problems [8].

With recent advances in nanotechnology, various types of metal and metal oxide nanoparticles with antimicrobial (microbiocidal or growth-inhibiting) activity have been synthesized. Several techniques like neutralization, adsorption, coagulation, flocculation, precipitation, ion exchange, sedimentation, biodegradation, reverse osmosis, membrane filtration, electrochemistry, photo-electrochemistry, advanced oxidation processes etc. have been used for the removal/degradation of pollutants from water. All these techniques have drawbacks or restrictions associated with them. Some of the techniques are ineffective for the removal of organic pollutants while in some cases more harmful substances are produced after degradation. In majority of techniques sophisticated instrumentation added with high input cost is needed. Adsorption techniques are widely used in industries for the removal of organic compounds, organic dyes, heavy metals, metalloids etc. For getting the desirable results specific pore size, surface area, chemical nature of adsorbate/adsorbent, temperature and pH of the medium, and presence of co-ions are to be considered as these factors affect the efficiency of the adsorption process [9].

1.1 Nanotechnology for Sustainable Development

Decreasing the size of any particle to the nano-scale (less than 100 nm) increases the proportion of surface and near surface atoms. Electronic and structural changes imparting new properties by decreasing the size of the particle is the basic principle of nanotechnology. These nano-particles possess a large surface area and 'surface to volume' ratio, high reactivity and catalytic potential and accordingly their other properties are changed [10]. According to Brundtland's Commission report social, economic and environment are the three basic requirements which must be fulfilled for achieving global sustainability and economic and societal development should be in such a manner that minimizes the impact of human activities on environment [11]. In brief, the environmentally sustainable solutions of nanotechnology are those which may successfully be used as efficient, cost-effective and environmentally friendly solutions for various applications. One major application of environmentally sustainable nanotechnology is in the removal of various toxic substances present in waste water such as dyes, heavy metals, inorganic anions, harmful microorganisms etc. [12].

2   Contaminations in Aqueous Media

2.1   Contamination of Dyes

Industries like textile, paper, printing, rubber, plastic, cosmetics, tanning, food processing etc. use complex organic compounds popularly known as 'Dyes' to colour the products [13]. Effluent coming out from these industries contains dyes which are potential hazards to the environment and human health. Presence of these dyes even at very low concentrations in the effluents is visible and they are resistant to chemical decomposition thus pose great danger to the aquatic life by hindering sunlight penetration which causes resistance to the photochemical reactions like photosynthesis. Waste products of industries during the dying process are toxic, mutagenic and carcinogenic and increase chemical oxygen demand (COD) and biochemical oxygen demand (BOD) levels and are thus very harmful for the aquatic life and human health [14]. Therefore, it is necessary to remove dyes from the industrial waste waters before it is released into the main stream [15].

2.2   Contamination of Heavy Metals

Heavy metals are the metals having specific gravity more than 5.0. In majority of the developing countries industries like metal plating and electro plating, fertilizer, tanning, batteries, paper and pesticides, etc., release their effluents containing heavy metal ions directly into the aquatic environment. In contrast to the organic pollutants metal ions are not biodegradable and most of them are toxic, carcinogenic which tend to accumulate in the living organisms and thus pose a great threat to the environment and the health of living beings [16, 17].

2.3   Contamination of Anions

In different parts of the world number of anionic pollutants such as fluoride, nitrate and nitrite, bromate, phosphate, sulphate etc. have been detected in surface and groundwater [18]. Both natural and anthropogenic factors are responsible for contamination of anionic pollutants in drinking water. For calcification of dental enamel fluoride is an essential micronutrient when present within the permissible limit (Table 2) of 1.0–1.5 mg/L. But at the same time when its concentration exceeds the permissible limit then it becomes the cause for fluorosis. Similarly, excess consumption of nitrates leads to two major problems in human beings, the 'Blue Baby Syndrome' (methemoglobinemia) especially in infants and the formation of carcinogenic nitrosamines [19].

Table 2 Permissible limit of toxic inorganic anionic contaminants in drinking water

Contaminants	Guideline values	
	US EPA (mg/L)	
Nitrate (measured as nitrogen)	10	11
Nitrite (measured as nitrogen)	1.0	WHO (mg/L)
Phosphate	0.1	–
Fluoride	4.0	–
Cyanide	0.2	1.5
Chloride	1.0	0.07
Bromate	0.01	–
Sulfate	0.02	0.01
		–

Phosphate levels greater than 1.0 mg/L may result in eutrophication, algal blooms and may interfere with coagulation in water treatment plants. Presence of excess of phosphate in water causes number of fatal health problems such as Addison's disease, severe heart and lung diseases, renal, thyroid and liver problems [20]. Therefore, the testing of these anions contamination in drinking waters is necessary. Some other anions, like chloride and sulfate, are considered secondary contaminants.

2.4 Microbial Contamination

Microscopic organisms called "microbes", are found almost everywhere on earth such as in air, water, soil and rock as well as, in plants, animals and the human body. Pathogens, germs or bugs are the microbes which are responsible for causing infectious diseases [21]. The microbial contaminants include pathogens like bacteria, viruses, fungi and parasites such as microscopic protozoa and worms. These living organisms can be spread by human and animal wastes knowingly or unknowingly [22]. Water pollution with pathogenic microorganisms is one of the serious threats to human health [23]. Microbial contamination generally spreads through drinking water. Food sources and poor sanitation are directly related to pathogen exposure. Poor water quality, sanitation and hygiene are basic causes of gastrointestinal diseases. Major pathogens include: *rotavirus*, *Campylobacter jejuni*, *Klebsiella* species, *Enterobacter* species, *Escherichia coli*, *Shigella* species, *Vibrio cholerae*, *Clostridium* species, *Salmonella typhae*, *Clostridium difficile*, *Cryptosporidium parvum* and Protozoa. These pathogens cause diarrhea, giardiasis, dysentery, and gastroenteritis [24]. Therefore, disinfection is necessary part of treatment of wastewater containing various microorganisms which affect human health.

3   Water Remediation and Nanotechnology

Nanotechnology has been applied successfully to the water/wastewater treatment and has emerged as a fast-developing promising field. In terms of wastewater treatment, nanotechnology is applicable in detection and removal of various pollutants in sustainable way. Various methods such as photocatalysis, nanofiltration, adsorption, and bioremediation are used for the remediation of water contaminants. Methods for the remediation of various water contaminants like heavy metals, dyes, anions and microbes are discussed bellow.

3.1   Remediation of Dyes

Nanomaterials have been extensively studied for the removal of dyes from wastewater. Adsorption method has been found to be an attractive technique for the removal of dyes from wastewater because of its flexibility and simplicity of design, cost effectiveness, eco-friendliness and high efficiency compared to other conventional methods. In addition, adsorption does not generate hazardous substances and avoids the secondary pollution. Adsorption process is a surface phenomenon in which the adsorbate is concentrated on the surface of porous adsorbent surface. The adsorption process is generally divided into two parts the physisorption in which an adsorbate is bound to the surface by weak physical forces like van der Waals forces, hydrophobicity, hydrogen bonding, polarity, static interactions, dipole interactions and π–π interactions. The second is chemisorption in which adsorbate is chemically bound to the surface of an adsorbent by the force of the electrons exchange or due to electrostatic attraction. Nanosized metal or metal oxides provide high surface area and specific affinity and thus are broadly used for the removal of dyes due to their minimal environmental impact, low solubility, and no secondary pollution [14].

Nanosized metal or metal oxide based materials and different types of carbon-based nanomaterials have been widely used for the removal of dyes in recent decades because these materials are found in abundance and are nontoxic can be prepared easily, have high surface area and porosity and thus have high sorption capacities. Activated carbon can be easily prepared from readily available carbonaceous precursors such as coal, wood, coconut shells and agricultural wastes. Due to its porous structure with high surface area, great adsorption capacity towards various kinds of pollutants has been widely used as adsorbents [12]. However, renewable and less expensive precursors for the synthesis of activated carbon have been reported by many researchers. Agricultural solid wastes are relatively cheap and are available in huge quantities. They can be used as adsorbents due to their physico-chemical properties. Agricultural by products and waste materials are the main sources for the production of activated carbon, such as banana and orange peel, bagasse, oil palm ash, spent tea leaves, degreased coffee bean etc. Coconut shell, a food solid waste, was successfully utilized as a low cost alternative adsorbent for the removal of hazardous textile dyes maxilon blue (GRL), and direct yellow (DY12). It was reported that acidic pH favoured the adsorption [25]. Preparation of activated carbon from the husk of coconut has been reported for the removal of methylene blue [26]. Many researchers reported modified agricultural waste as adsorbent [15] have utilized maize cob coated with magnetic ferric oxide nanoparticles as low-cost adsorbent for the adsorption of methylene blue (MB). In another study unmodified saw dust, a solid waste/by product produced in huge quantities at saw mills, has been used as an adsorbent for the removal of hazardous tartrazine from its aqueous solutions. In acidic medium (pH 3.0) maximum removal at 1 mg/L of tartrazine was reported to be 97% [27].

Inorganic nanomaterials like metal or metal oxide-based nanoparticles are widely used as adsorbents for the removal of heavy metal ions and dyes. These nanomaterials have high surface to volume ratio and create no secondary pollution [14]. Generally, nanoparticles can be readily produced using physical and chemical methods but these methods generate toxic byproducts in the synthesis protocol. So the use of biological resources available in nature including microorganisms and plants has received considerable attention for efficient and rapid synthesis of metal nanoparticles instead of using physical and chemical methods. Green methods for the synthesis of metal nanoparticles have possibility to minimize environmental problems, are cost effective, eco-friendly and energy efficient green alternatives. Synthesis of plant mediated nanosized metals or metal oxide nanoparticles has been broadly used for the removal of dyes e.g. copper nanoparticles synthesized from *Cassia Occidentalis* leaf extract, were used to remove bromo cresol green (BCG) [28].

Among such process, reductive degradation of hazardous dyes using nanocatalyst is a convenient degradation process because of their unique physiochemical and electronic properties which are not present in bulk materials. Many metal nanoparticles can act as good catalysts and hence catalyze many reduction reactions. Nanocatalysts are also widely used in water treatment as it increases the catalytic activity at the surface due to its special characteristics of having higher surface area with shape dependent properties. Silver nanoparticles of the size 15–50 nm prepared with the help of *Zanthoxylum armatum* (commonly known as toothache tree) leaves, were used in the degradation of hazardous dyes like safranine O, methyl red, methyl orange and methylene blue etc. [29]. Use of AgNPs prepared by the reduction of metal ion precursor with leaf extract of *Aegle marmelos* as nanocatalyst and nanoadsorbent for the reduction of malachite green dye has also been reported by Femila et al. [30].

In some cases it has been reported that incorporation of nanoparticles with other materials enhances their properties. The resulting material known as nanocomposite has recently gained increasing interest as the sustainable and efficient adsorbent for wastewater treatment. In one of the study soil was modified by coating the green synthesized AgNPs and the resulting nanocomposite was found (97.2% dye removal was obtained) to be an efficient adsorbent for the removal of the dye crystal violet. Neem (*Azadirachta indica*) leaf extract was used as an economical source to reduce the silver salt [31]. Chitosan, due to the presence of hydroxyl and amino groups on the polymer chains that can act as coordination and electrostatic interaction sites, is an effective material for the sorption of many organic compounds and dyes, except the basic dyes. Hydroxyapatite ($Ca_{10}(PO_4)_6(OH)_2$) (extracted from egg shell) incorporated with commercial chitosan improves its mechanical strength and adsorption capacity and has been used to almost completely remove the brilliant green dye (BG) [32].

Bioremediation is another approach for the removal of dyes in which microorganisms like algae, yeast, filamentous fungi and bacteria are for converting dye molecules into less harmful forms [33]. Heteropolysaccharides and lipids of the cell wall comprise different functional groups like amino, hydroxyl, carboxyl, phosphate and other charged groups which may cause strong attractive forces between the dyes and cell wall [34]. The ligninolytic enzymes produced by white rot fungi are found responsible for dye removal due to their nonspecific enzyme systems [14]. Normal and autoclaved hyphae of *A. oryzae* in aqueous solution have been shown to remove the dyes PV-H3R (Procion Violet H3R) and PR-HE7B (Procion Red HE7B) in acidic to near neutral pH range. The best pH for biosorption was found to be 2.50 [35]. Decolourization of Acid Red (AR) 151, Orange (Or) II, Sulfur Black (Sb) and Drimarene Blue (Db) K2RL dyes using *Pseudomonas aeruginosa, Pseudomonas putida* and *Bacillus cereus* at pH 7.0 has also been reported [36].

3.2 Remediation of Heavy Metals

Photocatalysis is one of the widely used methods for environmentally sustainable water remediation method which uses light to clean the waste water. Instead of separating as in other techniques photocatalysis actually degrades the pollutants. On exposure to UV light a catalyst, often titanium dioxide placed in polluted water starts a chemical reaction that degrades the pollutants [37]. Photocatalytic reduction of Cr(VI) under UV illumination has been reported by Rengaraj [38] who prepared neodymium (Nd) doped titanium dioxide (TiO_2) photocatalyst by sol–gel method. It has been described that the electrons accumulate on the site of deposited Nd(III) ions on the TiO_2 surface which enhances the photocatalytic reaction of Cr(VI) reduction. Separation of electrons and holes on modified TiO_2 surface allows efficient channeling of the charge carriers into redox reactions rather than recombination reactions. Electron donors such as formic acid enhance the photocatalytic reduction. The system reduces 98% Cr(VI) into Cr(III) within 2 h at pH 3. Barkat et al. studied the destruction of complex cyanide with a con-current removal of copper by using UV-irradiated TiO_2 suspension. Results revealed a complete removal of both copper and cyanide at a ratio of 10:1 in 3 h [39].

Adsorption is one of the most frequent and widely used technology for the treatment and removal of heavy metals and anions from wastewater. Nanoparticles as adsorbent have been used extensively for waste water treatment as compared to conventional adsorbents. Large surface area of the nanoparticles results in greater efficiency, short intra particle diffusion distance, tunable pore size and surface chemistry, fast rate of adsorption and easy separation after adsorption.

Extracts obtained from various plant parts like green tea leave, mango leaf, peels and bunch of banana, lemon, flower leave etc. have been used to prepare nanoparticles of metals, metal oxides, metal salts etc. These extracts consist of polyphenols which are nontoxic, act as reducing agent for the synthesis of nanoparticles and further protect nanoparticle from agglomeration. The benefit of using green synthesized nanoparticles is that it represents a "one-pot process". It is simple, environmentally sustainable and cheaper thereby representing a technology which degrades contaminants sustainably along with lowering the risk of release of toxic products and byproducts into the environment.

It has been reported that many plant extracts could be an alternative to existing chemical and physical techniques for the synthesis of nanoparticles [40]. Several plants extracts like green tea, oolong tea, and black tea [41], grape marc, black tea, and vine leaf extract [42], green tea and eucalyptus leaf [43] have been used for the synthesis of nanoparticles. Plant extracts are nontoxic to living organisms, renewable and being environmentally friendly and ecologically acceptable, when compared to other techniques of nanoparticles synthesis. Polyphenol/caffeine content of the plant extracts are probably responsible for the stabilization of NZVI and serve as the reducing and capping agent [40]. Most of the nanoadsorbents are non toxic and eco-friendly especially adsorbents synthesized from plants or plants extract. Fazlzadeh et al. [44] synthesized zero valent iron nanoparticles (NZVI) from green plants—Rosa damascene, Thymus vulgaris, and Urticadioica. Removal of 100% Cr(VI) at pH 2 in 30 min was reported. Kumar et al. [45] synthesized chitosan supported iron NPs using leaf extract of Mint (*Mentha spicata* L.). The composite was found to remove 98.79 and 99.65% of As(III) and As(V) respectively.

Depending on the size of the particle that can be retained, various types of membrane filtration such as ultrafiltration, nanofiltration and reverse osmosis can be employed for heavy metal removal from wastewater. Ultrafiltration (UF) utilizes permeable membrane to separate heavy metals, macromolecules and suspended solids from inorganic solution on the basis of the pore size (5–20 nm) and molecular weight of the separating compounds (1000–100,000 Da). Juang and Shiau [46], studied the removal of Cu(II) and Zn(II) ions from synthetic wastewater using chitosan-enhanced membrane filtration. Abu Qdaisa and Moussab [47], investigated application of both reverse osmosis (RO) and nanofiltration (NF) technologies for the treatment of wastewater containing copper and cadmium ions. 98 and 99% removal efficiency of copper and cadmium, respectively could be achieved by RO process. NF was capable of removing more than 90% of the copper ions existing in the feed water. Carboxyl methyl cellulose (CMC) [48], diethylaminoethyl cellulose [49], polyethyleneimine (PEI) [50, 51] were used as efficient water-soluble metal-binding polymers in combination with ultrafiltration (UF) for selective removal of heavy metals from water.

Nanofiltration, a process in between UF and RO, is a relatively recent membrane filtration process. Nanofiltration membranes (NF membranes) are widely used in removing pollutants from drinking water and air but it is also used for the desalination of salty water. Commercially available nanofilters in which carbon nanotubes or nanocapillaries are used while in reactive nanomembranes in which functionalize nanoparticles chemically convert impurities to give safe by products. NF membranes have been shown to remove turbidity, microorganisms, natural organic matter, biological contaminants, organic pollutants, inorganic anions, heavy metal from groundwater as well as surface water and in particular softening of groundwater. Nanofiltration is a promising technology for the rejection of heavy metal ions such as chromium [52], nickel [53], copper [54] and arsenic [55] from wastewater. Srivastava et al. [56] recently reported the successful fabrication of carbon nanotube filters. These new filtration membranes consist of hollow cylinders with radially aligned carbon nanotube walls. The carbon nanotube filters are readily cleaned by ultrasonication and autoclaving. Nanoceramic filters are a mixture of nanoalumina fiber and micro glass with high positive charge and can retain negatively charged particles. Nanoceramic filters have high efficiency for removing virus and bacteria. They have high capacity for particulates and less clogging and can chemisorb dissolved heavy metals [57].

3.3 Remediation of Anions

It is found that adsorption technique is a successful technique in removing various types of inorganic pollutants. Fluoride removal has been reported by using a titanium hydroxide-derived adsorbent [58]. The most commonly used treatment methods to remove/reduce NO_3^- include chemical denitrification using zero-valent iron (Fe^0), zero-valent magnesium (Mg^0), ion exchange, reverse osmosis, electrodialysis, catalytic denitrification and biological denitrification [59]. Iron oxides have been found to be effective to remove phosphate from aqueous solution [60]. Powdered activated carbons have been found to be good for bromate ion (BrO_3^-) removal from drinking water [61], while cationic surfactant-tailored activated carbons (ACs) have been found effective to remove perchlorate to below detection levels [62]. It is necessary to select suitable adsorbent material to remove specific type of anionic pollutants to achieve optimum removal rates.

3.4 Remediation of Microorganism

The commonly used chemical disinfectants (chlorine, chloramines, ozone etc.) applied in the treatment of drinking water can effectively control the microbial pathogens; however, these disinfectants can react with other constituents present in water to generate harmful disinfection by-products (DBPs). Thus it is desirable to search alternative sources of clean and renewable materials as well as novel processes to produce new materials including nanostructures. Many bacteria and fungi are able to produce nanoparticles of fairly good quality. Biological methods, besides being sustainable are capable to produce large quantities of nanoparticles and that too at a lower cost compared to other methods [63]. Recently many biological methods have been reported for the preparation of nanoparticles like by using microorganisms [64], enzymes [65], plants parts or the extract obtained from plant parts [66].

Since ancient times, silver has been known for its antimicrobial applications. Recently silver nanoparticles have attracted much attention due to their wide spread applications. Colloidal silver plasmon-resonant particles have been used as optical reporters in typical biological assays [67]. It has been reported that AgNPs increase the luminescent properties of europium complex with EDTA [68]. Strong antibacterial activity against gram negative bacteria of silver nanoparticles synthesized with marine red algae (*Porphyra vietnamensis*) [69], inhibition of the virus HIV-1 from binding to the host cells because of the preferential binding of silver nanoparticles to the gp120 glycoprotein knobs [70], enhancement of the chemiluminescence intensity of luminol–H_2O_2 system with AgNPs [71], use of AgNPs in catalysis [72], and in plant growth metabolism [73] have been reported by various workers. Based on the antibacterial properties of AgNPs, various nanosilver products, including nanosilver-coated wound dressings, contraceptive devices, surgical instruments, and implants have been developed [74]. Silver nanoparticles have been shown to be a powerful antibacterial agent against various pathogenic bacteria like *Staphylococcus aureus, Bacillus cereus, Escherichia coli, Pseudomonas aeruginosa* and *Klebsiella* species, when synthesized by using fungal species *Spirogyra varians* [75], against *E. coli, Staphylococcus* sps. and *Pseudomonas* sps. when synthesized with *Penciillium Citreonigum Dierck* and *Scopulaniopsos brumptii Salvanet–Duval* bacterial strains [76] and against *Staphylococcus aureus, Bacillus subtilis, Escherichia coli* and *Pseudomonas aeruginosa* when synthesized with the white rot fungi *Pycnoporus* sp (HE792771 [77].

Recently, the plant mediated nanomaterials have drawn more attention due to its vast application in various fields and their physico-chemical properties. Secondary metabolites such as phenolic acid, flavonoids, alkaloids and terpenoids present in crude extract of plant are mainly responsible for the reduction of ionic into bulk metallic nanoparticles formation [78].

Waste materials applications have given much attention in recent years. Use of plant waste materials like fruit peel, contains considerable amounts of minerals and vitamins. Previous studies have proven the presence of higher contents of phenolic, flavonoids, as well as other valuable secondary plant metabolites and essential oils in fruit peels. Various peel extracts have also been studied for the synthesis of silver, gold, platinum and titanium nanoparticles and their antimicrobial activities have been studied. For example, biosynthesis of titanium dioxide nanoparticles from the extracts of the peels of fruits (*Prunusdomestica* L. (Plum), *Prunus Persia* L. (Peach) and *Actinidia deliciosa* (Kiwi)) as agro wastes materials and their antibacterial activities have been reported [79]. In another study inactivation of E. coli with Ag-MgO nanocomposites has been reported in which *Citrus paradise* (grape fruit red) peel extracts were used for the synthesis of Ag-MgO nanocomposites [80]. Synthesis of iron oxide nanoparticles with the agro waste of *Anthocephalus cadamba* with its pesticidal activity against *Sitophilus granarius* and antimicrobial activity on the sequence of Gram-positive and Gram-negative bacteria have been reported by Sivapriya et al. [81].

The exact mechanisms of antibacterial or toxicity activities by these nanoparticles are still in investigation and a well debated topic. As in many studies it has been discussed that mechanism of antibacterial activity of iron nanoparticles might be mainly due to two reasons. Firstly the central mechanism causing the antibacterial activity by the nanoparticles might be through oxidative stress caused by Reactive Oxygen Species (ROS). ROS includes radicals like superoxide radicals (O^{-2}), hydroxyl radicals ($-OH$) and hydrogen peroxide (H_2O_2); that could be the reason for damaging the proteins and DNA in the bacterial cell and secondly the accumulation of nanoparticles, due to small size, in cytoplasmic region causes disintegration of the membrane through which leakage of cell constituents happens leading to bacterial death [82] (Fig. 1).

The fungicidal mechanism of biosynthesized metallic nanoparticles has been reported by many researchers. The membrane damage and damage in fungal intercellular components and finally destruction of cell function was reported in Candida species by plant derived silver nanoparticles [83]. Strong larvicidal and ovicidal

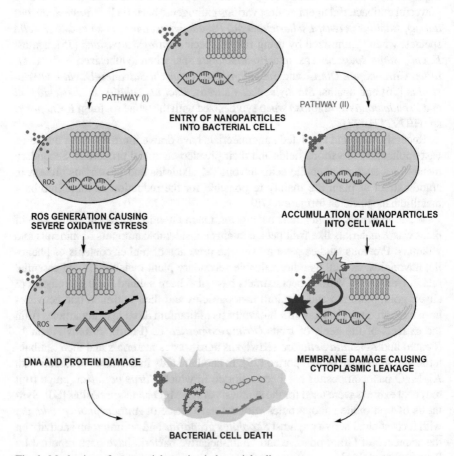

Fig. 1 Mechanism of nanoparticles action in bacterial cell

activities of ZnO NPs against *Aedes ageypti*, the common vector for causing dengue fever and the fungicidal effect of ZnONPs against fungal pathogens A. flavus (MTCC 873) and A. niger (MTCC 282) have been studied by Al-Dhabi et al. [84]. It was observed that the ZnONPs prepared with *Scadoxus multiflorus* leaf powder aqueous extract showed better results (96.4%) against *Aedes ageypti* compared to the activity of ZnONPs prepared with *Terminalia chebula* extracts (66%) mortality.

4 Benefits and Limitations of Nanotechnology

The concept of environmentally sustainable nanotechnology is the processes and technologies which are environmentally friendly, improved and processed in such a way so that it doesn't harm the environment and provides conservation to the natural resources. In brief, environmentally sustainable nanotechnology is referred as the environmentally green and clean technology. Benefits of environmentally sustainable nanotechnology include: minimization of the deterioration of environment, zero or low quantity of green house gas (GHG) emission, to remove other pollutants from the atmosphere, to conserve energy and natural resources and bring economic profits to certain areas.

With this brighter side of the environmentally sustainable nanotechnology there are some vivid apprehensions also. Detailed research work is needed to know the exact nature of the interaction of nanoparticles and biological systems and the response of living organisms to the presence of nanoparticles of varying size, shape, chemical composition and surface characteristics to understand the toxicity of nanoparticles. Although nanotechnology is likely to represent a beneficial replacement of current practices for site remediation, research into health and environmental effects of nanoparticles is the basic requirement which needs an urgent attention.

5 Conclusion

Hybrid technologies are always beneficial because there are limitations in every individual treatment technology. However, availability, selection, optimization, etc. are important for the best performances of the system. An increasing awareness towards green chemistry and the use of green routes for synthesis of metal nanoparticles lead a desire to develop environment-friendly techniques. Nanotechnology can make a considerable contribution to sustainable development through its combination of economic, ecological, and social benefits. On the basis of the available literature, it can be concluded that nanomaterials, apart from their enormous use in many fields, have the potential for efficiently removing various pollutants from water also. It is important to note that the processes of regeneration and reuse of nanomaterials are of high interest.

Thus, more research work should be focused on the cost-effective and feasible methods of the nanomaterial regeneration. In summary, the full-scale study of nano-materials for decontamination of wastewater is required for the practical applicability in future. It seems very plausible that nanomaterials may find wide commercial applications in wastewater treatment in the near future. Lastly, it must be mentioned that the future of the nanotechnology for water treatment is highly prosperous and we hope that one day it will fulfill the demand 'fresh water for everyone'.

References

1. Pappic, S., Koprivanac, N., Metes, A.: Optimizing polymer induced flocculation process to remove the active dyes from wastewater. Environ. Technol. **21**, 97–105 (2000)
2. Rajeshwari, S., Sivakumar, S., Senthilkumar, P., Subburam, V.: Carbon from cassava peel, an agricultural waste, as an adsorbent in the removal of dyes and metal ions from the aqueous solution. Bioresour. Technol. **80**, 233–235 (2001)
3. Yang, J., Hou, B., Wang, J., Tian, B., Bi, J., Wang, N., Li, X., Huang, X.: Nanomaterials for the removal of heavy metals from wastewater. Nanomaterials (2019). https://doi.org/10.3390/nano9030424
4. Munoz-Olivas, R., Camara, C.: Speciation related to human health. In: Ebdon, L., Pitts, L., Cornelis, R., Crews, H., Donard, O.F.X., Quevauviller, P. (eds.) Trace Element Speciation for Environment, pp. 331–353. The Royal Society of Chemistry, Food and Health (2001)
5. Celik, U., Oehlenschlager, J.: High contents of cadmium, lead, zinc and copper in popular fishery products sold in Turkish super markets. Food Control. **18**, 258–261 (2007)
6. Ortuzar, A., Escondrillas, I., Mijangos, F.: Inorganic anion removal from water using natural adsorbents. Int. Sch. Sci. Res. Innov. **12**, 284–287 (2018)
7. Inamori, Y., Fujimoto, N.: Water quality and standards—vol. II, microbial/biological contamination of water. Encycl. Life Support. Syst. (EOLSS). Available via dialog http://www.desware.net/Sample-Chapters/D16/E2-19-05-03.pdf (2009)
8. Nwachcuku, N., Gerba, C.P.: Emerging waterborne pathogens: can we kill them all? Curr. Opin. Biotechnol. **15**, 175–180 (2004)
9. Dabrowski, A., Podkoscielny, P., Hubicki, Z., Barczak, M.: Adsorption of phenolic compounds by activated carbon—a critical review. Chemosphere **58**, 1049–1070 (2005)
10. Hristovski, K., Baumgardner, A., Westerhoff, P.: Selecting metal oxide nanomaterials for arsenic removal in fixed bed columns: from nanopowders to aggregated nanoparticle media. J. Hazard. Mater. **147**, 265–274 (2007)
11. Diallo, M.S., Fromer, N.A., Jhon, M.S.: Nanotechnology for sustainable development: retrospective and outlook. J. Nanopart. Res. (2013). https://doi.org/10.1007/s11051-013-2044-0
12. Sadegh, H., Ali, G.A.M., Gupta, V.K., Makhlouf, A.S.H., Shahryari-ghoshekandi, R., Nadagouda, M.N., Sillanpa, M., Megiel, E.: The role of nanomaterials as effective adsorbents and their applications in wastewater treatment. J. Nanostructure Chem. **7**, 1–14 (2017)
13. Sathyaa, M., Kumarb, P.K., Santhib, M.: Dye removal from its aqueous solution by using nanoparticle as adsorbent—a review. IJASER **2**, 494–504 (2017)
14. Ruan, W., Hu, J., Qi, J., Hou, Y., Zhou, C., Wei, X.: Removal of dyes from wastewater by nanomaterials: a review. Adv. Mater. Lett. **10**, 09–20 (2019)
15. Tan, K.A., Morad, N., Teng, T.T., Norli, I., Panneerselvam, P.: Removal of cationic dye by magnetic nanoparticle (Fe_3O_4) impregnated onto activated maize cob powder and kinetic study of dye waste adsorption. APCBEE Procedia **1**, 83–89 (2012)
16. Fu, F., Wang, Q.: Removal of heavy metal ions from wastewaters: a review. J. Environ. Manage. **92**, 407–418 (2011)

17. Ihsanullah, Abbas A., Al-Amer, A.M., Laoui, T., Al-Marri, M., Nasser, M.S., Majeda, K., Atieh, M.A.: Heavy metal removal from aqueous solution by advanced carbon nanotubes: critical review of adsorption application. Sep. Purif. Technol. **157**, 141–161 (2016)
18. Kumar, E., Bhavnagar, A., Hogland, W., Marques, M., Sillanpaa, M.: Interaction of anionic pollutants with al-based adsorbents in aqueous media—a review. Chem. Eng. J. **241**, 443–456 (2014)
19. Srinivasan, R.: Advances in application of natural clay and its composites in removal of biological, organic, and inorganic contaminants from drinking water. Adv. Mater. Sci. Eng. (2011). https://doi.org/10.1155/2011/872531
20. Praveena, V.D., Kumar, K.V., Venkataraman, K.: Phosphate removal from aqueous solutions by a nano-structured Ag-chitosan film. J. Nanosci. Nanotechnol. **2**, 134–137 (2016)
21. Singh, S.R., Krishnamurthy, N.B., Baby, Mathew Blessy: A review on recent diseases caused by microbes. JAEM **2**, 106–115 (2014)
22. Sharma, S., Bhattacharya, A.: Drinking water contamination and treatment techniques. Appl. Water Sci. **7**, 1043–1067 (2017)
23. Nabeela, F., Azizullah, A., Bibi, R., Uzma, S., Murad, W., Shakir, S.K., Ullah, W., Qasim, M., Hader, D.P.: Microbial contamination of drinking water in Pakistan—a review. Environ. Sci. Pollut. Res. **21**, 13929–13942 (2014)
24. Ashbolt, N.J.: Microbial contamination of drinking water and disease outcomes in developing regions. Toxicology **198**, 229–238 (2004)
25. Aljeboree, A.M., Alshirifi, A.N., Alkaim, A.F.: Kinetics and equilibrium study for the adsorption of textile dyes on coconut shell activated carbon. Arab. J. Chem. **10**, 3381–3391 (2017)
26. Tan, I.A.W., Ahmad, A.L., Hameed, B.H.: Adsorption of basic dye on high-surface-area activated carbon prepared from coconut husk: equilibrium, kinetic and thermodynamic studies. J. Hazard. Mater. **154**, 337–346 (2008)
27. Banerjee, S., Chattopadhyaya, M.C.: Adsorption characteristics for the removal of a toxic dye, tartrazine from aqueous solutions by a low cost agricultural by-product. Arab. J. Chem. **10**, 1629–1638 (2017)
28. Raju, C.H.A.I., Nooruddin, S., Babu, K.S.: Studies on leaf extract mediated synthesis of copper nanoparticles for the removal of bromo cresol green dye from synthetic waste waters. IJSETR **6**, 1404–1411 (2017)
29. Jyoti, K., Singh, A.: Green synthesis of nanostructured silver particles and their catalytic application in dye degradation. J. Genet. Eng. Biotechnol. **14**, 311–317 (2016)
30. Eme, Femila, Srimathi, R., Charumathi, D.: Removal of malachite green using silver nanoparticles via adsorption and catalytic degradation. Int. J. Pharm. Pharm. Sci. **6**, 579–583 (2014)
31. Satapathy, M.K., Banerjee, P., Das, P.: Plant-mediated synthesis of silver-nanocomposite as novel effective azo dye adsorbent. Appl. Nanosci. **5**, 1–9 (2015)
32. Ragab, A., Ahmed, I., Bader, D.: The removal of brilliant green dye from aqueous solution using nano hydroxyapatite/chitosan composite as a sorbent. Molecules (2019). https://doi.org/10.3390/molecules24050847
33. McMullan, G., Meehan, C., Conneely, A., Kirby, N., Robinson, T., Nigam, P., Banat, I., Marchant, R., Smyth, W.: Microbial decolourisation and degradation of textile dyes. Appl. Microbiol. Biotechnol. **56**, 81–87 (2001)
34. Jafari, N., Soudi, M.R., Kasra-Kermanshahi, R.: Biodegradation perspectives of azo dyes by yeasts. Microbiology **83**, 484–497 (2014)
35. Corso, C.R.: Bioremediation of dyes in textile effluents by aspergillus oryzae. Microb. Ecol. **57**, 384–390 (2009)
36. Bayoumi, M.N., Al-Wasify, R.S., Hamed, S.R.: Bioremediation of textile wastewater dyes using local bacterial isolates. Int. J. Curr. Microbiol. **3**, 962–970 (2014)
37. Fulekar, M.H., Pathak, B., Kale, R.K.: Nanotechnology: perspective for environmental sustainability. In: Fulekar, M.H., Pathak, B., Kale, R.K. (eds.) Environment and Sustainable Development. https://doi.org/10.1007/978-81-322-1166-2 (2014)

38. Rengaraj, S., Venkataraj, S., Yeon, J.W., Kim, Y., Li, X.Z., Pang, G.K.H.: Preparation, characterization and application of Nd–TiO2 photocatalyst for the reduction of Cr(VI) under UV light illumination. Appl. Catal. B: Environ. **77**, 157–165 (2007)
39. Barakat, M.A.: New trends in removing heavy metals from industrial wastewater. Arab. J. Chem. **4**, 361–377 (2011)
40. Weng, X., Huang, L., Chen, Z., Megharaj, M., Naidu, R.: Synthesis of iron-based nanoparticles by green tea extract and their degradation of malachite. Ind. Crop. Prod. **51**, 342–347 (2013)
41. Tandon, P.K., Shukla, R.C., Singh, S.B.: Removal of arsenic(III) from water with clay supported iron nanoparticles synthesized with the help of tea liquor. Ind. Eng. Chem. Res. **52**, 10052–10058 (2013)
42. Machado, S., Stawinski, W., Slonina, P., Pinto, A.R., Grosso, J.P., Nouws, H.P., Albergaria, J.T., Delerue-Matos, C.: Application of green zero-valent iron nanoparticles to the remediation of soils contaminated with ibuprofen. Sci. Total. Environ. **461–462**, 323–329 (2013)
43. Wang, T., Lin, J., Chen, Z., Megharaj, M., Naidu, R.: Green synthesized iron nanoparticles by green tea and eucalyptus leaves extracts used for removal of nitrate in aqueous solution. J. Clean. Prod. **83**, 413–419 (2014)
44. Fazlzadeh, M., Rahmani, K., Zarei, A., Abdoallahzadeh, H., Nasiri, F., Khosravi, R.: A novel green synthesis of zero valent iron nanoparticles (NZVI) using three plant extracts and their efficient application for removal of Cr(VI) from aqueous solutions. Adv. Powder Technol. **28**, 122–130 (2017)
45. Prasad, K.S., Gandhi, P., Kaliaperumal, S.: Synthesis of green nano iron particles (Gnip) and their application in adsorptive removal of As(III) and As(V) from aqueous solution. Appl. Surf. Sci. **317**, 1052–1059 (2014)
46. Juang, R.S., Shiau, R.C.: Metal removal from aqueous solutions using chitosan-enhanced membrane filtration. J. Membr. Sci. **165**, 159–167 (2000)
47. Qdais, H.A., Moussa, H.: Removal of heavy metals from wastewater by membrane processes: a comparative study. Desalination **164**, 105–110 (2004)
48. Barakat, M.A.: Removal of Cu(II), Ni(II), and Cr(III) ions from wastewater using complexation—ultrafiltration technique. J. Environ. Sci. Technol. **1**, 151–156 (2008)
49. Trivunac, K., Stevanovic, S.: Removal of heavy metal ions from water by complexation-assisted ultrafiltration. Chemosphere **64**, 486–491 (2006)
50. Muslehiddinoglu, J., Uludag, Y., Ozbelge, H.O., Yilmaz, L.: Effect of operating parameters on selective separation of heavy metals from binary mixtures via polymer enhanced ultrafiltration. J. Membr. Sci. **140**, 251–266 (1998)
51. Canizares, P., Perez, A., Camarillo, R.: Recovery of heavy metals by means of ultrafiltration with water-soluble polymers: calculation of design parameters. Desalination **144**, 279–285 (2002)
52. Muthukrishnan, M., Guha, B.K.: Effect of pH on rejection of hexavalent chromium by nanofiltration. Desalination **219**, 171–178 (2008)
53. Murthy, Z.V.P., Chaudhari, L.B.: Separation of binary heavy metals from aqueous solutions by nanofiltration and characterization of the membrane using SpieglereKedem model. Chem. Eng. J. **150**, 181–187 (2009)
54. Csefalvay, E., Pauer, V., Mizsey, P.: Recovery of copper from process waters by nanofiltration and reverse osmosis. Desalination **240**, 132–142 (2009)
55. Figoli, A., Cassano, A., Criscuoli, A., Mozumder, M.S.I., Uddin, M.T., Islam, M.A., Drioli, E.: Influence of operating parameters on the arsenic removal by nanofiltration. Water Res. **44**, 97–104 (2010)
56. Srivastava, A., Srivastava, O.N., Talapatra, S., Vajtai, R., Ajayan, P.M.: Carbon nanotube filters. Nat. Mater. **3**, 610–614 (2004)
57. Shah, M.A., Ahmed, T.: Principles of Nanoscience and Nanotechnology, pp. 34–47. Narosa Publishing House, New Delhi, India (2011)
58. Wajima, T., Umeta, Y., Narita, S., Sugawara, K.: Adsorption behavior of fluoride ions using a titanium hydroxide-derived adsorbent. Desalination **249**, 323–330 (2009)

59. Bhatnagar, A., Sillanpaa, M.: A review of emerging adsorbents for nitrate removal from water. Chem. Eng. J. **168**, 493–504 (2011)
60. Ajmal, Z., Muhmood, A., Usman, M., Kizito, S., Lu, J., Dong, R., Wu, S.: Phosphate removal from aqueous solution using iron oxides: adsorption, desorption and regeneration characteristics. J. Colloid Interface Sci. **528**, 145–155 (2018)
61. Wang, L., Zhang, J., Liu, J., He, H., Yang, M., Yu, J., Ma, Z., Jiang, F.: Removal of bromate ion using powdered activated carbon. J. Environ. Sci. **22**, 1846–1853 (2010)
62. Robert, P., Cannon, F.S.: The removal of perchlorate from groundwater by activated carbon tailored with cationic surfactants. Water Res. **39**, 4020–4028 (2005)
63. Parikh, R.Y., Singh, S., Prasad, B.L., Patole, M.S., Sastry, M., Shouche, Y.S.: Extracellular synthesis of crystalline silver nanoparticles and molecular evidence of silver resistance from Morganella sp.: towards understanding biochemical synthesis mechanism. ChemBioChem **9**, 1415–1422 (2008)
64. Konishi, Y., Ohno, K., Saitoh, N., Nomura, T., Nagamine, S., Hishida, H., Takahashi, Y., Uruga, T.: Bioreductive deposition of platinum nanoparticles on the bacterium Shewanella algae. J. Biotechnol. **128**, 648–653 (2007)
65. Willner, I., Baron, R., Willner, B.: Growing metal nanoparticles by enzymes. Adv. Mater. **18**, 1109–1120 (2006)
66. Shankar, S.S., Rai, A., Ahmad, A., Sastry, M.: Rapid synthesis of Au, Ag, and bimetallic Au core Ag shell nanoparticles using Neem [Azadirachta indica] leaf broth. J. Colloid Interface Sci. **275**, 496–502 (2004)
67. Schultz, S., Smith, D.R., Mock, J.J., Schultz, D.A.: Single-target molecule detection with nonbleaching multicolor optical immunolabels. Proc. Natl. Acad. Sci. **97**, 996–1001 (2000)
68. Wang, Y.H., Zhou, J., Wang, T.: Enhanced luminescence from europium complex owing to surface plasmon resonance of silver nanoparticles. Mater. Lett. **62**, 1937–1940 (2008)
69. Venkatpurwar, V., Pokharkar, V.: Green synthesis of silver nanoparticles using marine polysaccharide: study of in vitro antibacterial activity. Mater. Lett. **65**, 999–1002 (2011)
70. Elechiguerra, J.L., Burt, J.L., Morones, J.R., Bragado, A.C., Gao, X., Lara, H.H., Yacaman, M.J.: Interaction of silver nanoparticles with HIV. J. Nanobiotechnol. (2005). https://doi.org/10.1186/1477-3155-3-6
71. Chen, H., Hao, F., He, R., Cui, D.X.: Chemiluminescence of luminol catalyzed by silver nanoparticles. J. Colloid Interface Sci. **315**(1), 58–163 (2007)
72. Crooks, R.M., Lemon, B.I., Sun, L., Yeung, L.K., Zhao, M.: Dendrimer-encapsulated metals and semiconductors: synthesis, characterization and application. Top. Curr. Chem. **212**, 82–135 (2001)
73. Krishnaraj, C., Jagan, E.G., Ramachandran, R., Abirami, S.M., Mohan, N., Kalaichelvan, P.T.: Effect of biologically synthesized silver nanoparticles on Bacopa monnieri (Linn.) Wettst. Plant growth metabolism. Process. Biochem. **47**, 651–658 (2012)
74. Lohse, S.E., Murphy, C.J.: Applications of colloidal inorganic nanoparticles: from medicine to energy. J. Am. Chem. Soc. **134**, 15607–15620 (2012)
75. Salari, Z., Danafar, F., Dabaghi, S., Ataei, S.A.: Sustainable synthesis of silver nanoparticles using macroalgae Spirogyra varians and analysis of their antibacterial activity. J. Saudi Chem. Soc. **20**, 459–464 (2016)
76. Moustafa, M.T.: Removal of pathogenic bacteria from wastewater using silver nanoparticles synthesized by two fungal species. Water Sci. **31**, 164–176 (2017)
77. Gudikandula, K., Maringanti, S.C.: Synthesis of silver nanoparticles by chemical and biological methods and their antimicrobial properties. J. Exp. Nanosci. **9**, 714–721 (2016)
78. Kuppusamy, P., Yusoff, M.M., Maniam, G.P., Govindan, N.: Biosynthesis of metallic nanoparticles using plant derivatives and their new avenues in pharmacological applications—An updated report. Saudi Pharm. J. **24**, 473–484 (2016)
79. Ajmal, N., Saraswat, K., Bakht, M.A., Riadi, Y., Ahsan, M.J., Noushad, M.: Cost-effective and eco-friendly synthesis of titanium dioxide (TiO_2) nanoparticles using fruit's peel agro-waste extracts: characterization, in vitro antibacterial, antioxidant activities. Green Chem. Lett. Rev. **12**, 244–254 (2019)

80. Ayinde, W.B., Gitari, M.W., Muchindu, M., Samie, A.: Biosynthesis of ultrasonically modified Ag-MgO nanocomposite and its potential for antimicrobial activity. J. Nanotechnol. (2018). https://doi.org/10.1155/2018/9537454

81. Sivapriya, V., Azeez, A.N., Deepa, S.V.: Phyto synthesis of iron oxide nano particles using the agro waste of *Anthocephalus cadamba* for pesticidal activity against *Sitophilus granaries*. J. Entomol. Zool. Stud. **6**, 1050–1057 (2018)

82. Lee, C., Kim, J.Y., Lee, W.I.I., Nelson, K.L., Yoon, J., Sedlak, D.L.: Bactericidal effect of zero-valent iron nanoparticles on *escherichia coli*. Environ. Sci. Technol. **42**, 4927–4933 (2008)

83. Logeswari, P., Silambarasan, S., Abraham, J.: Synthesis of silver nanoparticles using plant extracts and analysis of their antimicrobial activity. J. Saudi Chem. Soc. **4**, 23–45 (2012)

84. Al-Dhabi, N.A., Arasu, M.V.: Environmentally-friendly green approach for the production of zinc oxide nanoparticles and their anti-fungal, ovicidal, and larvicidal properties. Nanomaterials (2018). https://doi.org/10.3390/nano8070500

Nanostructured Membranes for Water Purification

Xin Li, Gomotsegang Fred Molelekwa, Meryem Khellouf
and Bart Van der Bruggen

Abstract Membrane based processes enjoy numerous industrial applications and have greatly enhanced our capabilities to restructure production processes, protect the environment and public health, and provide new technologies for water purification. The scope of membrane technology is still extending, stimulated by the developments of novel membrane materials and membranes with better properties, as well as by the decrease of capital and operation costs. Recent advances in nanomaterials empower next-generation multifunctional membrane processes with exceptional catalytic, adsorptive, optical and/or antibacterial abilities that enhance treatment cost-efficiency. This chapter reviews emerging opportunities and sustainable approaches for the design of nanostructured membranes for water purification. Potential development and implementation barriers are discussed along with research needs to overcome them for enhancing water purification.

Keywords Nanomaterials · Membrane · Water purification · Wastewater treatment

1 Introduction

Water is the lifeblood of ecosystems, vital to human health and well-being and a pre-condition for economic prosperity. For this reason, it is at the very core of the 2030

X. Li · B. Van der Bruggen (✉)
Department of Chemical Engineering, KU Leuven, Celestijnenlaan 200F, 3001 Leuven, Belgium
e-mail: Bart.VanderBruggen@kuleuven.be

G. F. Molelekwa
Department of Environmental Health, Tshwane University of Technology, Pretoria, South Africa

M. Khellouf
Central Direction of Research and Development, Sonatrach, Boumerdes, Algeria

B. Van der Bruggen
Faculty of Engineering and the Built Environment, Tshwane University of Technology, Private Bag X680, Pretoria 0001, South Africa

© Springer Nature Switzerland AG 2019
G. A. B. Gonçalves and P. Marques (eds.), *Nanostructured Materials for Treating Aquatic Pollution*, Engineering Materials,
https://doi.org/10.1007/978-3-030-33745-2_9

Agenda for Sustainable Development [1]. Although encouraging progress has been achieved in the field of water treatment in the 20th century, water scarcity is more than ever before the main challenge to provide clean, fresh water for the world's inhabitants. As estimated by the World Water Council (WWC), 3.9 billion people in the world will live in water-scarce regions by 2030 (WWC, Urban Urgency, Water Caucus Summary, 2007). In addition, water-associated vector-borne diseases continue to be a major public health problem in many countries. 2.1 billion people lack a safely managed drinking-water supply, and more than half of the 842,000 water-related die attributed to unsafe drinking-water (World Health Organization, Water, Sanitation and Hygiene strategy 2018–2025, 2018). Therefore, water resources, quality and treatment have become major topics of public and government.

Consolidated and emerging water purification strategies are pressingly acclaimed to offer compelling solutions to the water pollution issues and indeed, increasingly available approaches have surfaced. Initially, activated carbon was employed for wastewater purification, but has been replaced recently by more cost-effective membrane technologies [2]. Based on a range of synthetic membranes with various functions, membrane technologies offer the best options to 'drought proof' mankind on an increasingly thirsty planet by purifying seawater or wastewater [3]. Reverse osmosis (RO), which involves pushing water through a semipermeable membrane that blocks dissolved salts, is now considered the heart of conventional desalination plants. Other membrane processes, microfiltration (MF) and ultrafiltration (UF), have become major technologies for water treatment from non-saline sources, pretreatment for RO processes and in wastewater treatment in membrane bioreactors. Nanofiltration (NF) finds application in water softening and color removal, industrial wastewater treatment, water reuse, and desalination. Additionally, as emerging technology, forward osmosis (FO) has shown great promise in water supply, because it requires much less energy to induce a net flow of water across the membrane compared to traditional pressure-driven membrane processes such as RO. However, FO as a stand-alone process has to be integrated in an overall process scheme to be used as a desalination process. Although commercially available membranes perform well in water purification and desalination, current materials and fabrication methods for membranes are largely based on empirical approaches and lack molecular-level design [4]. The drive to protect existing water resources and produce new water resources demands membranes with improved productivity, selectivity, fouling resistance, and stability at lower cost and with fewer manufacturing defects.

The recent development of nanotechnology offers leapfrogging opportunities for developing innovative technologies for more efficient water purification and wastewater treatment processes. It is now a popular belief that many of the solutions to the existing and even future water-based challenges are most likely to come from nanotechnology and especially novel nanomaterials with increased affinity, capacity, and selectivity for water contaminants [5]. Particularly, the incorporation of these novel nanomaterials into the chemical composition of membranes for water purification has been the focus in this field, as illustrated by the number of publications and related citations (Fig. 1). The obtained nanostructured membranes can alter the membrane properties such as permeability, selectivity, mechanical resistance and

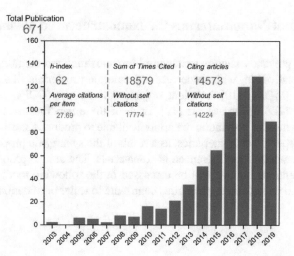

Fig. 1 The number of journal papers on nanostructured membrane for water purification since 2003. Data were obtained from Web of Science on July 2019. Search topics were 'membrane, nanoparticles and water purification'

hydrophilicity as compared to virgin polymers [6]. Nanomaterial properties desirable for nanostructured membranes also include a high surface area for adsorption, a high activity for (photo)catalysis and degradation, antibacterial properties for disinfection and biofouling control, and other unique optical and electronic properties that find use in novel treatment processes in water purification [7–9]. Since the use of nanostructured membranes in water purification is considered an achievement of high importance, they should be designed to maintain the surface reactivity, effective contact and uptake capacity. For this reason, the class of applied nanomaterials, and fabrication strategies of nanostructured membranes combining the respective advantages should be preferred for water purification.

This chapter presents promising nanostructured membranes and provides a broad view on how they could transform our water purification and wastewater treatment systems. The extraordinary properties of prevalent nanomaterials in nanostructured membranes, such as high surface area, photosensitivity, catalytic and antibacterial activity, and electrochemical, were introduced. The fabrication strategies and relevant applications of nanostructured membranes in water purification were critically reviewed based on their functions in unit operation processes. The potential impact of nanostructured membranes on ecosystems and economics as well as their technological capacity were also discussed.

2 Prevalent Nanomaterials for Nanostructured Membrane

Most of the prevalent nanomaterials used in nanostructured membranes are either metal or metal oxide, with some recently emerging materials like metal organic frameworks (MOFs) [10], covalent organic frameworks (COFs) [11], and two-dimensional (2D) nanomaterials [12]. Each of these nanomaterials can be incorporated with most of polymeric materials available to produce nanostructured membranes with specific characteristics, as a result of the synergistic properties between synthetic membranes and advanced nanomaterials. The unique properties of nanomaterials mentioned above will be discussed in the following sections. The focus will be placed on how these properties contribute to water purification in membrane processes.

2.1 Metal Nanoparticles

The challenge to achieve appropriate disinfection without forming harmful disinfection byproducts in point-of-use water treatment calls for new technologies for efficient disinfection and microbial control. Several natural and engineered nanomaterials have demonstrated strong antibacterial properties through diverse mechanisms. Among them, Ag nanoparticles have drawn special attention due to their excellent antibacterial and antifungal activity [13]. It is now well accepted that the antibacterial activity of Ag nanoparticles stems from the release of Ag^+, which can bind to thiol groups in vital proteins, resulting enzyme damage [7]. It has also been reported that Ag^+ can prevent DNA replication and induce structural changes in the cell envelope [14] (Fig. 2a). With these advantages, significant progress has been made in the development of Ag nanoparticles-based membranes for water purification. For example, Ag nanoparticles were impregnated in a polysulfone matrix to improve biofouling resistance and virus removal [15]. Biogenic Ag nanoparticles were also incorporated in polyethersulfone (PES) membranes for enhancing their antibiofouling capability in a membrane bioreactor [16]. Besides, Ag nanoparticles were embedded into woven fabric MF membranes. Such designed membranes were more hydrophilic, showing a higher water permeability and 100% removal of bacterial load from drinking water [17].

One challenge with the Ag nanoparticles inside the membrane matrices is their weak resistance to washing, as they can be released either as nanoparticles or dissolved ions form [18], so the antibacterial ability of membrane would be attenuated with time. A more advanced in situ reduction coating procedure is introduced to address this issue: Ag ions are firstly dissolved in the polymer/polyelectrolyte solution, followed by coating the mixtures and in situ synthesizing Ag nanoparticles onto a membrane for stable attachment. With this strategy, polydopamine (PDA) has been proposed to improve the long-term antibacterial efficiency of Ag nanoparticles-based membranes [19]. Huang et al. successfully immobilized Ag nanoparticles on

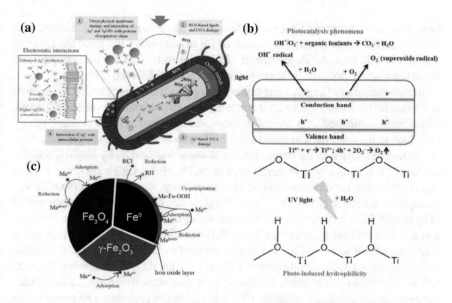

Fig. 2 **a** Proposed mechanism of Ag nanoparticles-related toxicity. Reprinted with the permission from Ref. [25]. Copyright 2014 The Royal Society of Chemistry. **b** Photocatalysis and photo-induced ultrahydrophilicity mechanism of TiO$_2$ nanoparticles. Reprinted with the permission from Ref. [26]. Copyright 2015 Elsevier. **c** Schematic model of magnetic nanoparticles. Reprinted with the permission from Ref. [27]. Copyright 2013 Elsevier

a PES UF membrane by PDA mediated in situ reduction of Ag(NH$_3$)$_2$OH. The static immersion test of the membrane exhibited a slight release of Ag$^+$ ions even after 12 days [20]. The inhibition zone study showed a clear halo, confirming its tremendous disinfection properties against *E. coli* and *B. subtilis*. Meanwhile, compared with a control membrane, the pure water flux of optimized membrane was increased from 248 to 336 L/m^{2}h. Similarly, tannic acid (TA) also has been used to enhance the stability of Ag nanoparticles due to its metal chelating and adhesive ability. Based on this strategy, an ultrathin precursor layer of TA-Fe-PEI complexes was deposited onto the membrane surface, followed by in situ reduction of Ag nanoparticles in the presence of polyvinylpyrrolidone. After contacting with *E. coli* and *B. subtilis* for 1.5 h, the bactericidal efficiency of TA-Fe-PEI/Ag-modified membrane could reach 100% [21]. Moreover, constructing nanocomposites by immobilizing Ag nanoparticles onto a larger-scale support can also tackle the instability problem. Promising carriers for Ag nanoparticles include carbon nanotubes (CNTs) [22], graphene oxide (GO) [23], and TiO$_2$ [24]. These carriers are either antibacterial agents themselves or possess the advantages of hydrophilic, antifouling, and photocatalytic properties, which can even lead to a higher antibacterial capability than that of pure Ag nanoparticles due to a synergistic enhancement effect.

Due to their considerably lower cost than silver, Cu nanoparticles were also explored in term of their antibacterial activity against various bacterial species. Although the exact antibacterial mechanisms remain elusive, substantial efforts have been devoted to designing Cu-based antibacterial membranes [28–30]. Similar with Ag nanoparticles, mitigating the release of Cu nanoparticles has been the focus of research in this area. An improvement of the pure water flux for Cu nanoparticles based membranes with the increase of Cu content was observed [28].

Beside bacteria or fungus, toxic organic compounds such as chlorinated aliphatic and aromatic molecules are another common class of contaminants in natural water, which were implicated with several diseases, chronic damages and carcinogenicity [31]. Due to the low cost and broad availability, zero-valent iron nanoparticles have drawn more attention in the detoxification of these compounds by electron transfer reactions [32]. Particularly, bimetals such as Pd/Fe, Ni/Fe, Cu/Fe are found to have higher efficiency for dichlorination of chlorinated organic compounds (COCs). For example, in the Pd/Fe bimetallic system, the corrosion of iron leads to the emergence of hydrogen. Pd is known to have the unique ability to absorb hydrogen into its lattice and acts as a catalyst to accelerate hydrodechlorination [33]. During the past years, many theoretical and experimental studies focused on designing Fe-based catalytic membranes for COCs dichlorination [33, 34]. In parallel, a continuous effort to enhance the stability of bimetals nanoparticles in membrane structures and to evaluate the degradation efficiency of target COCs is still under progress. The preservation and regeneration of Fe-based catalytic membranes are also an important issue for their prolonged applications.

2.2 Metal Oxide Based Nanomaterials

Metal oxides nanomaterials are widely explored as highly efficient adsorbents and catalysts for contaminants removal from water/wastewater, due to their easy and well-developed synthesis method, cost effectiveness and functionalization to nanostructured membranes. There has been bountiful literature published on the application of TiO_2 nanoparticles in membranes for water purification [35]. This could be attributed to two main reasons—its excellent photocatalysis that decomposes/disintegrates organic foulants and microbes, and photo-induced ultrahydrophilicity that repels hydrophobic foulants and forms hydration layer of water on the membrane surface, thereby enhancing fouling control [36, 37] (Fig. 2b). TiO_2 based photocatalytic membranes exhibit a more obvious effect on dye degradation [38], humic substances removal [39], and oil-water emulsion separation [40], as compared with virgin membranes. In particular, TiO_2 free-standing membranes have been reported, providing higher photocatalytic surface area and mass transfer rate than TiO_2 nanoparticles coated and blended membranes. For example, TiO_2 nanotube membranes shown a satisfying permeability and antifouling performance due to the rejection and photodegradation of foulants on the surface of membrane under UV irradiation [41]. Despite of these advantages, UV-responsive TiO_2 based membranes still face challenges. TiO_2 based catalysts suffer from the limited photocatalytic performance due to fast recombination of the photogenerated charges, and exhibit poor photoactivity in the visible light region [42]. Moreover, the membrane structure may suffer from severe destruction by both UV light and reactive oxygen species [43]. Recently, modification of TiO_2, by metal or nonmetal doping, co-doping, coupling with another semiconductor or developing composites with carbon materials, has led to a significant enhancement of its photocatalytic performance under visible light irradiation. With this progress, the attempt to extend the modified TiO_2 photocatalysts into the design of visible-light responsive membranes has become a topic of great interest. A timely review by Shi et al. discussed the development of these novel TiO_2 based membranes for water purification [44].

The presence of heavy metals in aqueous systems is considered a major worldwide problem related to many harmful effects on the health of humans and other life forms. Iron-based nanoparticles are the widely applied materials for the uptake of heavy metals in water [45]. During the water treatment, zero-valent iron is oxidized forming in situ oxy-hydroxides and oxides with enhanced surface area, charge density and reactivity (Fig. 2c). Iron oxide nanoparticles may surpass the drawback of soluble iron release. In addition, the magnetic behavior of Fe_3O_4 and γ-Fe_2O_3 facilitates their recovery after application. Cr (VI) removal is the early studies case of iron oxide nanoparticles based membranes application in water purification, based on the electrostatic adsorption and redox reaction [46, 47]. Similarly, the adsorptive removal of Cu (II) was remarkably improved using an Fe_3O_4 nanoparticles embedded PES membrane [48]. Recently, for improving stability and controlling aggregation, a metformin/GO/Fe_3O_4 hybrid was introduced into nanofiltration for effective removal of dye and Cu (II) [49].

Several metal oxide nanoparticles, namely CuO and ZnO were also studied for inclusion in antibacterial membranes. CuO nanoparticles presented enhanced bactericidal behavior against *E. coli* and *P. aeruginosa* [50]. The proposed antibacterial mechanism of ZnO nanoparticles is closely correlated with the photocatalytic formation of reactive oxygen species or the release of Zn^{2+} ions, both of which can effectively inhibit the bacteria. A recent work reported the development of CuO or ZnO decorated polyacrylonitrile (PAN) nanofibers membranes for drinking water purification [51]. The hybrid membranes exhibited an excellent antibacterial potential against both *E. coli* and *S. aureus*. The developments and breakthroughs on antibacterial membranes with metal oxide nanoparticles for water treatment have been well described by Zhu et al. [18]. Besides, ZnO nanoparticles have shown more obvious advantages related to membrane fouling control based on their hydrophilic nature. The reduced flux drop, extended operational lifespan or higher flux recovery rate for ZnO nanoparticles based membranes have been experimentally/theoretically confirmed when tested with model foulants (e.g., oleic acid solution [52], sodium alginate solution [53]) and river water [54]. More recently, surface modified ZnO nanoparticles were also immobilized on the polyvinylidene difluoride (PVDF) membrane surface for oily wastewater treatment [55].

In the same direction, SiO_2, ZrO_2, and Al_2O_3 nanoparticles have been proven efficient for designing antifouling membranes in water purification. The addition of these nanoparticles to membranes enhanced the surface hydrophilicity, water permeability, or fouling resistance, along with the mechanical and thermal stability [7]. In the light of molecular-level design, in situ preparation, permitting the in situ generation of nanoparticles into/on the membrane matrix, has pushed the design of metal oxide based nanostructured membranes to the next stage [56]. With the in situ formation process (i.e., the formation of SiO_2, ZrO_2, and Al species via sol-gel process), the deposition of nanoparticles can be preciously controlled on the membrane surface or around membrane pores, which triggers the maximum exposure of hydrogen accepter groups. Research has been undertaken to compare the influence of in situ and ex situ created SiO_2 nanoparticles on the structure and performance of PES membranes [57]. In recent research PVDF MF/UF membranes were modified via in situ biomimetic silicification for oily emulsion and protein wastewater treatment [58] (Fig. 3a). Similarly, a successful design was achieved with ultrathin ZrO_2 film as selective layer to prepare a novel NF membrane using hydrolysis of zirconium sulfate [59] (Fig. 3b). By this stage, researchers realized that the chemistry and the functions of metal oxide nanoparticles could be deliberately pre-designed for specific water treatment purposes, while integrating with a suitable membrane formation process.

2.3   Metal Organic Framework/Covalent Organic Framework

The development of crystalline materials extended their applications for designing next-generation nanostructured membranes in water purification. Among these materials, porous inorganic solids are highly attractive due to their porous structure and high surface area, which can increase the overall porosity and separation ability of membranes. MOFs/COFs have emerged as porous solids of superior new types [60]. These solids have a regular and highly tunable pore structure, along with an enormous variability in secondary building units connected by multitopic organic ligands and linker topology, connectivity, and chemical functionality, which are preferred over conventional porous materials such as zeolites and carbon based materials. Both continuous MOFs/COFs and MOFs/COFs based composite membranes have been developed as multifunctional materials for membrane separation beyond the traditional uses for dye removal and fouling-resistance, and even contribute to catalytic degradation, heavy metal adsorption, and oil/water separation. For example, trace nickel (II) (ppm level) could be effectively removed from high-salinity wastewater ($[Na^+] = 15,000$ mg/L) with a ZIF-8 modified PVDF UF membrane under the protection of a polyacrylic acid layer [61]. The hybrid membrane could be regenerated by a HCl-NaCl solution (pH = 5) for repeated use. Recently, COFs based mixed matrix membranes with much higher permselectivities were obtained by embedding carboxyl-functionalized COFs in PAN UF membrane [62]. The best performing membrane displayed a relatively high pure water flux of 940 L/m^2 h bar and an enhanced fouling resistance, maintaining a 99.4% rejection of γ-globulin. Furthermore, novel composites, such as UiO-66-$(COOH)_2$/prGO [63] and TpBD@Fe_3O_4 [64], were introduced into membrane for water treatment, which represented the latest trends for MOFs/COFs based membranes. The recent progress of MOFs/COFs based membranes for water purification has been discussed in published reviews [10, 11].

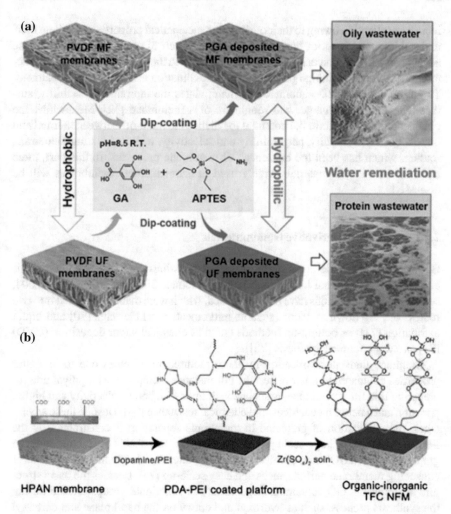

Fig. 3 Scheme of (**a**) co-deposition of gallic acid (GA) and γ-aminopropyltriethoxysilane (APTES) on the PVDF MF/UF membranes for application in water remediation on oily and protein wastewater. PGA refers to the abbreviation of the hydrophilic hybrid network derived from polymerization of GA and APTES-derived polysiloxane. Reprinted with the permission from Ref. [58]. Copyright 2018 American Chemical Society. **b** Preparation process and mechanism for the organic–inorganic TFC membranes with ultrathin ZrO_2 film as the selective layer on the PDA-PEI coated PAN membranes. Reprinted with the permission from Ref. [59]. Copyright 2015 Elsevier

2.4 Two-Dimensional Materials

With the development in materials science, the dimension has become a significant parameter of materials classification. 2D nanomaterials, such as graphene and GO [65], exfoliated nanosheets of MOFs [66], zeolite nanosheets [67], transition metal dichalcogenides (TMDs) [68], and 2D MXene nanosheets [69], have attracted increasing attention owing to their outstanding mechanical properties, excellent thermal stability, and superior flexibility that surpass their 3D counterparts. With these advantages, 2D nanomaterials are rapidly emerging in the development of nanostructured membranes with high performance in desalination and wastewater treatment. The introduction of 2D nanomaterials manipulates the morphology, stability, surface hydrophilicity, charge, and roughness of a membrane [12]. Meanwhile, the obtained membranes have the potential to balance the trade-off between thermal and physicochemical stability, permeability and selectivity, while mitigating membrane fouling, which has been the bottleneck of membrane processes. In this part, three main families of 2D materials widely used in nanostructured membranes will be discussed.

2.4.1 Graphene Derivative Nanomaterials

Graphene is a 2D single-layered nanomaterial consisting of sp^2-related carbon atoms arranged in a hexagonal honeycomb lattice structure. Since its discovery in 2004, several synthesis methods have been proposed, which were mainly classified into two routes, i.e., top-down methods (such as micromechanical cleavage [70] and liquid exfoliation [71]) or bottom-up methods (such as chemical vapor deposition (CVD) [72] and wet-chemical synthesis [73]).

Graphene finds its application in nanostructured membranes due to its better properties compared to other materials. For instance, compared with polyamide for preparing thin film membrane, graphene is thinner with better selectivity and higher porosity, and a better mechanical, and chemical resistance [74]. Despite these advantages, the application of graphene in membrane separation is constrained by the limited potential for large scale production. Carbon nanosheets of GO and reduced graphene oxide (rGO) have received increasing attention due to their potentials for enhancing membrane performances and easy scale-up [75]. Besides the basic structure of graphene, GO contains a large variety of functional groups created during the synthesis process, such as hydroxyl and epoxy on the basal plane and carboxyl groups at the edges [76]. These functional groups facilitate the adjustment of the adjacent interlayer distance of the stacked GO sheets by different methods so that the stability, the interaction with the charged particles or the channels size of the laminated GO membrane were modified. Moreover, the oxygen groups of a GO nanosheet ensure the interlayers hydrogen bonding interactions that play an important role in assembling laminated GO membranes [77].

Figure 4 illustrates the investigations about tuning interlayer spacing GO sheets to create laminates GO and change the surface charge of modified GO based membranes [78]. By intercalating various nanoparticles, such as TiO_2, SiO_2 and Si_3N_4, GO based composite nanomaterials have been developed to design novel GO based nanostructured membranes with high rejection against specific water pollutant and desired water flux. For example, an intercalated GO base membrane with Si_3N_4 nanoparticles provided approximatively a doubled water flux compared with a pristine membrane, while maintaining a high rejection of methylene blue and methyl

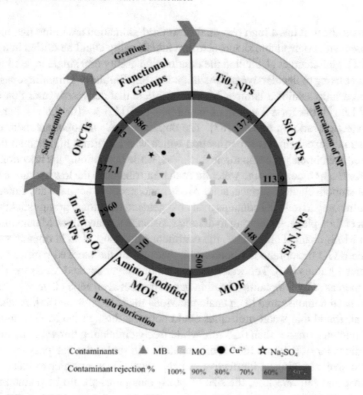

Fig. 4 Effect of GO intercalation with different nanoparticles and grafting of functional groups on membrane's performances (contaminant rejection % and water flux L/m² h MPa)

orange of 88.9% and 99.3%, respectively [79]. Moreover, rGO nanosheets were doped with oxidized CNTs to enlarge the GO interlayer spacing using layer-by-layer self-assembly technique, which instantly increased the water permeability to 5.65 L/m² h bar with a Na_2SO_4 rejection of 80.0% [80].

2.4.2 TMDs Materials

2D nanomaterials with semiconducting character, such as TMDs, were also received extraordinary research attention due to their electrical, physicochemical, mechanical, and biological properties [68], which make them promising candidates to fabricate TMDs based membranes in water purification.

TMDs are a kind of 2D nanoporous materials of three-atom thickness with the formula MX_2, where M is a transition metal atom (like Mo or W) and X is the chalcogen atom (mainly S, Se or Te); showing a 'sandwich' structure with two chalcogen atomic layers separated by a transition metal atomic layer [68]. Molybdenum disulfide (MoS_2) is one of the most extensively studied TMDs materials in membrane separation. Although there exist few in-plane nanopores within a single

MoS$_2$ nanosheet, it has a high potential in water desalination assuming that in-plan nanopores with controllable shape, size and functionality could be drilled in a large scale [81]. Heiranian et al. studied the desalination ability of a single layered MoS$_2$ membrane using molecular dynamics simulation [82]. Similarly, Hirunpinyopas et al. prepared a functionalized laminar MoS$_2$ membrane that delivered water flux of 40 $\times$ 10^{-3} L/m^2 h bar for a 3 μm-thick membrane and 11.6 $\times$ 10^{-3} L/m^2 h bar, with a salt rejection ratio up to 97% [83]. The obtained MoS$_2$ nanoporous membranes demonstrated excellent water permeation which was 2–5 times higher than that of GO based membranes as comparable thickness, while maintaining the rejection ratio (89%) for Evans blue. The high water flux can be attributed to the low hydraulic resistance of smooth channel surface, since MoS$_2$ nanosheets do not have any functional groups. In comparison, GO contains many oxygenated functional groups sticking out from its carbon plane, generating hydraulic resistance to water flow. Meanwhile, the smooth MoS$_2$ channel also allows the continuous transport of light organic vapors, to which a GO membrane is impermeable because of the blocked pathway by its oxygenated groups [84]. The water permeance of a layer stacked MoS$_2$ membrane can be increased by templating ultrathin nanowires between MoS$_2$ layers, a strategy that has been demonstrated in an analogue MoS$_2$ made membrane [85]. A study by Sun et al. found that water molecules could not pass through the MoS$_2$ nanopores with a diameter smaller than 0.23 nm, while freely circulating throw the nanopores with a diameter of 0.44 nm [86]. Unfortunately, salt molecules can pass through at this pore size, resulting in ineffective salt rejection. Thus, in order to balance the water flux and salt rejection, the size of MoS$_2$ nanopores should be maintained in the range of 0.23 nm.

2.4.3   2D MXene Nanosheets

MAX phases are a new class of ternary nanolayered, hexagonal ceramics with the general formula M$_{n+1}$AX$_n$, where: M is an early transition metal, A is a group of element mainly Al, Ga, In, Tl, Si, Ge, Sn and Pb, X is carbon or nitrogen with $n = 1$–3. These materials are attracting increasing scientific attention as candidate materials for diverse applications [87]. They have a unique combination of metallic and ceramic properties, which has advantages in low density, good machinability, high strength and high Young's modulus with excellent thermal and chemical resistance [88]. The excitement about the mass production of 2D materials from 3D layered one using exfoliation and sonication, was shared by the finding of a new family derived from MAX phases called MXene with the general formula M$_{n+1}$X$_n$T$_z$ (T is the functional group edging the surface, such as O^{2-}, OH$^-$, F$^-$, NH$_3$, NH^{4+}). Till now, over 70 compounds of MXene were reported in both experimental syntheses and numerical simulation, which has been shown in Fig. 5 [89].

Due to its excellent exfoliation capability, Ti$_3$C$_2$T$_x$ is the most reported MXene material for designing nanostructured membranes. Gogotsi et al. prepared Ti$_3$C$_2$T$_x$ MXene based membranes via assembling freestanding or supported membranes for

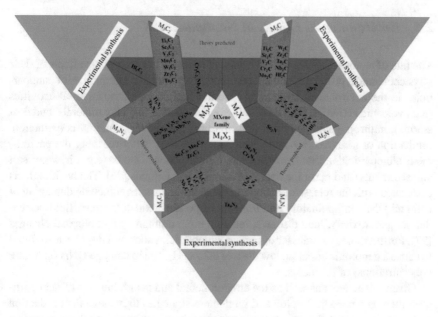

Fig. 5 Schematic classification of MXene compounds with examples

the charge- and size-selective rejection [90]. The micrometer-thick MXene membranes was found to have an ultrafast water flux, which was attributed to the hydrophilic nature and the presence of H_2O molecule layers between the wet $Ti_3C_2T_x$ layers. It also showed a high selectivity towards metal and dye cations with different charge and size. Another nano-sized MXene sheets based membrane was prepared by Wang et al. who intercalated Fe_3O_4 nanoparticles followed by mild acid etching as pillars to achieve larger interlayer distance and create more nanochannels [91]. These nanochannels exhibited a water flux of over 10,000 L/m² h MPa while achieving a 90% rejection rate for particles larger than 2.5 nm.

Recently, for further improving the performance of MXene based membranes, composite MXene based nanomaterials, such as $Ti_3C_2T_x$-GO [92] and $Ti_3C_2T_x$-PEI [93], were introduced to overcome defects in the structure and selectivity of nanochannels for pure MXene membranes. The post-modification of MXene materials was also explored, by grafting multiple functional groups including $-NH_2$ and $-COOR$ [94]. Moreover, improving the hydrophilicity of MXene based nanomaterials may extend their applications in water treatment.

2.5 Environmental Effects of Nanomaterials

The fate of nanoparticles in the environment relies on the synergistic effects of their physicochemical properties and interactions with pollutants. The sources of nanoparticles in the environment include, but are not limited to, various natural activities (e.g., volcanic activities, veld fires, soil erosion, weathering, clay minerals, and dust storms); anthropogenic activities (e.g., burning fossil fuels, mining, construction, production of nanoparticles and waste materials) [95]. After reaching the environment, nanoparticles accumulate in different environmental media (e.g., air, water, soil and sediments) and harm the health of humans and animals [96]. The health effects associated with the release of nanoparticles to the environment include damaging of cell and DNA, inflammatory and immune responses, oxidative stress, lipid peroxidation, genotoxicity, lung diseases, inflammation, pulmonary pathological changes [97]. Furthermore, the transfer of certain nanoparticles such as TiO_2 was also found to cause a genotoxic effect (at low dose of 0.25 mM), and to damage DNA (at higher concentrations) of the plants.

Given the benefits as well as the environmental and health impacts of nanoparticles, there is a need to develop and implement strategies to manage the production and use of nanoparticles in a manner that would prevent or reduce their impact on human health and the environment. To this effect, Patil et al. suggested the use of innovative greener routes for nanoparticle synthesis; advanced engineering ways for manufacturing smarter and more degradable nanoparticles; and governing local and international legislation to monitor nanoparticles released into the environment [98].

3 Rational Design Strategies

3.1 Blending

Blending is the classical membrane elaboration method which was mainly applied in fabricating nanomaterials based mixed matrix membranes (MMMs), using either organic or organic-inorganic materials. A large number of studies has been dedicated to investigating the effect of blending polymeric materials with mineral nanoparticles including silica, ZrO_2 and TiO_2 etc. [26], which have been considered as convenient operation with mild conditions and stable performances. Besides, blending newly emerging nanomaterials, such as GO, MOFs, COFs, and 2D materials, into membrane matrix were also intensively investigated [10–12, 75]. The major inconvenience of the blending method was the absence of evenly dispersed inorganic interconnected networks compared with other methods such as the sol-gel process. In the area of nanostructured membranes, the blending procedure is assisted by phase inversion. Typically, nanomaterials were directly dispersed into a polymer dope (e.g., PSF, PES, PVDF) or in situ formed within the dope in the presence of solvent [26]. The blending procedure in presence of solvent conventionally involves in three main steps, (i) mixing the nanomaterials and the polymer with a solvent, (ii) casting the mixed solution on a porous support, (iii) drying and thermal treatment to eliminate the solvent.

The blended nanomaterials influence the pore size and pore size distribution of MMMs. Meanwhile, the performances of MMMs depend upon the size and loading rate of blended nanomaterials, as well as their interaction with the polymer matrix. It has been reported that nanomaterials blended UF membrane display a higher permeability and hydrophilicity, improved photocatalytic activity, antifouling and antimicrobial properties, and lower compaction under pressure [26].

3.2 Interfacial Polymerization

The interfacial polymerization process is a classical method, where two monomers react on the surface of the porous support. The developed membranes to date are integrally composed of an ultrathin polyamide (PA) layer containing functional nanomaterials and a porous polymer support [99]. The addition of these nanomaterials to the barrier layer endows the membranes with various attractive properties such as high permeability, increased mechanical strength, desirable contaminant selectivity, and antibacterial ability etc. Taking silica as an example, a representative fabrication step for silica nanoparticles based thin film RO membrane was reported by Peyki et al. [100]. The substrate was dipped in an aqueous amine solution containing different contents of silica nanoparticles. Then the membrane was immersed into the hexane solution of acid chloride, which formed a PA thin film with silica nanoparticles. In another study, NF membrane active layers of polyimine COF were synthesized via the interfacial polymerization of terephthalaldehyde and tris (4-aminophenyl) benzene monomers on top of a PES UF membrane support [101]. The rejection efficiencies of the COF NF membrane for a model organic compound, Rhodamine WT, and a background electrolyte, NaCl, were higher than those of the PES support without the COF film. The changes in properties of nanocomposite membranes are strongly influenced by the chemical properties, type, size and concentration of the nanomaterial used. Based on rational design strategies, intentionally selected nanomaterials were employed to modify PA thin film composite (TFC) membranes for enhancing forward osmosis performances. A comprehensive review summarized the recent advances of nanostructured TFC membrane in this area [102].

The control of the molecular and structural characteristics of the PA layer are of great importance in improving the permeance and selectivity. Porous nanomaterials can be ideal fillers embedded in the PA layer for preparing TFN membranes, which behave like water channels to facilitate water permeability while blocking hydrated cations, leading to an enhanced permeance without adversely affecting the selectivity. However, a significant change of water transport properties in the PA-based active layers is difficult to achieve because water transport is heavily restricted by the highly cross-linked polymer chains in a tortuous manner. Recently, Wang et al. fabricated a TFC NF membrane with a crumpled PA layer, which provided a simple avenue to overcome this challenge [103]. Interfacial polymerization was performed in a single-walled carbon nanotubes/polyether sulfone composited support loaded with ZIF-8 as a sacrificial templating material. ZIF-8 can be removed by water dissolution, which facilitates the formation of a rough PA active layer with crumpled nanostructure. The resultant membrane exhibits an unprecedented permeance up to 53.2 L/m^2 h bar while maintaining a high rejection of 95.2% for Na$_2$SO$_4$.

3.3 In Situ Preparation

In situ preparation has been considered a promising strategy to fabricate next-generation nanostructured membranes with rational design. Under this in situ design platform, the functionalized elaboration for nanomaterials and sophisticated modification for membrane matrix/surface with nanomaterials can be judiciously merged, which farthest exploits the advantages of nanomaterials in disinfection, adsorption and catalytic degradation for water treatment. Based on the general preparation process for membranes, in situ formation of nanomaterials takes place (i) within the casting solution, (ii) during the solidification of casting solution, or (iii) after the solidification of the casting solution [56].

In the past two decades, two kinds of main in situ design strategies have been developed: (i) sol-gel process. Through the controlled hydrolysis and polycondensation reaction of the inorganic precursors, accompanying the polymer solidification, a nanostructured membrane is obtained during the in situ generation of metal oxide nanoparticles on/within the membrane matrix. For example, the classical in situ preparation of TiO$_2$ based nanostructured membranes is moderately controlling the synchronous formation of TiO$_2$ nanoparticles (derived from the sol-gel process of titanium (IV) butoxide [104] or titanium tetraisopropoxide [105]) and a polymeric membrane. Furthermore, SiO$_2$ [106] and ZnO [107] nanoparticles were also introduced into membrane via the in situ sol-gel process for water treatment. Recently, biomimetic mineralization opens a new avenue to design nanostructured membranes with complex structures [108]. The mineralization inducers act as catalysts to promote the in situ nucleation and growth of inorganics within the confined spaces among polymer chains, which favor the uniform dispersion of inorganics in a molecule level. (ii) In situ reduction. Inorganic precursors are reduced by a reducing agent; metal nanoparticles are in situ synthesized in the casting solution or on the membrane surface/pore. The fabrication strategy was attractive particularly for preparing membranes functionalized with Ag [109], Fe [34], Cu [30] nanoparticles. Two methods have been reported to fabricate nanostructured membranes via in situ reduction [56]. The classical method is the 'generating before' method, in which the precursor is in situ reduced in the casting solution before the formation of the membrane matrix. The 'generating after' method allows the in situ reduction and localized loading of nanoparticles on the membrane exterior surface or membrane pore.

3.4 Surface Modification

It is well known that surfaces/interfaces play a crucial role in the fabrication and application of nanostructured membranes in water purification owing to its simplicity and sustainability [110]. Compared with blending, surface modification minimized the effects of nanomaterials on the internal structure of membranes, which offers opportunities for modification of commercial membranes. In addition, the functionality of nanomaterials could be fully utilized.

A simple method is to modify the membrane surface with active moieties to 'capture' nanomaterials from suspension via physical interactions. Some researchers grafted hydroxyl or sulfonic acid groups onto the membrane surfaces to capture TiO_2 nanoparticles [111, 112]. To further improve the nanomaterials coverage and their stability on the membrane surface, strong interactions, such as electrostatic interaction, hydrogen bonding, coordination, or covalent bonding, were applied between the membrane and nanomaterials. For example, Choi et al. attempted to modify the TFC membrane surface by depositing multilayers of oppositely charged GO and aminated GO nanosheets using layer-by-layer technique [113]. Similarly, to develop a low-fouling antibacterial membrane, azide-functionalized GO was covalently anchored onto commercial flat sheet membrane surfaces via azide photochemistry [23]. Alternatively, in another development, a commercial RO membrane was functionalized by a spray- and spin-assisted layer-by-layer (SSLbL) method through the alternate coating of polyethyleneimine modified Cu nanoparticles and poly(acrylic) acid [29]. Cu nanoparticles can be potentially regenerated on the membrane surface via the same SSLbL method. Besides, as introduced in Sect. 3.3, biomimetic mineralization, defined as a mineral formation process regulated by bio-macromolecules, was also involved for sophisticated surface modification of membranes. The mineralized membranes are composed of three major layers: polymer membrane (as the skeleton), intermediate layer (as the connection, e.g., polydopamine/polyethyleneimine intermediate layer) and mineral coating (e.g., $CaCO_3$, SiO_2, ZrO_2 nanoparticles). The relevant progress on fabrications of nanostructured membrane for water purification have been extensively reviewed by other researchers [9].

Polymer/nanomaterials composites were considered as promising candidate for membrane surface modification in the past few years. Especially in the field of membrane fouling, these composites imparted synergistic effects of nanoparticles and polymers to membrane surfaces, where nanoparticles and polymers provide microbial and fouling resistant properties, respectively. Recently, a hybrid of GO-poly L-lysine (GO/PLL-H) was covalently bonded to the PA layer of TFC FO membranes [114]. The membrane surface grafted with GO/PLL-H had improved surface morphology, smoothness, antibacterial and hydrophilic properties.

3.5 3D Printing

Three-dimensional   (3D)   printing   or   additive   manufacturing
is a game changer in manufacturing space due to its capability to design complex customized products. Specifically, this technology presented the capability to
design advanced materials, such as nanocomposites by introducing nanoparticles
into the matrix [115]. It is different from conventional manufacturing in that it is
the closest to the 'bottom up' manufacturing where a structure can be built into its
designed shape using a 'layer-by-layer' approach rather than casting or forming
by forging or machining. 3D printing is versatile, flexible, highly customizable
and, as such, can suit most sectors of industrial production. Again, a wide range
of raw materials can be used, including metallic, ceramic and polymeric materials
along with combinations in the form of composites, hybrid, or functionally graded
materials [116]. Generally, all the 3D printing methods follow a basic process of
additive manufacturing [115]: (1) 3D modeling through CAD software, 3D scanner,
or a photogrammetry procedure; (2) digitalization of the 3D model by converting it
into a STL file; (3) converting the STL file data into a G-code file which contains the
geometrical information of each 2D layer slicing from the 3D model; (4) printing
the materials in a layer-by-layer manner.

3D printing has been applied in various fields, including membranes. For instance,
3D printing was used by Thomas et al. to create different feed channel spacer designs
aimed at enhancing the spacer performance specifically for membrane distillation
(MD) application [117]. The triply periodic minimal surfaces (TPMS) were used
as feed spacers. The best performing TPMS spacer topology exhibited 60% higher
water flux and 63% higher overall film heat transfer coefficient than the commercial
spacer. Al-Shimmery et al. fabricated 3D printed composite membranes by depositing a thin PES selective layer onto ABS-like 3D printed flat and wavy structured
supports [118]. The resultant flat and wavy composite membranes were tested in
terms of permeance, rejection, and cleanability by filtering oil-in-water emulsions.
Results showed that the pure water permeance through a wavy membrane is 30%
higher than of a flat membrane. The wavy 3D printed membrane had a 52% higher
permeance recovery ratio compared to the flat one after the first filtration cycle,
with both membranes having an oil rejection of 96% ± 3% [118]. Besides, Tsai
et al. designed three turbulence promoters with different configurations (circular,
diamond and elliptic) using 3D printing technology for cross-flow MF [119]. The
elliptic type of promoter with a hydraulic angle of 90° displays the flux enhancement by approximately 30–64% under 20 kPa compared to the normal type of MF,
whereas the diamond type of promoter with a hydraulic angle of 60° shows a lower
flux enhancement by approximately 7–16%.

Although the applications of 3D printing in membrane technology are increasing, some challenges still need attention, including void formation, poor adhesion of
fillers and matrix, blockage due to filler inclusion, increased curing time, and limited

resolution [120], production of high performance membranes (e.g., highly perms-elective membranes). These could be addressed by conducting extensive research using both experimental and simulation approaches.

4 Applications of Nanostructured Membrane for Water Purification

4.1 Desalination

Desalination has been touted as one of the promising technologies that could provide a solution for global water crisis. It offers effective removal of minerals and salts from brackish or seawater to achieve pure water for industrial, domestic, agriculture and human consumption [121]. The Global Water Intelligence and International Desalination Association revealed that the installed desalination plants around the world have the capacity to produce more than 74.8 billion liter freshwater per day and there is an increase in demand for membrane-based desalination market throughout the globe due its advantage and advancement in the production of low-cost pure water.

With the progress of membranes in desalination, nanostructured membranes have been an area of interest for many researchers to further improve the desalination performances. For instance, Shi et al. prepared GO-cellulose acetate (CA) nanocomposite membranes for high-flux desalination [122]. The GO-CA membrane with 0.01 wt.% GO showed the optimal water flux (16.82 L/m² h) due to its best hydrophilicity and highly porous structure. Similarly, the effect of the loading rate of candle soot (CS) nanoparticles on the desalination process using CA/polyethylene glycol (PEG) membranes was investigated [123]. The results indicated that CS nanoparticles contributed in improving the NaCl rejection with a slight reduction in the water permeability. Perera et al. fabricated a novel TFN membrane by interfacial polymerization to enhance FO membrane desalination performance [124]. They incorporated fullerenol ($C_{60}(OH)_n$, $n = 24$–28) nanomaterials into the active layer. The TFN membrane with a fullerenol loading of 400 ppm exhibited a water flux of 26.1 L/m² h, which is 83.03% higher than that of TFC membrane with a specific reverse salt flux of 0.18 g/L using 1 M NaCl draw solution against deionized water in FO mode. The incorporation of fullerenol also contributed to a decreased fouling propensity. Besides, Nemati et al. examined the desalination and heavy metal removal abilities using $NiFe_2O_4$/hydrogel based on 2-acrylamido-2-methyl propane sulfonic acid (HAMPS) nanocomposite modified ion exchange membrane [125].

Given their unique structural and morphological features, nanomaterials have gained considerable attention for their applications in membrane desalination. The progress in enhancing water flux and salt rejection for desalination has been witnessed and well-reviewed [126, 127]. Future studies should be directed at more long-term performance data using real feed solutions.

4.2 Contaminant Removal

In the last two decades, the advancement in membrane technology permitted the development of a variety of nanostructured membranes to effectively remove multiple water contaminants ranging in size and nature from yeast and virus to humic acids and salts. Table 1 gives some examples on recently reported nanostructured membranes (mainly in NF and RO process) elaborated for contaminants removal. Contaminants can be classified in four main classes: (i) emergent contaminants and pharmaceutical compounds that can be eliminated by TFC membrane, (ii) textile dyes, (iii) metal ions, and (iv) salts. Overall, the ability of nanostructured membranes for the removal of contaminants has been well confirmed.

Table 1 Nanomaterial based membrane preparation methods and their contaminant/solute removal efficiency

Nanomaterial	Preparation method	Applied process	Contaminant	Efficiency (%)	Ref.
Polyamide thin-film composite	Commercial	RO	Caffeine, theobromine, theophylline, amoxicillin, and penicillin G	100	[128]
PVDF/APTES-functionalized halloysites	Blending	NF	Dye, Cr (VI)	94.9, 52.3	[129]
Ultrathin GO framework	Layer-by-layer self-assembly	NF	Pb (II)	95	[130]
SAPO-34 nanoparticles	Wet phase inversion	NF	Methyl violet 6B, reactive blue 4, and acid blue 193	100	[131]
Multiple hierarchic structures on porous polypropylene	Surface modification	NF	Methylene blue, nano sized oil droplet	$\leq$95, $\leq$99	[132]
rGO–CNT-AAO	Vacuum assisted filtration method, surface modification	NF	Na_2SO_4, NaCl	84, 42	[133]
$Ti_3C_2T_x$ (MXene)	Vacuum assisted filtration method	NF	MB+	~99	[92]
GO-TFC	IP	RO	NaCl	99	[134]

4.3   Fouling Mitigation

The main issue of membrane processes in water purification is its high-energy consumption derived from fouling in long term operation [26]. Membrane fouling occurs during an increase in trans membrane pressure to maintain a particular flux or during a decrease in flux when the system is operated at constant pressure. Membrane fouling can be classified as reversible fouling and irreversible fouling, which was mainly determined by pore size, surface charge, roughness, and hydrophilicity/hydrophobicity of membranes [135].

Because of the adverse effects that fouling has on membrane processes, the literature on developing nanostructured membranes in improving fouling resistance is vast, and readers are referred to comprehensive reviews [26, 136]. Incorporating metal oxide nanoparticles by blending or surface modification is one of the most promising and versatile antifouling strategies for nanostructured membranes. This favorable effect was attributed to the hydrophilic nature of most metal oxide nanoparticles. For example, the self-assembly of TiO_2 nanoparticles on a PES UF membrane via a synergistic analysis method for enhancing the antifouling behavior was demonstrated [137]. Li et al. studied the synthesis and incorporation of hollow mesoporous SiO_2 spheres (HMSS) to improve the antifouling property of PES membrane [138]. Especially, the irreversible fouling resistance was improved for a membrane with 1.5% HMSS compared with the pristine PES membrane using bovine serum albumin as the model foulant. In addition, a significant number of research papers indicates that membranes incorporated with GO show enhanced hydrophilic performance and improved water flux [139]. Lee et al. reported that a GO membrane had a high water permeability and antibiofouling capability due to its hydrophilicity and electrostatic repulsion characteristics [140]. Similarly, the negatively charged GO-incorporated nanocomposite membranes were responsible for a gradual decline in the adsorbed amount of humic acid [141].

Nanomaterials such as zeolite, multiwalled CNTs and TiO_2 nanoparticles have also been incorporated into the PA layer to enhance the membrane anti-scaling properties [142]. For instance, Zhao et al. incorporated multiwalled CNTs into PA polymer membrane [143]. The flux decline of the PA/multiwalled CNTs membrane was remarkably lower than that of neat PA membrane, demonstrating that the PA/MWNTs membrane could effectively reduce inorganic fouling. The improved antifouling property against $Ca(HCO_3)_2$ was mainly attributed to the enhanced electrostatic repulsion between charged membrane surface and the inorganic ions. The excitement and great benefits harnessing from the nanomaterials on antifouling ability of TFC membrane have spurred great interest in the desalination industry. Several key technical constraints ranging from economic feasibility to the synthesis process have to be addressed. Furthermore, research needs to continue to discover alternative means for fabricating such membranes on a large scale.

5 Technological Capacity and Economic Impact of Nanostructured Membranes

Figure 6 displays a graphic classification of the most used membrane technologies for aquatic depollution and water treatment based on pore size, operating pressure, contaminant rejection, and process cost investment. Meanwhile, the respective rate of nanocomposite membrane is illustrated. The nanostructured membrane applied in UF process represents 45% of total nanostructured membranes whereas their application on NF and RO processes represents 32% and 6%, respectively.

The incorporation of nanomaterials in membranes improved their water permeability and contaminant removal ability. The current state-of-the-art of hydrophilic commercial and hydrophobic laboratory membranes performances has been the research focus [144]. All reported data agree that commercial hydrophilic dense membranes were mainly applied in RO process with relatively low permeability ranging from 0.01 to 0.12 L/m^2 h bar, while exhibiting high selectivity toward pollutants. Conversely, laboratory-made nanostructured membranes showed an excellent permeability with relatively low contaminant rejection/selectivity.

Figure 7 compares the commercial and laboratory nanostructured membranes according to their technical capacities (pollutant removal rate and water permeability). Commercial nanostructured membranes from different manufacturers (e.g., DOW, Microdyn Nadir and Trisep) exhibited the highest pollutant removal ability (>99%) with relatively low water permeability [145]. Taking MXene nanomaterial as example, the laboratory made MXene based membranes presented an unpreceded

Fig. 6 Schematic classification of nanostructured membranes in water treatment

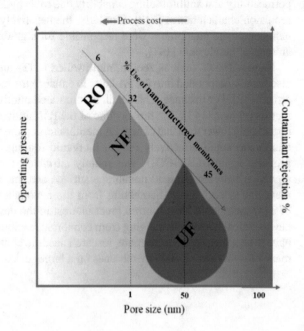

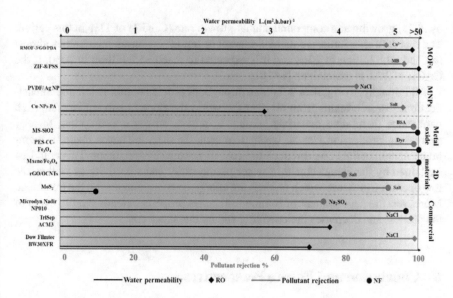

Fig. 7 Technical capacities of commercial and lab-made nanostructured membranes

ability for water permeation [91]. Correspondingly, metal/metal oxide nanoparticles based nanostructured membranes showed a limited improvement in water permeability and fouling resistance, owing to the lack of functional groups on their surface. In this issue, Li et al. reported the incorporation of carboxylated nanodiamonds into PVDF membrane during phase inversion [145]. Carboxylate group minimized the aggregation of nanomaterials compared with the raw nanodiamonds, at the same time improving the porosity and surface pore size, water permeability and hydrophilicity, and antifouling ability of obtained membrane.

For a membrane based configuration, three main parameters are considered to be crucial for an economic analysis, (i) cost investment of general installations (e.g., pumps, pipe, control and monitoring devices, energy consumption and efficiency), (ii) physical and chemical properties of the feed (i.e., salinity, pollutant concentration, feed flowrate, and target quality of treated water according to the final reuse purpose), (iii) membrane performances (e.g., water flux, pollutant hydraulic permeability, structural properties, and cost). Today, RO and NF membranes are the leading technologies for desalination installations and water treatment plants. However, current commercial membranes performances are less than optimal, resulting in inefficient and energy-intensive separation processes.

The economic impact and cost assessment of membrane based processes have been analyzed in many papers. They focused on those membrane based separation processes, especially NF and RO, which continue to be used worldwide for water treatment, industrial and municipal wastewater reuse, and desalination [146]. In 2017, BCC Research claimed that the global market for RO system components for water treatment reached nearly $6.6 billion in 2016 and it should reach over $11.0 billion by 2021, growing at a compound annual growth rate (CAGR) of 11% in this period. The report further indicated that US market of liquid separation membranes almost reached \$3.5 billion in 2018 and it is expected to reach \$5.0 billion in 2030, if the CAGR of 30% is maintained [147].

Additionally, to engineer solutions throughout process combination and intensification, research teams from all over the world developed nano-engineered, high permeance membrane materials, aiming to reduce energy consumption, thereby decreasing production cost in water treatment. To date, advances made in membrane nanomaterials could reduce the cost of wastewater treatment. This could be explained by the enhancement of water permeability and selectivity as well as the low operation pressure which reduces the energy consumption during the treatment. Moreover, the development of fouling mitigation property extends the life cycle of the process and decreases the operating and maintenance cost.

6 Conclusion and Future Perspectives

In the present chapter, various nanomaterials used to design nanostructured membrane for water purification are reviewed. 2D materials and MOFs/COFs with tunable surface chemistry are qualified to optimize membrane properties or add additional functionalities. A special emphasis has been given to the most adopted preparation techniques of nanostructured membranes, along with their applications in water purification. Specifically, 3D printing, expected to have a significant future impact on designing nanostructured membranes, is highlighted.

The toxicity of nanomaterials was also reviewed with a brief discussion of their economic impact on water purification. Nanostructured membranes have huge potential in developing next-generation membrane technology by removing microorganisms, heavy metal, salt ions, organic compounds, and emerging micropollutants. They would have leapfrogging potential for futurist multifunctional membranes. Nevertheless, several challenges for industrial application still exist, including (i) difficulty/cost for large scale (ii) recyclability/reusability of the membranes at the end of cycle, (iii) lack of comprehension on their behavior during the treatment of complex municipal and industrial effluents. Meanwhile, a continue working is necessary on understanding and improving the surface physicochemical properties of nanostructured membrane, such as hydrophilicity and surface charge, to gain wider social and economic acceptability.

References

1. Progress on Safe Treatment and Use of Wastewater: Piloting the Monitoring Methodology and Initial Findings for SDG Indicator 6.3.1. World Health Organization and UN-HABITAT, Geneva (2018)
2. Saraswathi, S.A., Sundaram, M., Nagendran, A., Rana, D.: Tailored polymer nanocomposite membranes based on carbon, metal oxide and silicon nanomaterials: a review. J. Mater. Chem. A **7**(15), 8723–8745 (2019)
3. Fane, A.G., Wang, R., Hu, M.X.: Synthetic membranes for water purification: status and future. Angew. Chem. Int. Ed. **54**(11), 3368–3386 (2015)
4. Werber, J.R., Osuji, C.O., Elimelech, M.: Materials for next-generation desalination and water purification membranes. Nat. Rev. Mater. 16018 (2016)
5. Alvarez, P.J., Chan, C.K., Elimelech, M., Halas, N.J., Villagrán, D.: Emerging opportunities for nanotechnology to enhance water security. Nat. Nanotechnol. **13**(8), 634 (2018)
6. Pendergast, M.M., Hoek, E.M.V.: A review of water treatment membrane nanotechnologies. Energy Environ. Sci. **4**(6), 1946–1971 (2011)
7. Qu, X., Alvarez, P.J.J., Li, Q.: Applications of nanotechnology in water and wastewater treatment. Water Res. **47**(12), 3931–3946 (2013)
8. Li, R., Zhang, L., Wang, P.: Rational design of nanomaterials for water treatment. Nanoscale **7**(41), 17167–17194 (2015)
9. Yang, H.-C., Hou, J., Chen, V., Xu, Z.-K.: Surface and interface engineering for organic-inorganic composite membranes. J. Mater. Chem. A **4**, 9716–9729 (2016)
10. Li, X., Liu, Y., Wang, J., Gascon, J., Li, J., Van der Bruggen, B.: Metal-organic frameworks based membranes for liquid separation. Chem. Soc. Rev. **46**(23), 7124–7144 (2017)
11. Yuan, S., Li, X., Zhu, J., Zhang, G., Van Puyvelde, P., Van der Bruggen, B.: Covalent organic frameworks for membrane separation. Chem. Soc. Rev. **48**(10), 2665–2681 (2019)
12. Zhu, J., Hou, J., Uliana, A., Zhang, Y., Tian, M., Van der Bruggen, B.: The rapid emergence of two-dimensional nanomaterials for high-performance separation membranes. J. Mater. Chem. A **6**(9), 3773–3792 (2018)
13. Chernousova, S., Epple, M.: Silver as antibacterial agent: ion, nanoparticle, and metal. Angew. Chem. Int. Ed. **52**(6), 1636–1653 (2013)
14. Wang, Z., Xia, T., Liu, S.: Mechanisms of nanosilver-induced toxicological effects: more attention should be paid to its sublethal effects. Nanoscale **7**(17), 7470–7481 (2015)
15. Zodrow, K., Brunet, L., Mahendra, S., Li, D., Zhang, A., Li, Q., Alvarez, P.J.J.: Polysulfone ultrafiltration membranes impregnated with silver nanoparticles show improved biofouling resistance and virus removal. Water Res. **43**(3), 715–723 (2009)
16. Zhang, M., Zhang, K., De Gusseme, B., Verstraete, W.: Biogenic silver nanoparticles (bio-Ag^0) decrease biofouling of bio-Ag^0/PES nanocomposite membranes. Water Res. **46**(7), 2077–2087 (2012)
17. Mecha, C., Pillay, V.L.: Development and evaluation of woven fabric microfiltration membranes impregnated with silver nanoparticles for potable water treatment. J. Membr. Sci. **458**, 149–156 (2014)
18. Zhu, J., Hou, J., Zhang, Y., Tian, M., He, T., Liu, J., Chen, V.: Polymeric antimicrobial membranes enabled by nanomaterials for water treatment. J. Membr. Sci. **550**, 173–197 (2018)
19. Mukherjee, M., De, S.: Antibacterial polymeric membranes: a short review. Environ Sci: Water Res. Technol. **4**(8), 1078–1104 (2018)
20. Huang, L., Zhao, S., Wang, Z., Wu, J., Wang, J., Wang, S.: In situ immobilization of silver nanoparticles for improving permeability, antifouling and anti-bacterial properties of ultrafiltration membrane. J. Membr. Sci. **499**, 269–281 (2016)
21. Dong, C., Wang, Z., Wu, J., Wang, Y., Wang, J., Wang, S.: A green strategy to immobilize silver nanoparticles onto reverse osmosis membrane for enhanced anti-biofouling property. Desalination **401**, 32–41 (2017)

22. Akhavan, O., Abdolahad, M., Abdi, Y., Mohajerzadeh, S.: Silver nanoparticles within verti-
    cally aligned multi-wall carbon nanotubes with open tips for antibacterial purposes. J. Mater.
    Chem. **21**(2), 387–393 (2011)
23. Huang, X., Marsh, K.L., McVerry, B.T., Hoek, E.M., Kaner, R.B.: Low-fouling antibacte-
    rial reverse osmosis membranes via surface grafting of graphene oxide. ACS Appl. Mater.
    Interface **8**(23), 14334–14338 (2016)
24. Habib, Z., Khan, S.J., Ahmad, N.M., Shahzad, H.M.A., Jamal, Y., Hashmi, I.: Anti-
    bacterial behavior of surface modified composite polyamide nanofiltration (NF) membrane
    by immobilizing Ag doped TiO_2 nanoparticles. Environ. Technol. 1–48 (2019, in press)
25. Rizzello, L., Pompa, P.P.: Nanosilver-based antibacterial drugs and devices: mechanisms,
    methodological drawbacks, and guidelines. Chem. Soc. Rev. **43**(5), 1501–1518 (2014)
26. Jhaveri, J.H., Murthy, Z.V.P.: A comprehensive review on anti-fouling nanocomposite mem-
    branes for pressure driven membrane separation processes. Desalination **379**, 137–154
    (2016)
27. Tang, S.C., Lo, I.M.: Magnetic nanoparticles: essential factors for sustainable environmental
    applications. Water Res. **47**(8), 2613–2632 (2013)
28. Duan, L., Zhao, Q., Liu, J., Zhang, Y.: Antibacterial behavior of halloysite nanotubes decorated
    with copper nanoparticles in a novel mixed matrix membrane for water purification. Environ.
    Sci.: Water Res. Technol. **1**(6), 874–881 (2015)
29. Ma, W., Soroush, A., Luong, T.V.A., Brennan, G., Rahaman, M.S., Asadishad, B., Tufenkji,
    N.: Spray- and spin-assisted layer-by-layer assembly of copper nanoparticles on thin-film
    composite reverse osmosis membrane for biofouling mitigation. Water Res. **99**, 188–199
    (2016)
30. Zhu, J., Wang, J., Uliana, A.A., Tian, M., Zhang, Y., Zhang, Y., Volodin, A., Simoens, K.,
    Yuan, S., Li, J.: Mussel-inspired architecture of high-flux loose nanofiltration membrane
    functionalized with antibacterial reduced graphene oxide–copper nanocomposites. ACS Appl.
    Mater. Interface **9**(34), 28990–29001 (2017)
31. Simeonidis, K., Mourdikoudis, S., Kaprara, E., Mitrakas, M., Polavarapu, L.: Inorganic engi-
    neered nanoparticles in drinking water treatment: a critical review. Environ. Sci.: Water Res.
    Technol. **2**(1), 43–70 (2016)
32. Lowry, G.V., Johnson, K.M.: Congener-specific dechlorination of dissolved PCBs by
    microscale and nanoscale zerovalent iron in a water/methanol solution. Environ. Sci. Technol.
    38(19), 5208–5216 (2004)
33. Wang, X., Chen, C., Liu, H., Ma, J.: Preparation and characterization of PAA/PVDF
    membrane-immobilized Pd/Fe nanoparticles for dechlorination of trichloroacetic acid. Water
    Res. **42**(18), 4656–4664 (2008)
34. Wan, H., Briot, N.J., Saad, A., Ormsbee, L., Bhattacharyya, D.: Pore functionalized PVDF
    membranes with in-situ synthesized metal nanoparticles: material characterization, and toxic
    organic degradation. J. Membr. Sci. **530**, 147–157 (2017)
35. Bet-Moushoul, E., Mansourpanah, Y., Farhadi, K., Tabatabaei, M.: TiO_2 nanocomposite based
    polymeric membranes: a review on performance improvement for various applications in
    chemical engineering processes. Chem. Eng. J. **283**, 29–46 (2016)
36. Liu, K., Cao, M., Fujishima, A., Jiang, L.: Bio-inspired titanium dioxide materials with special
    wettability and their applications. Chem. Rev. **114**(19), 10044–10094 (2014)
37. Chen, X., Mao, S.S.: Titanium dioxide nanomaterials: synthesis, properties, modifications,
    and applications. Chem. Rev. **107**(7), 2891–2959 (2007)
38. Ngang, H., Ooi, B., Ahmad, A., Lai, S.: Preparation of PVDF-TiO_2 mixed-matrix membrane
    and its evaluation on dye adsorption and UV-cleaning properties. Chem. Eng. J. **197**, 359–367
    (2012)
39. Rajesh, S., Senthilkumar, S., Jayalakshmi, A., Nirmala, M., Ismail, A., Mohan, D.: Prepara-
    tion and performance evaluation of poly (amide-imide) and TiO_2 nanoparticles impregnated
    polysulfone nanofiltration membranes in the removal of humic substances. Colloid Surf. A:
    Physicochem. Eng. Asp. **418**, 92–104 (2013)

40. Wang, Y., Li, Y., Yang, H., Xu, Z-l: Super-wetting, photoactive TiO_2 coating on amino-silane modified PAN nanofiber membranes for high efficient oil-water emulsion separation application. J. Membr. Sci. **580**, 40–48 (2019)
41. Zhang, X., Du, A.J., Lee, P., Sun, D.D., Leckie, J.O.: Grafted multifunctional titanium dioxide nanotube membrane: separation and photodegradation of aquatic pollutant. Appl. Catal. B: Environ. **84**(1–2), 262–267 (2008)
42. Kuvarega, A.T., Khumalo, N., Dlamini, D., Mamba, B.B.: Polysulfone/N, Pd co-doped TiO_2 composite membranes for photocatalytic dye degradation. Sep. Purif. Technol. **191**, 122–133 (2018)
43. Chin, S.S., Chiang, K., Fane, A.G.: The stability of polymeric membranes in a TiO_2 photocatalysis process. J. Membr. Sci. **275**(1–2), 202–211 (2006)
44. Shi, Y., Huang, J., Zeng, G., Cheng, W., Hu, J.: Photocatalytic membrane in water purification: is it stepping closer to be driven by visible light? J. Membr. Sci. **584**, 364–392 (2019)
45. Li, L., Fan, M., Brown, R.C., Van Leeuwen, J., Wang, J., Wang, W., Song, Y., Zhang, P.: Synthesis, properties, and environmental applications of nanoscale iron-based materials: a review. Crit. Rev. Environ. Sci. Technol. **36**(5), 405–431 (2006)
46. Xu, G.-R., Wang, J.-N., Li, C.-J.: Preparation of hierarchically nanofibrous membrane and its high adaptability in hexavalent chromium removal from water. Chem. Eng. J. **198**, 310–317 (2012)
47. Li, C.-J., Li, Y.-J., Wang, J.-N., Cheng, J.: PA6@FexOy nanofibrous membrane preparation and its strong Cr (VI)-removal performance. Chem. Eng. J. **220**, 294–301 (2013)
48. Daraei, P., Madaeni, S.S., Ghaemi, N., Salehi, E., Khadivi, M.A., Moradian, R., Astinchap, B.: Novel polyethersulfone nanocomposite membrane prepared by $PANI/Fe_3O_4$ nanoparticles with enhanced performance for Cu (II) removal from water. J. Membr. Sci. **415**, 250–259 (2012)
49. Abdi, G., Alizadeh, A., Zinadini, S., Moradi, G.: Removal of dye and heavy metal ion using a novel synthetic polyethersulfone nanofiltration membrane modified by magnetic graphene oxide/metformin hybrid. J. Membr. Sci. **552**, 326–335 (2018)
50. Das, D., Nath, B.C., Phukon, P., Dolui, S.K.: Synthesis and evaluation of antioxidant and antibacterial behavior of CuO nanoparticles. Colloids Surf. B **101**, 430–433 (2013)
51. Shalaby, T., Hamad, H., Ibrahim, E., Mahmoud, O., Al-Oufy, A.: Electrospun nanofibers hybrid composites membranes for highly efficient antibacterial activity. Ecotoxicol. Environ. Saf. **162**, 354–364 (2018)
52. Leo, C., Lee, W.C., Ahmad, A., Mohammad, A.W.: Polysulfone membranes blended with ZnO nanoparticles for reducing fouling by oleic acid. Sep. Purif. Technol. **89**, 51–56 (2012)
53. Li, N., Zhang, J., Tian, Y., Zhao, J., Zhang, J., Zuo, W.: Anti-fouling potential evaluation of PVDF membranes modified with ZnO against polysaccharide. Chem. Eng. J. **304**, 165–174 (2016)
54. Ahmad, A., Abdulkarim, A., Shafie, Z.M., Ooi, B.: Fouling evaluation of PES/ZnO mixed matrix hollow fiber membrane. Desalination **403**, 53–63 (2017)
55. Chen, X., Huang, G., An, C., Feng, R., Wu, Y., Huang, C.: Plasma-induced PAA-ZnO coated PVDF membrane for oily wastewater treatment: preparation, optimization, and characterization through Taguchi OA design and synchrotron-based X-ray analysis. J. Membr. Sci. **582**, 70–82 (2019)
56. Li, X., Sotto, A., Li, J., Van der Bruggen, B.: Progress and perspectives for synthesis of sustainable antifouling composite membranes containing in situ generated nanoparticles. J. Membr. Sci. **524**, 502–528 (2017)
57. Ananth, A., Arthanareeswaran, G., Mok, Y.S.: Effects of in situ and ex situ formations of silica nanoparticles on polyethersulfone membranes. Polym. Bull. **71**, 2851–2861 (2014)
58. Yang, X., Sun, H., Pal, A., Bai, Y., Shao, L.: Biomimetic silicification on membrane surface for highly efficient treatments of both oil-in-water emulsion and protein wastewater. ACS Appl. Mater. Interface **10**(35), 29982–29991 (2018)
59. Lv, Y., Yang, H.-C., Liang, H.-Q., Wan, L.-S., Xu, Z.-K.: Novel nanofiltration membrane with ultrathin zirconia film as selective layer. J. Membr. Sci. **500**, 265–271 (2016)

60. Slater, A.G., Cooper, A.I.: Function-led design of new porous materials. Science **348**(6238), 8075 (2015)

61. Li, T., Zhang, W., Zhai, S., Gao, G., Ding, J., Zhang, W., Liu, Y., Zhao, X., Pan, B., Lv, L.: Efficient removal of nickel (II) from high salinity wastewater by a novel PAA/ZIF-8/PVDF hybrid ultrafiltration membrane. Water Res. **143**, 87–98 (2018)

62. Duong, P.H., Kuehl, V.A., Mastorovich, B., Hoberg, J.O., Parkinson, B.A., Li-Oakey, K.D.: Carboxyl-functionalized covalent organic framework as a two-dimensional nanofiller for mixed-matrix ultrafiltration membranes. J. Membr. Sci. **574**, 338–348 (2019)

63. Zhang, P., Gong, J.-L., Zeng, G.-M., Song, B., Liu, H.-Y., Huan, S.-Y., Li, J.: Ultrathin reduced graphene oxide/MOF nanofiltration membrane with improved purification performance at low pressure. Chemosphere **204**, 378–389 (2018)

64. Yang, H., Cheng, X., Cheng, X., Pan, F., Wu, H., Liu, G., Song, Y., Cao, X., Jiang, Z.: Highly water-selective membranes based on hollow covalent organic frameworks with fast transport pathways. J. Membr. Sci. **565**, 331–341 (2018)

65. Shen, J., Liu, G., Huang, K., Jin, W., Lee, K.R., Xu, N.: Membranes with fast and selective gas-transport channels of laminar graphene oxide for efficient CO_2 capture. Angew. Chem. **127**(2), 588–592 (2015)

66. Liu, G., Cadiau, A., Liu, Y., Adil, K., Chernikova, V., Carja, I.D., Belmabkhout, Y., Karunakaran, M., Shekhah, O., Zhang, C.: Enabling fluorinated MOF-based membranes for simultaneous removal of H_2S and CO_2 from natural gas. Angew. Chem. Int. Ed. **57**(45), 14811–14816 (2018)

67. Varoon, K., Zhang, X., Elyassi, B., Brewer, D.D., Gettel, M., Kumar, S., Lee, J.A., Maheshwari, S., Mittal, A., Sung, C.-Y.: Dispersible exfoliated zeolite nanosheets and their application as a selective membrane. Science **334**(6052), 72–75 (2011)

68. Manzeli, S., Ovchinnikov, D., Pasquier, D., Yazyev, O.V., Kis, A.: 2D transition metal dichalcogenides. Nat. Rev. Mater. **2**(8), 17033 (2017)

69. Zheng, Z., Grünker, R., Feng, X.: Synthetic two-dimensional materials: a new paradigm of membranes for ultimate separation. Adv. Mater. **28**(31), 6529–6545 (2016)

70. Novoselov, K.S., Jiang, D., Schedin, F., Booth, T., Khotkevich, V., Morozov, S., Geim, A.K.: Two-dimensional atomic crystals. Proc. Natl. Acad. Sci. USA **102**(30), 10451–10453 (2005)

71. Edwards, R.S., Coleman, K.S.: Graphene synthesis: relationship to applications. Nanoscale **5**(1), 38–51 (2013)

72. Yu, J., Li, J., Zhang, W., Chang, H.: Synthesis of high quality two-dimensional materials via chemical vapor deposition. Chem. Sci. **6**(12), 6705–6716 (2015)

73. Tan, C., Zhang, H.: Wet-chemical synthesis and applications of non-layer structured two-dimensional nanomaterials. Nat. Commun. **6**, 7873 (2015)

74. Cohen-Tanugi, D., Grossman, J.C.: Nanoporous graphene as a reverse osmosis membrane: recent insights from theory and simulation. Desalination **366**, 59–70 (2015)

75. Song, N., Gao, X., Ma, Z., Wang, X., Wei, Y., Gao, C.: A review of graphene-based separation membrane: materials, characteristics, preparation and applications. Desalination **437**, 59–72 (2018)

76. Dreyer, D.R., Park, S., Bielawski, C.W., Ruoff, R.S.: The chemistry of graphene oxide. Chem. Soc. Rev. **39**(1), 228–240 (2010)

77. Sun, P., Wang, K., Zhu, H.: Recent developments in graphene-based membranes: structure, mass-transport mechanism and potential applications. Adv. Mater. **28**(12), 2287–2310 (2016)

78. Zhang, N., Qi, W., Huang, L., Jiang, E., Bao, J., Zhang, X., An, B., He, G.: Review on structural control and modification of graphene oxide-based membranes in water treatment: From separation performance to robust operation. Chinese J. Chem. Eng. (2019)

79. Chen, L., Li, N., Wen, Z., Zhang, L., Chen, Q., Chen, L., Si, P., Feng, J., Li, Y., Lou, J.: Graphene oxide based membrane intercalated by nanoparticles for high performance nanofiltration application. Chem. Eng. J. **347**, 12–18 (2018)

80. Zhang, H., Quan, X., Chen, S., Fan, X., Wei, G., Yu, H.: Combined effects of surface charge and pore size on co-enhanced permeability and ion selectivity through RGO-OCNT nanofiltration membranes. Environ. Sci. Technol. **52**(8), 4827–4834 (2018)

81. Azamat, J., Khataee, A.: Improving the performance of heavy metal separation from water using MoS_2 membrane: molecular dynamics simulation. Comput. Mater. Sci. **137**, 201–207 (2017)

82. Heiranian, M., Farimani, A.B., Aluru, N.R.: Water desalination with a single-layer MoS_2 nanopore. Nat. Commun. **6**, 8616 (2015)

83. Hirunpinyopas, W., Prestat, E., Worrall, S.D., Haigh, S.J., Dryfe, R.A., Bissett, M.A.: Desalination and nanofiltration through functionalized laminar MoS_2 membranes. ACS Nano **11**(11), 11082–11090 (2017)

84. Huang, L., Zhang, M., Li, C., Shi, G.: Graphene-based membranes for molecular separation. J. Phys. Chem. Lett. **6**(14), 2806–2815 (2015)

85. Deng, M., Kwac, K., Li, M., Jung, Y., Park, H.G.: Stability, molecular sieving, and ion diffusion selectivity of a lamellar membrane from two-dimensional molybdenum disulfide. Nano Lett. **17**(4), 2342–2348 (2017)

86. Sun, L., Ying, Y., Huang, H., Song, Z., Mao, Y., Xu, Z., Peng, X.: Ultrafast molecule separation through layered WS_2 nanosheet membranes. ACS Nano **8**(6), 6304–6311 (2014)

87. Low, I.M.: Advances in ceramic matrix composites: introduction. In: Advances in Ceramic Matrix Composites, pp. 1–7. Elsevier (2018)

88. Pang, W.K., Low, I.M.: Understanding and improving the thermal stability of layered ternary carbides in ceramic matrix composites. In: Advances in Ceramic Matrix Composites, pp. 340–368. Elsevier (2014)

89. Chakraborty, P., Das, T., Saha-Dasgupta, T.: MXene: a new trend in 2D materials science. Comprehensive Nanoscience and Nanotechnology, 2nd edn., pp. 319–330 (2018)

90. Ren, C.E., Hatzell, K.B., Alhabeb, M., Ling, Z., Mahmoud, K.A., Gogotsi, Y.: Charge-and size-selective ion sieving through $Ti_3C_2T_x$ MXene membranes. J. Phys. Chem. Lett. **6**(20), 4026–4031 (2015)

91. Ding, L., Wei, Y., Wang, Y., Chen, H., Caro, J., Wang, H.: A two-dimensional lamellar membrane: MXene nanosheet stacks. Angew. Chem. Int. Ed. **56**(7), 1825–1829 (2017)

92. Kang, K.M., Kim, D.W., Ren, C.E., Cho, K.M., Kim, S.J., Choi, J.H., Nam, Y.T., Gogotsi, Y., Jung, H.-T.: Selective molecular separation on $Ti_3C_2T_x$-graphene oxide membranes during pressure-driven filtration: comparison with graphene oxide and mXenes. ACS Appl. Mater. Interface **9**(51), 44687–44694 (2017)

93. Wu, X., Hao, L., Zhang, J., Zhang, X., Wang, J., Liu, J.: Polymer-$Ti_3C_2T_x$ composite membranes to overcome the trade-off in solvent resistant nanofiltration for alcohol-based system. J. Membr. Sci. **515**, 175–188 (2016)

94. Hao, L., Zhang, H., Wu, X., Zhang, J., Wang, J., Li, Y.: Novel thin-film nanocomposite membranes filled with multi-functional $Ti_3C_2T_x$ nanosheets for task-specific solvent transport. Compos. A: Appl. Sci. Manuf. **100**, 139–149 (2017)

95. Kabir, E., Kumar, V., Kim, K.-H., Yip, A.C.K., Sohn, J.R.: Environmental impacts of nanomaterials. J. Environ. Manag. **225**, 261–271 (2018)

96. Colvin, V.L.: The potential environmental impact of engineered nanomaterials. Nat. Biotechnol. **21**(10), 1166 (2003)

97. Lee, J., Mahendra, S., Alvarez, P.J.: Nanomaterials in the construction industry: a review of their applications and environmental health and safety considerations. ACS Nano **4**(7), 3580–3590 (2010)

98. Patil, S.S., Shedbalkar, U.U., Truskewycz, A., Chopade, B.A., Ball, A.S.: Nanoparticles for environmental clean-up: a review of potential risks and emerging solutions. Environ. Technol. Innov. **5**, 10–21 (2016)

99. Paul, M., Jons, S.D.: Chemistry and fabrication of polymeric nanofiltration membranes: a review. Polymer **103**, 417–456 (2016)

100. Peyki, A., Rahimpour, A., Jahanshahi, M.: Preparation and characterization of thin film composite reverse osmosis membranes incorporated with hydrophilic SiO_2 nanoparticles. Desalination **368**, 152–158 (2015)

101. Valentino, L., Matsumoto, M., Dichtel, W.R., Mariñas, B.J.: Development and performance characterization of a polyimine covalent organic framework thin-film composite nanofiltration membrane. Environ. Sci. Technol. **51**(24), 14352–14359 (2017)

102. Akther, N., Phuntsho, S., Chen, Y., Ghaffour, N., Shon, H.K.: Recent advances in nanomaterial-modified polyamide thin-film composite membranes for forward osmosis processes. J. Membr. Sci. **584**, 20–45 (2019)
103. Wang, Z., Wang, Z., Lin, S., Jin, H., Gao, S., Zhu, Y., Jin, J.: Nanoparticle-templated nanofiltration membranes for ultrahigh performance desalination. Nat. Chem. **9**(1), 2004 (2018)
104. Li, X., Fang, X., Pang, R., Li, J., Sun, X., Shen, J., Han, W., Wang, L.: Self-assembly of TiO$_2$ nanoparticles around the pores of PES ultrafiltration membrane for mitigating organic fouling. J. Membr. Sci. **467**, 226–235 (2014)
105. Fischer, K., Grimm, M., Meyers, J., Dietrich, C., Gläser, R., Schulze, A.: Photoactive microfiltration membranes via directed synthesis of TiO$_2$ nanoparticles on the polymer surface for removal of drugs from water. J. Membr. Sci. **478**, 49–57 (2015)
106. Hu, Y., Lü, Z., Wei, C., Yu, S., Liu, M., Gao, C.: Separation and antifouling properties of hydrolyzed PAN hybrid membranes prepared via in-situ sol-gel SiO$_2$ nanoparticles growth. J. Membr. Sci. **545**, 250–258 (2018)
107. Li, X., Li, J., Van der Bruggen, B., Sun, X., Shen, J., Han, W., Wang, L.: Fouling behavior of polyethersulfone ultrafiltration membranes functionalized with sol-gel formed ZnO nanoparticles. RSC Adv. **5**(63), 50711–50719 (2015)
108. Zhao, X., Jia, N., Cheng, L., Liu, L., Gao, C.: Dopamine-induced biomimetic mineralization for in situ developing antifouling hybrid membrane. J. Membr. Sci. **560**, 47–57 (2018)
109. Li, X., Pang, R., Li, J., Sun, X., Shen, J., Han, W., Wang, L.: In situ formation of Ag nanoparticles in PVDF ultrafiltration membrane to mitigate organic and bacterial fouling. Desalination **324**, 48–56 (2013)
110. Miller, D.J., Dreyer, D.R., Bielawski, C.W., Paul, D.R., Freeman, B.D.: Surface modification of water purification membranes. Angew. Chem. Int. Ed. **56**(17), 4662–4711 (2017)
111. Zhang, R.-X., Braeken, L., Luis, P., Wang, X.-L., Van der Bruggen, B.: Novel binding procedure of TiO$_2$ nanoparticles to thin film composite membranes via self-polymerized polydopamine. J. Membr. Sci. **437**, 179–188 (2013)
112. Ren, S., Boo, C., Guo, N., Wang, S., Elimelech, M., Wang, Y.: Photocatalytic reactive ultrafiltration membrane for removal of antibiotic resistant bacteria and antibiotic resistance genes from wastewater effluent. Environ. Sci. Technol. **52**(15), 8666–8673 (2018)
113. Choi, W., Choi, J., Bang, J., Lee, J.-H.: Layer-by-layer assembly of graphene oxide nanosheets on polyamide membranes for durable reverse-osmosis applications. ACS Appl. Mater. Interface **5**(23), 12510–12519 (2013)
114. Hegab, H.M., ElMekawy, A., Barclay, T.G., Michelmore, A., Zou, L., Saint, C.P., Ginic-Markovic, M.: Fine-tuning the surface of forward osmosis membranes via grafting graphene oxide: performance patterns and biofouling propensity. ACS Appl. Mater. Interface **7**(32), 18004–18016 (2015)
115. Guo, H., Lv, R., Bai, S.: Recent advances on 3D printing graphene-based composites. Nano Mater. Sci. **1**(2), 101–115 (2019)
116. Tofail, S.A., Koumoulos, E.P., Bandyopadhyay, A., Bose, S., O'Donoghue, L., Charitidis, C.: Additive manufacturing: scientific and technological challenges, market uptake and opportunities. Mater. Today **21**(1), 22–37 (2018)
117. Thomas, N., Sreedhar, N., Al-Ketan, O., Rowshan, R., Al-Rub, R.K.A., Arafat, H.: 3D printed triply periodic minimal surfaces as spacers for enhanced heat and mass transfer in membrane distillation. Desalination **443**, 256–271 (2018)
118. Al-Shimmery, A., Mazinani, S., Ji, J., Chew, Y.J., Mattia, D.: 3D printed composite membranes with enhanced anti-fouling behaviour. J. Membr. Sci. **574**, 76–85 (2019)
119. Tsai, H.-Y., Huang, A., Luo, Y.-L., Hsu, T.-Y., Chen, C.-H., Hwang, K.-J., Ho, C.-D., Tung, K.-L.: 3D printing design of turbulence promoters in a cross-flow microfiltration system for fine particles removal. J. Membr. Sci. **573**, 647–656 (2019)
120. Parandoush, P., Lin, D.: A review on additive manufacturing of polymer-fiber composites. Compos. Struct. **182**, 36–53 (2017)

121. Chandrashekara, M., Yadav, A.: Water desalination system using solar heat: a review. Renew. Sustain. Energy Rev. **67**, 1308–1330 (2017)
122. Shi, Y., Li, C., He, D., Shen, L., Bao, N.: Preparation of graphene oxide-cellulose acetate nanocomposite membrane for high-flux desalination. J. Membr. Sci. **52**(22), 13296–13306 (2017)
123. Abdelhamid, A.E., Khalil, A.M.: Polymeric membranes based on cellulose acetate loaded with candle soot nanoparticles for water desalination. J. Macromol. Sci. A **56**(2), 153–161 (2019)
124. Perera, M.G.N., Galagedara, Y.R., Ren, Y., Jayaweera, M., Zhao, Y., Weerasooriya, R.: Fabrication of fullerenol-incorporated thin-film nanocomposite forward osmosis membranes for improved desalination performances. J. Polym. Res. **25**(9), 199 (2018)
125. Nemati, M., Hosseini, S., Parvizian, F., Rafiei, N., Van der Bruggen, B.: Desalination and heavy metal ion removal from water by new ion exchange membrane modified by synthesized $NiFe_2O_4$/HAMPS nanocomposite. Ionics 1–11 (2019)
126. Daer, S., Kharraz, J., Giwa, A., Hasan, S.W.: Recent applications of nanomaterials in water desalination: a critical review and future opportunities. Desalination **367**, 37–48 (2015)
127. Teow, Y.H., Mohammad, A.W.: New generation nanomaterials for water desalination: a review. Desalination **451**, 2–17 (2019)
128. Lopera, A.E.-C., Ruiz, S.G., Alonso, J.M.Q.: Removal of emerging contaminants from wastewater using reverse osmosis for its subsequent reuse: pilot plant. J. Water Process. Eng. **29**, 100800 (2019)
129. Zeng, G., He, Y., Zhan, Y., Zhang, L., Pan, Y., Zhang, C., Yu, Z.: Novel polyvinylidene fluoride nanofiltration membrane blended with functionalized halloysite nanotubes for dye and heavy metal ions removal. J. Hazard. Mater. **317**, 60–72 (2016)
130. Zhang, Y., Zhang, S., Gao, J., Chung, T.-S.: Layer-by-layer construction of graphene oxide (GO) framework composite membranes for highly efficient heavy metal removal. J. Membr. Sci. **515**, 230–237 (2016)
131. Ghaemi, N., Safari, P.: Nano-porous SAPO-34 enhanced thin-film nanocomposite polymeric membrane: simultaneously high water permeation and complete removal of cationic/anionic dyes from water. J. Hazard. Mater. **358**, 376–388 (2018)
132. Wang, Z., Ji, S., Zhang, J., He, F., Xu, Z., Peng, S., Li, Y.: Dual functional membrane with multiple hierarchical structures (MHS) for simultaneous and high-efficiency removal of dye and nano-sized oil droplets in water under high flux. J. Membr. Sci. **564**, 317–327 (2018)
133. Chen, X., Qiu, M., Ding, H., Fu, K., Fan, Y.: A reduced graphene oxide nanofiltration membrane intercalated by well-dispersed carbon nanotubes for drinking water purification. Nanoscale **8**(10), 5696–5705 (2016)
134. Chae, H.-R., Lee, J., Lee, C.-H., Kim, I.-C., Park, P.-K.: Graphene oxide-embedded thin-film composite reverse osmosis membrane with high flux, anti-biofouling, and chlorine resistance. J. Membr. Sci. **483**, 128–135 (2015)
135. Qu, F., Liang, H., Zhou, J., Nan, J., Shao, S., Zhang, J., Li, G.: Ultrafiltration membrane fouling caused by extracellular organic matter (EOM) from *Microcystis aeruginosa*: effects of membrane pore size and surface hydrophobicity. J. Membr. Sci. **449**, 58–66 (2014)
136. Kim, J., Van der Bruggen, B.: The use of nanoparticles in polymeric and ceramic membrane structures: review of manufacturing procedures and performance improvement for water treatment. Environ. Pollut. **158**(7), 2335–2349 (2010)
137. Li, X., Li, J., Fang, X., Bakzhan, K., Wang, L., Van der Bruggen, B.: A synergetic analysis method for antifouling behavior investigation on PES ultrafiltration membrane with self-assembled TiO_2 nanoparticles. J. Colloid Interface Sci. **469**, 164–176 (2016)
138. Li, Q., Pan, S., Li, X., Liu, C., Li, J., Sun, X., Shen, J., Han, W., Wang, L.: Hollow mesoporous silica spheres/polyethersulfone composite ultrafiltration membrane with enhanced antifouling property. Colloid Surf. A **487**, 180–189 (2015)
139. Jiang, Y., Biswas, P., Fortner, J.D.: A review of recent developments in graphene-enabled membranes for water treatment. Environ. Sci.: Water Res. Technol. **2**(6), 915–922 (2016)

140. Lee, J., Chae, H.-R., Won, Y.J., Lee, K., Lee, C.-H., Lee, H.H., Kim, I.-C., Lee, J-m: Graphene oxide nanoplatelets composite membrane with hydrophilic and antifouling properties for wastewater treatment. J. Membr. Sci. **448**, 223–230 (2013)

141. Shao, J., Hou, J., Song, H.: Comparison of humic acid rejection and flux decline during filtration with negatively charged and uncharged ultrafiltration membranes. Water Res. **45**(2), 473–482 (2011)

142. Ong, C., Goh, P., Lau, W., Misdan, N., Ismail, A.: Nanomaterials for biofouling and scaling mitigation of thin film composite membrane: a review. Desalination **393**, 2–15 (2016)

143. Zhao, H., Qiu, S., Wu, L., Zhang, L., Chen, H., Gao, C.: Improving the performance of polyamide reverse osmosis membrane by incorporation of modified multi-walled carbon nanotubes. J. Membr. Sci. **450**, 249–256 (2014)

144. Sterlitech (2019) https://www.sterlitech.com/flat-sheet-membranes.html

145. Li, Y., Huang, S., Zhou, S., Fane, A.G., Zhang, Y., Zhao, S.: Enhancing water permeability and fouling resistance of polyvinylidene fluoride membranes with carboxylated nanodiamonds. J. Membr. Sci. **556**, 154–163 (2018)

146. Osipi, S.R., Secchi, A.R., Borges, C.P.: Cost assessment and retro-techno-economic analysis of desalination technologies in onshore produced water treatment. Desalination **430**, 107–119 (2018)

147. BCC Research LLC MST041J Membrane Technology for Liquid and Gas Separations (2019). https://www.bccresearch.com/market-research/membrane-and-separation-technology/membrane-separation-technology-research-review.html

Nanostructured Polymer Composites for Water Remediation

Michael Ovbare Akharame, Ogheneochuko Utieyin Oputu, Omoniyi Pereao, Bamidele Oladapo Fagbayigbo, Lovasoa Christine Razanamahandry, Beatrice Olutoyin Opeolu and Olalekan Siyanbola Fatoki

Abstract Nano-based advanced processes and technologies for water treatment operations provide innovative and progressive approaches with the capability to solve water scarcity, and contamination challenges across the globe. The incorporation of the numerous nanomaterials employed in this field into polymeric matrices bequeaths safer and more environmentally friendly materials-nanostructured polymer composites, for water treatment processes. In addition, the various functionalities of the polymers in combination with those of the incorporated nanomaterials often create synergistic treatment efficiencies in the hybrid composites. In this chapter, we focused on the complete process overview involved in the utilization of nanostructured polymer composites in water remediation and monitoring operations; this includes the different types of polymer matrices and nanomaterials in use, the design and synthesis methods, and characterisation techniques employed to elucidate their morphological, composite structural and functional properties. Scrutiny of their application potentials and efficiencies was done based on the different mode of applications which includes catalysis, adsorption, membrane filtration, microbial control, and monitoring and detection programs. This contribution harped on some areas of future concern

M. O. Akharame (✉) · O. U. Oputu · O. S. Fatoki
Department of Chemistry, Cape Peninsula University of Technology,
Cape Town 8000, South Africa
e-mail: michael.akharame@uniben.edu

M. O. Akharame
Department of Environmental Management and Toxicology, University of Benin,
Benin-City 3002, Nigeria

O. Pereao · B. O. Fagbayigbo · B. O. Opeolu
Department of Environmental and Occupational Health, Cape Peninsula University of
Technology, Cape Town 8000, South Africa

L. C. Razanamahandry
Nanoscience's/Nanotechnology Laboratories (U2AC2N), College of Graduate Studies,
University of South Africa, Pretoria, South Africa

Material Research Department, iThemba LABS, Nanosciences African Network
(NANOAFNET), Somerset West 7129, South Africa

© Springer Nature Switzerland AG 2019
G. A. B. Gonçalves and P. Marques (eds.), *Nanostructured Materials
for Treating Aquatic Pollution*, Engineering Materials,
https://doi.org/10.1007/978-3-030-33745-2_10

for their sustainability as a treatment option such as the development of multifunctional hybrid composites, the composites life span or cycle, and the safe disposal of the spent nanostructured polymer composites and their regeneration medium or solution.

Keywords Nanotechnology · Nanomaterials and nanoparticles · Nanostructured polymer composites · Water treatment and monitoring · Regeneration and reuse

1 Introduction

The world's freshwater sources are being depleted and contaminated with a wide array of pollutants due to increasing industrialization, urbanization and population growth [1, 2]. More so, natural phenomena such as severe droughts, earthquakes, land/mudslides, typhoons and tsunamis due to climate change (mostly the offshoot of the aforementioned issues) have increased cases of and/or worsened the scarcity of this vital and fundamental natural resource needed by humans and all living organisms [3]. These challenges have necessitated the need to seek alternative water sources such as wastewaters, and to develop advanced technologies with capabilities to effect remediation to acceptable treatment levels for possible reuse. In this trend, the use of nanoparticles, nanomaterials and their composites is receiving much interest due to the inherent potentials derivable from the application of nanotechnology in water treatment operations [4].

Technological advancements come at a price. While we race off to a world of cleaner waters produced by advanced water treatment technologies based on nanomaterials, there is a significant knowledge gap regarding the behaviours of these nanomaterials when they enter environmental matrices. Whether we be water treatment experts or environmental toxicologist, one thing is certain: the presence of these nanomaterials in the natural environment is undesirable. There is therefore a need to design effective and safe procedures for the utilization of nanoparticles or nanomaterials and this has led to increased interest in the use of polymeric matrices as support materials for their incorporation [5]. This is coming on the verge of the challenges experienced by the use of bare nanomaterials which include: loss of the nanoparticles during treatment operations and difficulty in their recovery processes, the tendency of the nanoparticles to aggregate which leads to reduced active surface area, coupled with possible health implications that may arise due to their release into the environment [6–12].

The surge in the use of polymer-based materials as supports for the incorporation of nanomaterials is encouraged by their high thermal, mechanical and environmental stability. Furthermore, their excellent pore-forming and easy processability characteristics, the huge number of functional groups they possess, and relatively low cost contribute to this trend [13]. Operational gains derived from the incorporation of the nanomaterials into polymer matrices include ease of recovery, prevention of nanomaterial loses and aggregation during process operations and the subsequent safeguarding of the environment. In effect, applying a nanomaterial in a heterogeneous continuous flow setup would require the nanomaterial is stabilized on supports during use. While several supports (many of which are based on other nanomaterials) have been tested in literature, only supports based on polymers may easily be adapted to continuous flow set-ups that could be upscaled for industrial use. This is because these polymer composites lend themselves to structural manipulations and may be tailor made to reactor designs either as membranes or coated on reactor walls [14]. Also, improved treatment and monitoring capabilities due to synergistic effect by both starting materials are often obtained as compared to the use of bare nanomaterials [15, 16].

Nanostructured polymers composites (NPCs) are fabricated by the incorporation of nanomaterial(s) into and/or unto an appropriate polymer matrix, polymer blends or grafts [17]. The process of incorporation is usually designed to provide high interaction between both materials for best results. This is very important as the final properties of the composites are dependent on the size scale of its components and the degree of interaction between the two phases [18]. Nanostructured polymer composites have been reported to possess enhanced properties ranging from improved fouling resistance and membrane permeability, sorption and catalytic capabilities to higher thermal and mechanical stability [19]. These intrinsic properties are resultant from the characteristic properties of the nanomaterials and those of the polymer matrices, each contributing to the final hybrid properties. The enhanced properties contribute to the versatility in their remediation and monitoring functionalities in water treatment processes as presented in Table 1.

2 Composition of Nanostructured Polymer Composites

2.1 Polymer Matrices

The use of polymer matrices to provide support for the incorporation of numerous (and still emerging) nanomaterials has opened a floodgate of possibilities for their use in water treatment operations. This stems from the fact that the polymeric materials possess some unique and useful properties such as their numerous chemical functionalities, and hydrophobic/hydrophilic characteristics, relatively low cost and easy processability which creates more advantages and spread for their possible areas of application [4]. Added to the aforementioned properties, is the derived advantage of their low weight and often ductile nature [90]. Their selection for specific application and usage is guided by factors bordering on the desired properties which may include thermal, mechanical, opto-electronic, magnetic, and chemical stability [91]. More importantly, the compatibility between the polymer support and the nanomaterial should be considered in order to obtain desirable and functional nanostructured composites that will suit the intended purpose.

Table 1 Application potential of NPCs for water treatment [20]

Polymer matrix	Incorporated nanomaterials	Reported application area	Target pollutant/application	References
Polyamide	ZnO	Photocatalysis	MB dye	Ummartyotin and Pechyen [21]
	Graphene oxide	Membrane filtration	Desalination (NaCl/Na$_2$SO$_4$)	Yin et al. [22]
	Carbon nanotubes (CNT)	Membrane filtration	Desalination (NaCl/Na$_2$SO$_4$)	Kim et al. [23]
	Carbon nanotubes	Membrane filtration	Desalination (NaCl)	de Lannoy et al. [24]
	TiO$_2$	Photocatalysis and microbial control	Congo red; *E. coli*	Pant et al. [25]
Polystyrene	Graphene oxide	Adsorption	PB(II)	Ravishankar et al. [26]
	Fe$_3$O$_4$	Demulsification	Oil in water	Reddy et al. [27], Yu et al. [28]
	TiO$_2$	Photocatalysis	MB dye	Vaiano et al. [29]
	Au	Sensing/detection	17β-estradiol	Pule et al. [30]
	CNT-CuO/AgO	Sensing/detection	COD detection/measurement	Gutierrez-Capitan et al. [31]
Polyacrylamide	Zr(IV) vanadophosphate	Photocatalysis and microbial control	Congo red; *E. coli and S. aureus* Congo red and MO dyes	Sharma et al. [32]
	ZnS	Photocatalysis	MG and rhodamine B dyes	Pathania et al. [33]
	Ni$_{0.02}$Zn$_{0.98}$O	Adsorption and photocatalysis	Dewatering of sludge	Kumar et al. [34]
	Montmorillonite	Flocculation	Cu(II)	Huang and Ye [35]
	Bentonite	Adsorption	Cu(II)	Zhao et al. [36]
	Attapulgite	Adsorption		Chen and Wang [37]
Polysulfone	Graphene oxide	Membrane filtration	Antifouling	Lee et al. [38]
	FeOOH (hydrous ferric oxide)	Adsorption	Pb(II)	Abdullah et al. [39]
	CNT	Microbial control	*E. coli*	Schiffman and Elimelech [40]

(continued)

Table 1 (continued)

Polymer matrix	Incorporated nanomaterials	Reported application area	Target pollutant/application	References
Polypyrrole	TiO_2	Photocatalysis and sensing/detection	Rhodamine B dye	He et al. [41]
	Fe_3O_4		Cr (VI)	Yao et al. [42]
	Montmorillonite	Adsorption	Cr (VI)	Setshedi et al. [43]
	Multi-walled carbon nanotubes (MWCNT)	Adsorption	Pb(II)	Sahmetlioglu et al. [44]
	Magnetic zeolite	Adsorption	Vanadium	Mthombeni et al. [45]
Polydimethyl Siloxane	ZnSe	Ion-exchange	Pb(II) and Hg(II)	Chavan et al. [46]
	Au	Adsorption	Toluene, sulfide, thiophenol and thioether	Scott et al. [47]
	Carbon nanofiber	Absorption	Oils in water	Abdulhussein et al. [48]
Polyvinyl alcohol	TiO_2/graphene-MWCNT	Photocatalysis	MB dye	Jung and Kim [49]
	TiO_2	Sensing/detection	Potassium ferricyanide	Mondal et al. [50]
	Ag	Microbial control and	E. coli, P. aeruginosa, B. cereus and S. aureus; Cd(II)	Devi and Umadevi [51]
	Laponite/Fe_3O_4	Sensing/detection	MB dye	Mahdavinia et al. [52]
	Laponite/Fe_3O_4	Adsorption	Bovine serum albumin	Mahdavinia et al. [53]
		Adsorption		
Polyacrylonitrile	Reduced graphene oxide	Membrane filtration	MO, Acid brilliant blue and chrome blue-black R dyes	Liang et al. [54]
	Ag	Microbial control (filtration process)	S. aureus and E. coli	Zhang et al. [55]
Polyethersulfone	TiO_2	Photocatalysis	MO dye	Hir et al. [56]
	α-FeOOH (goethite)	Membrane filtration	Direct red 16 and Antifuoling	Rahimi et al. [57]
	ZnO/MWCNT	Membrane filtration	Antifouling	Zinadini et al. [58]
	Fe–NiO	Membrane filtration	Desalination ($NaCl/Na_2SO_4$)	Bagheripour et al. [59]
	ZnO	Microbial control	E. coli and S. aureaus	Jo et al. [60]
	Ag	Microbial control	E. coli and S. aureaus	Toroghi et al. [61]
	Ag	Microbial control	E. coli and S. aureaus	Basri et al. [62]

(continued)

Table 1 (continued)

Polymer matrix	Incorporated nanomaterials	Reported application area	Target pollutant/application	References
Polyvinylidene fluoride	TiO_2	Photocatalysis	Nonylphenol	Dzinun et al. [63]
	Fe_2O_3/MWCNT	Catalysis	Organic contaminants	Alpatova et al. [64]
	Al_2O_3	Membrane filtration	Oil in water	Yan et al. [65]
	SiO_2	Membrane filtration	Desalination (NaCl)	Obaid et al. [66]
Polyaniline	β-FeOOH (akaganéite)	Adsorption	Cr(VI)	Ebrahim et al. [7]
	Zr(IV) silicophosphate	Microbial control and photocatalysis	E. coli and	Pathania et al. [67]
	Fe	Photocatalysis	MB dye	Haspulat et al. [68]
	CdO	Photocatalysis	MB, MG, MO and MR dyes	Gülce et al. [69]
	Graphene/WO_3	Photocatalysis	MB and MG dyes	Tovide et al. [70]
	$VOPO_4$	Photocatalysis	Phenanthrene	Khan et al. [71]
	TiO_2	Sensing/detection	Pb(II)	Zheng et al. [72]
	Th(IV) tungstomolybdophosphate	Sensing/detection	Trimethylamine	Sharma et al. [73]
	Polygorskite/Ag	Ion-exchange and microbial control	Cu(II) and Pb(II); S. aureus	Tian et al. [74]
	Montmorillonite/vermiculite/Fe_3O_4	Catalysis Adsorption	4-nitrophenol and CR dye MB, BG and CR dyes	Mu et al. [75]
Epoxy	MWCNT; MWCNT	Electrocatalysis and sensing/detection	Chlorine; Ibuprofen	Muñoz et al. [76] Motoc et al. [77]
	TiO_2	Adsorption and Photocatalysis	Pb(II), Cd(II) and Cr(II); MB dye	Benjwal and Kar [78]
Poly acrylic acid	Fe_3O_4	Adsorption	Rb(I) and Cs(I)	Zhu et al. [79, 80]
	Silane modified Fe_3O_4	Adsorption	Cu(II) and Pb(II)	Zhu et al. [81]
	Attapulgite	Adsorption	MB dye	Zhu et al. [82]
	Bentonite	Adsorption	Pb(II)	Rafiei et al. [83]
	Attapulgite	Adsorption	Pb(II)	Liu et al. [84, 85]
	Vermiculite	Adsorption	MB dye	Liu et al. [86]
	Halloysite	Adsorption	Ammonium	Zheng and Wang [87]
	Rectorite	Adsorption	Ammonium	Zheng and Wang [88]
	Biotite	Adsorption	MB Dye	Liu et al. [89]

Polymeric matrices for nanostructured polymer composites comprise mainly of the thermosetting and thermoplastic polymers. The main difference between the two types of polymers is their interaction or reaction to heat. Thermosetting polymers, also referred to as thermosets are irreversibly hardened polymers, which undergo degradation without melting or going through the fluid state, while the thermoplastic polymers become soft or molten and pliable when heat is applied to them [92]. The versatility of both groups of polymers has seen them being employed in several areas of water treatment operations which covers catalysis, adsorption, membrane filtration, disinfection and microbial control, and for detection and monitoring purposes. In this regards, the recent application of thermosetting polymers as supports include their utilization by Wang et al. [93] to synthesize biomorphic MgAl layered double oxide/acrylic ester resins, which were employed for oil remediation in contaminated water. The PNCs reportedly showed high absorption capabilities, with excellent composite stability recorded after over 5 cycles of use and regeneration without a significant decrease in the oil absorption. Also, Sun and co-workers [94] prepared magnetic triethylene tetramine-graphene oxide ($CoFe_2O_4$-TETA-GO) ternary nanocomposites which were applied for the removal of Cr(VI) from aqueous solution. They reported the saturation adsorption capacity of Cr(VI) to be about 180.12 mg g^{-1} on the $CoFe_2O_4$-TETA-GO composite at pH 2. The PNCs equally showed good recycling and regeneration potential as 78% of the adsorptive capacity was retained after 5 cycles of use.

Benjwal and Kar [78] reported the fabrication of titania/activated carbon/carbonized epoxy nanocomposite (TiO_2/AC/CE) which was applied for simultaneous photocatalytic and adsorptive removal of methylene blue dye and Pb(II) ions from aqueous solution. The as-synthesized TiO_2/AC/CE composite showed high activity for the removal of both pollutants with about 90% degradation of methylene blue recorded, while 97% of the Pb(II) ions were removed from the aqueous solution. This is an excellent case of a composite combining the unrelated photocatalytic property of titania and adsorption property of the reinforced polymer. Furthermore, Motoc et al. [77] reported the fabrication of two types of epoxy PNCs electrodes in their research namely; multi-walled carbon nanotubes (MWCNT)-epoxy and silver-modified zeolite-MWCNT-epoxy (AgZMWCNT). The electrode composites were applied for electrochemical detection and degradation of ibuprofen from water. Results obtained showed that electrochemical determination of ibuprofen was achieved using AgZMWCNT by cyclic voltammetry, differential-pulsed voltammetry, square-wave voltammetry and chronoamperometry. Also, both electrode composites exhibited degradation capabilities for ibuprofen under controlled electrolysis at 1.2 and 1.75 V versus Ag/AgCl. However, the AgZMWCNT composite uniquely showed dual character, allowing a double application in ibuprofen degradation and its detection.

For the thermoplastics, they have been utilised as support matrices for numerous nanomaterials employed for various water treatment operations. For instance, electrically conductive polymer nanocomposites assemblage with thin membrane fabricated from polyamide-carbon nanotubes were synthesized by de Lannoy and co-researchers [24]. The PNCs were reported to possess high electrical conductivity

(~400 S/m), good NaCl rejection (>95%), and ability to resist fouling, even under extreme bacteria and organic material loadings. Similarly, Lee et al. [38] synthesized PNCs using polysulfone and nanoplatelets of graphene oxide (GO) with the resultant hybrid material showing impressive antifouling capabilities. PNCs applied as adsorbents were synthesized by the assembly of Fe_3O_4 and GO on the surface of polystyrene. The composite formed was reported to possess a spontaneous and excellent adsorption capacity (73.52 mg g^{-1}) for the removal of Pb (II) ions, with a maximum removal efficiency of 93.78% [26]. Thermoplastic PNCs have also been employed as photocatalyst for the degradation of recalcitrant methyl orange and Congo red dyes from aqueous systems. The PNCs synthesized from aqueous solution of chitosan-g-poly(acrylamide)/ZnS through microwaves radiation were reported to degrade Congo red dye by 70% and methyl orange 69%, after 2 and 4 h of irradiation, respectively [33].

In all scenarios above, both groups of polymers showed excellent potentials when utilised as supports for the different nanomaterials. Essentially, the thermosets possess high thermal, mechanical, and environmental stability due to their chemical structures and crosslinking densities, aided by processing conditions such as temperature and pressure [95]. However, their irreversibility once cured and susceptibility to degradation without going through the fluid state pose major processability challenge [96]. Examples of thermosetting polymers include the epoxies, phenolics, polyesters, polyimides and polyurethanes [97]. On the other hand, thermoplastics offer good mechanical and durability properties, as well as processability advantages. Their long-chain entanglement with temporary crosslinking effects and the ability to slip past each other when local stress is applied contributes to their versatility during process operations [98]. This class of polymers include polyamide, polycarbonate, polybutylene terephthalate, polystyrene, polyethylene, polypropylene, poly(vinyl chloride), poly(methyl methacrylate), poly(ρ-phenylene oxide), acrylonitrile butadiene and styrene acrylonitrile [99].

2.2 Inorganic Nanomaterials

There is a plethora of nanomaterials employed for the fabrication of nanostructured polymer composites. These include metals or mixed metals, metal oxides, clay/silicate-based, carbon-based nanoparticles, and a host of other hybrid nanomaterials with mixed compositions as depicted in Fig. 1. Generally, the nanomaterials are structured components with less than 100 nm in at least one dimension [100]. The nano-metre scale size translates to increased surface area and quantum effects [101], which leads to enhanced intrinsic and functional properties such as reactivity, magnetic, mechanical/strength, electrical, optical characteristics [101, 102]. Other exhibited properties are superparamagnetism, localized surface plasmon resonance and quantum confinement effects [19]. These nanomaterials are usually of different sizes, shapes and dimensions such as nanoparticles, nanocrystals, nanosheets, nanoshells, nanowires, nanorods, fullerenes and quantum dots [103].

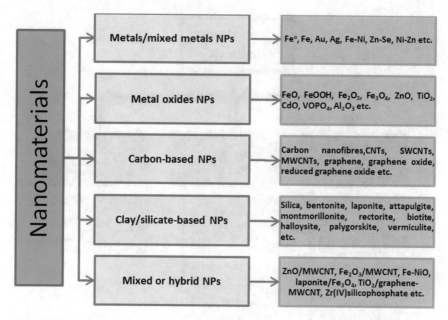

Fig. 1 Classification of nanomaterials

The incorporation of various nanomaterials into or unto polymer composites create improve structural properties and functionalities in the hybrid materials. For example, the embedding of certain nanoparticles in membrane filters brings about improvement in their water permeability, and pore sizes of the membranes [2]. This phenomenon comes from the unique surface chemistry of the nanomaterials, with added advantages of possible functionalization or grafting with functional groups for more efficiency or effectiveness in their application area [104]. In most scenarios, the enhanced properties of the nanomaterials are combined with those of the selected polymer matrix or matrices to yield hybrid composites with desirable properties required for specific water remediation and monitoring operations. Hence, the choice of nanomaterials, as well as, the polymer support is of critical importance when designing nanostructured polymer composites for water treatment applications.

There are numerous methods and techniques currently being employed in the designing and synthesis of nanomaterials. Broadly, they fall under chemical, green and mechanical synthesis methods as presented in Fig. 2. The chemical methods tend to be more popular and widely use for the synthesizing of a range of nanomaterials as it offers versatility in terms of functionalization and tuning of intrinsic properties. The green synthesis approach, which involves the use of microorganisms and/or plants extracts, has also received much attention in recent times due to the perceived environmental friendliness of the technique. Mechanical methods were mostly used during the inception of this nanotechnology as it provides a simple, economical and high yield route for the synthesis of nanomaterials [105, 106]. However, the challenge of inherent agglomeration and the limited possibility of functionalization

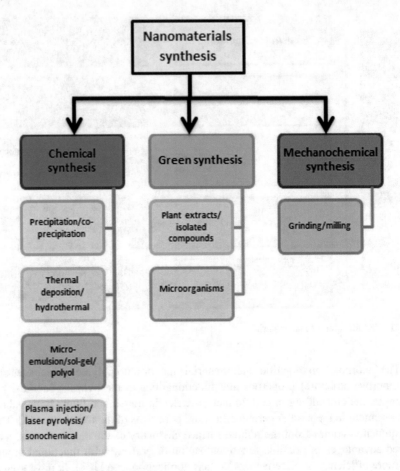

Fig. 2 Nanomaterials synthesis routes

of the nanomaterials is causing a decline in the usage of the synthesis route for nanomaterials employed for water treatment processes. Also, the method poses the challenge of possible significant contamination of the nanomaterials from the milling media or atmosphere [106]. The different synthesis routes present opportunities for the preparation of nanomaterials with different functionalities. This is achievable due to changes obtained in the intrinsic properties of nanomaterials with tuning or variations in the synthesis parameters or processes.

3 Synthesis Procedures for NPCs

Various developed techniques exist for the preparation of NPCs [5, 107], which could be divided into two broad categories depending on the formation processes of the

Fig. 3 Synthesis of
polymeric nanocomposites

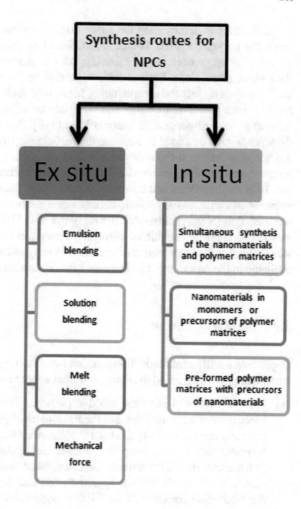

nanocomposites. These are ex situ (blending or direct compounding) and in situ
synthesis [11] as shown in Fig. 3.

3.1 Ex Situ Synthesis

The ex situ synthesis also referred to as blending or direct compounding technique
is used to fabricate PNCs when mass production is desired and is comparatively low
cost, and operational convenient. Polymer supporters and nanofillers are initially sep-
arately prepared and thereafter compounded by emulsion, solution, melt blending
or mechanical forces [108, 109]. However, for most systems, only limited success
is achieved by using direct compounding of polymers with nanofillers because of

the difficulties in determining the space distribution parameter of nanoparticles in or on the polymer matrix. The nanoparticles often demonstrate great tendency to form bigger aggregates during blending, which greatly reduced the advantages of their small dimensions. Several surface treatments have been adopted in the synthesis procedure, and the compounding conditions such as the configuration of the reactor, shear force, temperature and time, can be adjusted in polymer matrices to achieve a good dispersion of nanoparticles [110]. Appropriate compatibilizers or dispersants may be added to improve the particle dispersion and/or miscibility and the adhesion between the nanoparticles and the matrix [111]. Razzaz et al. [108] reported chitosan/TiO$_2$ composite nanofibrous adsorbents by two techniques including TiO$_2$ nanoparticles coated chitosan nanofibers (coating method) and electrospinning of chitosan/TiO$_2$ solutions (entrapped method), while Musico et al. [112] reported Poly(N-vinylcarbazole) (PVK) blend with GO to form a PVK–GO polymer nanocomposite capable of adsorbing heavy metal from aqueous solutions. It was found that the increasing concentration of oxygen-containing functional groups available in the nanocomposite increased the adsorption capacity.

3.2 In-Situ Synthesis

Depending on different fabrication processes and starting materials, in situ synthesis may be largely categorised into three different approaches:

(1) Nanoparticles and polymers could be prepared simultaneously by blending the precursors of nanoparticles and the monomers of polymers with an initiator in proper solvent [113, 114]. Li et al. [115] modified titanium dioxide nanoparticles by polyaniline (PANI) using 'in situ' chemical oxidative polymerization method in hydrochloric acid solutions. Results showed that deposited TiO$_2$ nanoparticles in PANI matrix did not agglomeration and the modification did not alter the crystalline structure of the TiO$_2$ nanoparticles from the X-ray diffraction patterns.

(2) Another comparable method used the polymeric hosts monomers and the nanofillers targeted as the starting materials [116, 117]. Typically, the nanoparticles are first dispersed into the monomers or precursors of the polymeric hosts, and the mixture is then polymerized under desirable conditions with the addition of a suitable catalyst. Consideration for this method increased because it permits the synthesis of nanocomposites with tailored physical properties. A well-directed dispersion of the nanoparticles into the liquid monomers or precursors will avoid their agglomeration in the polymer matrix and subsequently increase the interfacial connection between both phases. Wang et al. [118] prepared nano-zinc oxide and nanoalumina/EPDM composites with good performance, thermal conductivity and mechanical properties; in-situ modifications with the silane-coupling agent bis-(3-triethoxy silylpropyl)-tetrasulfide (Si69)

enhanced the interfacial interaction between nano-particles and rubber matrix remarkably, and therefore contributed to the better dispersion of filler.

(3) Preloaded metal ions within polymer matrix serve as nanoparticle precursors, where the ions are expected to dispense evenly. Thereafter, the precursors are exposed to the corresponding gas or liquid having S^{2-}, OH^-, or S^{2-} to "in situ synthesize" the target nanoparticles [119]. Luo et al. [120] developed and used Lactic acid (LA) to modify TiO_2 surface by the Ti-carboxylic coordination bonds, and LA chemically bonded to TiO_2 nanoparticles to form functionalized oligomeric-poly(lactic acid)-grafted TiO_2 nanoparticles (g-TiO_2) which were added to the poly(lactic acid) (PLA) matrix to prepare PLA/TiO_2 nanocomposites via melting processing. The result showed that functionalized nanoparticles played an important role in improving the mechanical properties.

3.3 Emerging Synthesis Methods

Several other methods utilized for preparing PNCs recently include electrospinning, self-assembly, template synthesis, phase separation. Electrospinning is a unique synthetic method used for spinning nanofibers with the assistance of electrostatic forces to produce nanofibers in the range of 40–200 nm diameters [121, 122]. Electrospinning can be carried out by applying high voltage to a capillary tube filled with polymer fluid, which is ejected out toward a collector serving as the counter electrode [123]. The production of nanofibers by the electrospinning procedure is influenced by the viscoelastic behaviour of the polymer and the electrostatic forces [124, 125]. Electrospun nanofibers have attracted attention because of the special properties such as high permeability, low basis weight, layer thinness, surface area per unit mass, cost-effectiveness, high porosity, flexibility in surface functionalities, and high superior directional strength [126].

Self-assembly technique is a molecule-mediated technique used for the construction of various nanocomposite films with desirable thicknesses [127]. Instead of strong chemical bonds, nanoparticles are often linked by weak van der Waals, hydrogen bond and electric/magnetic dipole interactions [128]. Szabó et al. [129] reacted natural graphite with a strong oxidizing agent to form hydrophilic, negatively charged graphite oxide colloids dispersed in water which allowed the deposition of thin graphite oxide/cationic polymer (poly(diallyldimethylammoniumchloride, PDDA)) multilayer films on a glass substrate using wet-chemical self-assembly. Lu et al. [130] demonstrated the use of self-assembly of conjugated polymer/silica nanocomposite films with hexagonal, cubic or lamellar mesoscopic order using polymerizable amphiphilic diacetylene molecules as both structure-directing agents and monomers. The phase separation involves nano-fibrous matrices preparation from polymer solutions by processes involving thermally induced gelation, solvent exchange and freeze-drying [131]. Zhou and Wu [132] prepared transparent ZrO_2 dispersions in organic solvents by functionalizing ZrO_2 nanoparticles with 3-methacryloxypropyltrimethoxysilane (MPS) and subsequently blended the composite using UV-curable polyurethane (PU) coatings by ultrasonication.

The template method for preparing nanomaterials entails synthesis of monodisperse tubular and fibrillar nanostructures within the pores of a membrane or other nanoporous solid [133]. The most significant feature of this method remains the nanometer fibrils and tubules of several raw materials such as metals, semiconductors, electronically conducting polymers, and carbons which can be fabricated [11, 134]. Hulteen et al. [135] prepared arrays of carbon nanotubes, capped nanotubes and nanofibrils with a nominal outer diameter of 200 nm and length of 60 μm. This was accomplished via the polymerization of acrylonitrile within the pores of aluminium oxide membranes. Wang et al. [136] prepared micro/nanoscale magnesium silicate hollow spheres synthesized by using silica colloidal spheres as a chemical template in one pot. The as-synthesized magnesium silicate hollow spheres with large specific surface area showed availability for the removal of organic and heavy-metal ions efficiently from waste water.

The aforementioned emerging techniques for NPCs synthesis ultimately fall into the broad categories of classification. For example, the electrospinning techniques mostly involve the use of pre-formed polymer matrices and appropriate nanomaterials dissolved in suitable solvent solutions before the electrospinning operation which makes it an ex situ synthesis method. However, scenarios which involve the electrospinning of the NPCs with the pre-form polymer and the precursor to the nanomaterials (in situ method) is also in use [6]. In the template method, precursors of nanomaterials are loaded into the pores of polymeric membranes or other porous support where they are converted to the desired nanomaterials, and this is akin to the in situ synthesis preparation route.

4 Characterization Techniques for NPCs

Nanostructure polymer composites is a combination of the nanomaterials multilayers with a polymer matrix as support [137]. Therefore, the characterisation involves various criteria on the NPCs facet [5]. Firstly, the NPCs nanometre scale dimensions should be verified [138]; and the hybrid properties between the nanomaterial and the polymer need to be described as well. Mittal [139] reported properties elucidated in NPCs to include: (i) the nanomaterials distribution and dispersion level within the polymer support, (ii) the nanomaterials and polymer interaction through the functional groups, (iii) the NPCs structural changes as occasioned by its various synthesis routes, and (iv) the wide range of the NPCs functionalities for its potential application.

Several methods and techniques are utilized to analyse NPCs properties, which can be grouped into optical, structural, morphological, elemental and functional analysis. The NPCs optical properties are detected by using two techniques, namely UV-Visible spectroscopy and Fourier transmission infrared spectroscopy (FT-IR). The UV-Visible defines the NPCs spectrum absorption and concentration [140],

Fig. 4 FTIR spectra of ZIF-8, PSF membrane, and ZIF-8@PSF membrane. Reprinted with permission from [141]. Copyright (2015) American Chemical Society

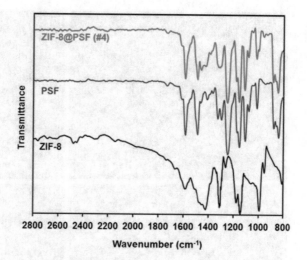

while the FT-IR identifies chemical bonding and the functional groups (see Fig. 4), and confirms the dispersion of the nanomaterials into the polymer matrix through the additional peak spectra [5]. X-ray diffraction (XRD) and X-ray photon spectroscopy (XPS) techniques are used to study the NPCs structure, i.e. the NPCs crystallinity and crystallography, while XPS reveals composition uniformity and binding energy the NPCs [139].

Several techniques are available to detect the NPCs morphology. NPCs size, topology and morphology could be determined by running Scanning electron microscopy (SEM) or Transmission electron microscopy (TEM, HRTEM) or Atomic force microscopy (AFM), Scanning tunnelling microscopy (STM) or Dynamic light scattering (DLS) [142]. By choosing the AFM and STM techniques, the surface texture and the chemical composition could be known, respectively, as additional characterisations [142]. Nevertheless, TEM and HRTEM resolution are highly appreciated and there is a possibility to get the selected area electron diffraction (SAED) microstructure of NPCs through the NPCs crystallographic structure [142]. Figure 5a–d give typical morphological microstructures of NPCs view using the SEM and TEM techniques.

The Energy dispersive X-ray (EDX), the X-ray absorption spectroscopy (XAS) and X-ray fluorescence (XRF) techniques are employed to determine the NPCs elemental analysis [5]. However, the EDX analysis clarifies the NPCs purity. Specifically, the XAS and XRF techniques identify any materials geometry and NPCs coating thickness, respectively [138]. The NPCs surface analyse is possible by running Zeta potential to confirm the NPCs colloidal stability and its surface charge, with the nitrogen adsorption-desorption analysis playing a useful role for surface area and pore size distribution studies [138].

Some NPCs have magnetic properties, which could be analysed using vibrating sample magnetometer (VSM) or superconducting quantum interference device (SQUID) or magnetic force microscopy (MFM). VSM, SQUID and MFM reveal the

Fig. 5 SEM micrographs of electrospun pristine nylon-6 mat (**a**) and TiO$_2$/nylon-6 mat (**b**). TEM images of TiO$_2$/nylon-6 nanofibers (**c**) and TEM images of TiO$_2$/nylon-6 nanofibers (**d**). Reprinted with permission from [25]. Copyright (2011) Elsevier

NPCs magnetic behaviour, the NPCs magnetisation and the NPCs surface magnetic properties, respectively [138]. Thermal studies of NPCs is possible using differential scanning calorimetry (DSC) to verify the NPCs amorphous content and polymorphism and using the thermal gravimetric analysis (TGA), as well as, the dynamic mechanical thermal analysis (DMTA) and thermal mechanical analysis (TMA) for NPCs kinetic, physical, and chemical properties elucidation [5]. Table 2 summarises the NPCs characterisations techniques.

5   Application of NPCs for Treatment Processes

Nanostructured polymer composites have been utilized in various capacities for water treatment and monitoring purposes. This section covers their application as catalyst, adsorbents, membrane filters, monitoring and detecting devices, and microbial control agents. In general, the choice of the nanomaterials and polymer used for the fabrication of the nanostructured polymer composites plays a vital role in the area

Table 2 NPCs characterisation techniques adapted from Mourdikoudis et al. [142] and Nasrollahzadeh et al. [138]

NPCs properties	NPCs analysis types	Characterisation technique
Optical	Concentration, size, agglomeration state, hints on NPCs shape	UV-Visible spectroscopy
	Surface composition, ligand binding, functional groups	Fourier transmission infrared spectroscopy (FT-IR)
Structural	Crystal composition and structure	X-ray diffraction (XRD)
	Crystallography, elemental analysis, energy binding	X-ray photon spectroscopy (XPS)
Morphological	Morphology, size, dispersion, topology, crystallography	Scanning electron microscopy (SEM)
	Shape, size, quantification, Topology, crystallographic structure	Transmission electron microscopy (TEM, HRTEM)
	Elemental composition, dispersion, morphology, surface roughness, size, shape, texture	Atomic force microscopy (AFM)
	Texture, topology, elemental analysis	Scanning tunnelling microscopy (STM)
	Size, agglomeration, dispersion	Dynamic light scattering (DLS)
Elemental	Elemental composition, purity	Energy dispersive X-ray (EDX)
	Chemical elements, materials geometry	X-ray absorption spectroscopy (XAS)
	Elemental composition, Thickness, concentration	X-ray fluorescence (XRF)
Surface	Surface charge, colloidal stability	Zeta potential
	Surface area active	Brunauer-Emmett-Teller (BET)
Magnetic	Magnetic behaviour	Vibrating sample magnetometer (VSM)
	NPCs magnetisation	Superconducting quantum interference device (SQUID)
	NPCs surface magnetic properties	Magnetic force microscopy (MFM)
Thermal	Amorphous status	Differential scanning calorimetry (DSC)
	Mass and composition of stabilizers	Thermal gravimetric analysis (TGA)
	Viscoelastic behaviour	Dynamic mechanical thermal analysis (DMTA)
	Kinetic, physical and chemical properties	Thermal mechanical analysis (TMA)

of application and effectiveness or degree of treatment derivable from the remediation processes. For example, TiO_2 is an established photo-catalyst due to its high photonic band structure or band-gap, while polypyrrole is a widely applied conducting polymer. These are desirable properties that are considered and harnessed when designing and synthesizing NPCs. Hence, review of the properties and functionalities of the nanomaterials and polymer matrices is important to obtain the best results for specific treatment operations. Additionally, the various methods of preparation and synthesis of NPCs could influence their performance during application for water treatment.

5.1 NPCs as Photocatalysts

Application of NPCs as catalysts during the wastewater treatment process has received enormous attention among researchers in the field of nanomaterials in recent time. Studies have proffered insight into the development of various synthesis methods, characterization, chemical stability, applications, regeneration potential, among other research areas. NPCs have been discovered to have great potential as catalysts due to the unique characteristics such as superior influence on the rate of chemical reactions, specific surface area, and electronic properties inherent in the hybrid materials. Studies have revealed that NPCs can function as photocatalysts in chemical reactions when light is absorbed which brings about excitation into higher energy level [143]. The suitability of NPCs as efficient photocatalysts for water treatment purposes is based on the quality of nanocomposite material used and their ability to irradiate light source [144]. Matrix of different metals such as gold, copper, and silver among others with their corresponding polymeric materials gave disparate properties during applications, while radiation source provides activation energy with sufficient excitation for degradation reaction to occur during water treatment processes [145]. Furthermore, the application of NPCs as photocatalysts has involved ZnO, ZnS, graphene/GO, perovskites, MoS_2, WO_3, Fe_2O_3 composite; however, TiO_2 remains the most commonly used photocatalyst due to its relatively low cost, non-toxicity and high oxidising ability [145]. Photocatalysis have been employed for the degradation and containment of water contaminants such as 1,4 dioxane, n-nitrosodimethylamine (NDMA), tris-2-chloroethyl phosphate (TCEP), gemfibrozil, and 17β estradiol [146]. Other contaminants that have been remediated using catalytic NPCs include 2-nitrophenol and 4-nitrophenols [147], nonylphenol [63], nitrobenzene [148], lindane [149], O-nitroaniline [150]; while organic dyes are Congo red [32], Rhodamine B [41, 151], methyl orange [33, 56], methyl green [34, 69], methyl red [68], methylene blue [152], Eosine Y [147], amongst others.

5.2 NPCs as Adsorbents

Adsorption has been identified as one of the simplest and effective approaches of removing contaminants from wastewater/water with NPCs widely explored for this purpose [153, 154]. Adsorption is described as the adherence or accumulation of a substance such as gaseous, liquid or chemical species into a solid surface [155]. The process of adsorption involves the mass transfer of adsorbate molecule from solution onto the available binding sites on the adsorbent. Several sorbents have been reported in the literature for removal of various contaminants such as activated carbons sourced from different precursors, clay, zeolites among others potential sorbents [156]. The aforementioned sorbents have provided encouraging results during water treatment operations, however, the effectiveness and efficiency of some of these sorbents are limited due to their intrinsic properties [157]. The application of nano-sorbents in wastewater remediation process is an alternative to improve the removal of some the contaminants that are difficult to eliminate during wastewater treatment. Blending of a small amount of nano-sized inorganic material to the organic polymer can enhance the performance of adsorbents [158, 159]. The use of NPCs as adsorbents have been reported in previous studies, which reveals their efficiency for removal of contaminants; heavy metals, pesticides, pharmaceuticals, perfluoroalkyl acids, PAHs, amongst other ubiquitous pollutants from aqueous phase [160]. Recent studies showed that enhanced physicochemical characteristic including surface functional morphology, pore volume and sizes, regeneration capacities and other properties of the PNCs greatly influence their removal efficiency for contaminants [161]. Pollutants are trapped by the hybrid composites as a result of their intrinsic characteristics such as hydrophobic interaction, electrostatic attraction, hydrogen bonding, micro-porosity and complexity of chemical formations with contaminants. Optimized variables such as solution pH, contact time, agitation speed, adsorbent load, temperature, the effect of initial contamination concentration dose play vital roles during the adsorption process. Rekos et al. [153] reported the application of magnetic nano-adsorbents fabricated from GO with polymer supports for bisphenol removal; adsorption was found to be dependent on the oxygen-containing functional groups of graphene nanosheets and the amino groups present on the polymers used. The GO-based magnetic nanocomposites presented sufficient adsorption capacities for BPA. In another study, iron–oxide nanoparticles (IONPs) based NPCs synthesized using *Terminalia chebula* leaf extract for enhanced adsorption of arsenic (V) from water was investigated. The study revealed that PVA–alginate-supported IONPs were found to be more effective than the other adsorbents in terms of adsorption, stability, and reusability [162].

5.3 NPCs as Filtration Membranes

Membranes and indeed filtration membranes are designed specifically to exclude certain pre-determined molecules or species. By the inclusion of nanomaterial on a membrane, additional functions such as "active site reaction" on the membrane can be achieved. In such a scenario, the polymeric membrane serves as a porous support for an active catalyst with the catalyst improving the mechanical strength/character of the polymer. Application of filtration membrane techniques for remediation has been widely used especially for water treatment due to its adaptable nature. Efficient filtration membrane for wastewater treatment scheme is expected to possess some inherent qualities such as high sorption capacity, environmentally friendly, cost-effective and among other qualities [163]. The operations of some filtration membrane is also similar to what is obtained in adsorption techniques with separations brought about due to electrostatic, hydrophilic and hydrophobic interactions between the membrane and pollutants. Meanwhile, physical separation of large particles of pollutant (sieving) is a distinctive characteristic of filtration membranes [5]. Performance of filtration membrane is largely dependent on the quality of membrane. Some of the polymeric materials in use include polyvinylidene fluoride, polyacrylonitrile, polypropylene, polysulfone, and polyethersulfone. The sort-after surface characteristic of these materials include hydrophobicity, surface charge, pH and oxidant tolerance, stability etc. [163]. Application of NPCs as filtration membranes has been greatly embraced due to their ability to overcome some of the reoccurring challenges of conventional membrane such as permeability, fouling and selectivity. NPCs have demonstrated resistance to fouling, low permeability, contaminant selectivity, anti-microbial activities, oil resistance, catalytic activities among others [164]. Evidence of improved filtration during wastewater treatment process have been reported in literature with better efficiency and stability using various nanocomposites materials as membrane filters [19]. NPCs filter membranes have been successfully and efficiently applied for water remediation processes. In this trend, several fabricated sorbents with promising remediation capabilities has been synthesized and utilized for the removal of oils [65], dyes [54, 57], solutes/salts [22, 59, 66], and heavy metals [39, 165]. Additionally, NPCs have been employed for antimicrobial [60], antifouling [58, 166], flocculation/dewatering of sludge [35], decolourization of distillery effluent [167], removal of chlorinated byproducts [168], and natural organic matter (NOM) rejection or elimination purposes [169]. The incorporation of the nanomaterials also leads to improved water fluxes of the filtration membranes [170].

5.4 NPCs as Monitoring and Detection Devices

The increasing awareness and deployment of NPCs for water treatment purposes are partly due to their multi-functionality. This versatility has seen NPCs being used as monitoring and detection devices in the recent time. NPCs have been incorporated as monitoring and detecting devices based on pH sensitivity, affinities for both organic and inorganic ions and responses to microorganism [171]. The ability of the NPCs to carry out the functions of sensing and detection is derived from combined intrinsic properties of the nanomaterials and the selected polymer matrices. Hence, NPCs utilized in this area of water remediation require careful selection of both materials. In this regards, the nanomaterials employed mostly possess properties such as magnetic, optical and electrochemical functionalities, which aid the sensitivity and rate of detection [100]. Also, conducting polymers or the appropriate polymeric support contributes to the final improved conductive, dielectric, electronic, ion-selective, optically active, and/or molecular recognition ability of the fabricated NPCs [172]. NPCs has been utilized for the detection of contaminants such as heavy metals, trace organic and inorganic contaminants, persistent organic pollutants, and pathogens in aqueous solution as evident in the literature [170]. Additionally, some NPCs have been used for monitoring pH [173], dissolved oxygen [174] and chemical oxygen demand [31] of water systems.

5.5 NPCs as Antimicrobial Agents

The removal of microbial contaminants such as *Escherichia coli*, *Staphylococcus aureus*, et cetera from water is among the several challenges encountered during water treatment operations. On another hand, the use of microorganisms to reduce pollution through biological route (bioremediation) converts pollutants to non-harmful or environmentally friendly non-toxic substances. In both scenarios, the microbial content of the in-process treated water needs to be completely eliminated to obtain the final treated water. There have been a great number of studies that have successfully used nanomaterials for their anti-microbial properties [175]. Application of nano-based technology in the area of disinfection has received enormous attention in recent times due to the associated challenges with other disinfection methods, which include resistance, the release of bye-product, and incomplete extermination of microorganisms amongst others. Anti-microbial tendency of some nanomaterials such as nano-Ag, nano-Au, nano-Cu, nano-ZnO, nano-TiO$_2$ and nano-CeCO$_4$ has been explored with reports indicating minimal oxidation and formation of toxic by-product when compared to conventional antimicrobial agents [176]. Apart from the nanomaterials, certain polymeric matrices contribute to the antimicrobial activity of the NPCs [177]. The nano-based disinfection and microbial control procedure is expected to receive more attention in the coming years as a viable alternative in water treatment operations. However, this technique still requires more investigation to eliminate associated challenges encountered during application processes. For example, most NPCs tend to give encouraging treatment efficiencies with simulated contaminated water assays, but exhibit drastic decline in their capabilities when utilized for the remediation of waters from natural sources [178].

6 Summary and Future Outlook

The use of nanostructured polymer composites offers great potentials for the safe utilization of nanomaterials in water treatment and monitoring operations. They present a wide array of possibilities for the deployment of nanotechnology in resolving water scarcity and contamination challenges through effective water reclamation and management opportunities as evident from the numerous literatures. The current pace and development in this area should be channelled to cover identified loopholes and challenges that may limit the growth and wide-scale application of the nanostructured polymer composites. One of these challenges is the loss/leaching of the nanomaterials during water treatment operations [179]. Apart from the decline in performance and efficiency occasioned by these losses [180], the released nanomaterials may pose adverse health effects to organisms in the ecosystem. Their nanoscale size allows them to easily enter cells, tissues, organelles and functional biomolecular structures with dire risk to the organisms [181]. To this end, the compatibility of the nanomaterials and polymeric matrices should be considered in designing and synthesizing the nanostructured polymer composites. This entails a careful selection of both materials, considering the functionalities and functional groups available in both materials, coupled with the application of the best suited synthesis route. Also, the optimum loading capacity of the polymer matrix in use should be considered as exceeding this limit may create higher possibilities of nanomaterial losses.

The development and sustainability of any water treatment technology depends on several factors with the economic value of such operations playing a vital role. Hence, the fabrication of nanostructured polymer composites with huge potential for reusability will be a leading opening for their large scale application [20]. Researches bordering on careful design and fabrication of nanostructured polymer composites with the ability to undergo several reuse cycles with no or minimal decline in their treatment capabilities will be the next daunting challenge to overcome with this promising nano-based water treatment technology. The regeneration processes for these nanostructured polymer composites should equally be such that they are simple, fast, and cheap to align with the economy of scale. Moreover, solutions to challenges such as the safe disposal of the spent composites and the regeneration medium are needed for a holistic and completely safe treatment technology to be developed. Information on this critical aspect is scanty or absent from most investigations as evident in literature [26].

Greater sustainability and large-scale application of nanostructured polymer composites may be boosted by the fabrication and utilization of hybrid multifunctional composites. The ability of the composites to remove multiple contaminants or have combined functionalities as a catalyst, adsorbent or microbial control agent will be the next level of their development. In this regard, a few reported investigations give some insight to the possibility of this concept [32, 34, 41, 51, 67, 74, 78]. More studies on the development of multifaceted nanostructured polymer composites will position the technology at the forefront of water treatment options.

References

1. Gorza, F.D.S., Pedro, G.C., da Silva, R.J., Medina-Llamas, J.C., Alcaraz-Espinoza, J.J., Chávez-Guajardo, A.E., de Melo, C.P.: Electrospun polystyrene-(emeraldine base) mats as high-performance materials for dye removal from aqueous media. J. Taiwan Inst. Che. Eng. **82**, 300–311 (2018)
2. Yin, J., Deng, B.: Polymer-matrix nanocomposite membranes for water treatment. J. Membr. Sci. **479**, 256–275 (2015)
3. Zhu, J., Hou, J., Zhang, Y., Tian, M., He, T., Liu, J., Chen, V.: Polymeric antimicrobial membranes enabled by nanomaterials for water treatment. J. Membr. Sci. **550**, 173–197 (2018)
4. Cantarella, M., Sanz, R., Buccheri, M.A., Romano, L., Privitera, V.: PMMA/TiO$_2$ nanotubes composites for photocatalytic removal of organic compounds and bacteria from water. Mater. Sci. Semicond. Process. **42**, 58–61 (2016)
5. Akharame, M.O., Fatoki, O.S., Opeolu, B.O., Olorunfemi, D.I., Oputu, O.U.: Polymeric nanocomposites (PNCs) for wastewater remediation: an overview. Polym.-Plast. Technol. Eng. **57**(17), 1801–1827 (2018)
6. Li, Y., Zhao, H., Yang, M.: TiO$_2$ nanoparticles supported on PMMA nanofibers for photocatalytic degradation of methyl orange. J. Colloid Interface Sci. **508**, 500–507 (2017)
7. Ebrahim, S., Shokry, A., Ibrahim, H., Soliman, M.: Polyaniline/akaganéite nanocomposite for detoxification of noxious Cr(VI) from aquatic environment. J. Polym. Res. **23**(4), 79 (2016)
8. Muliwa, A.M., Leswifi, T.Y., Onyango, M.S., Maity, A.: Magnetic adsorption separation (MAS) process: An alternative method of extracting Cr(VI) from aqueous solution using polypyrrole coated Fe$_3$O$_4$ nanocomposites. Sep. Purif. Technol. **158**, 250–258 (2016)
9. Wu, J., Yu, C., Li, Q.: Regenerable antimicrobial activity in polyamide thin film nanocomposite membranes. J. Membr. Sci. **476**, 119–127 (2015)
10. Samiey, B., Cheng, C.-H., Wu, J.: Organic-inorganic hybrid polymers as adsorbents for removal of heavy metal ions from solutions: a review. Materials **7**(2), 673–726 (2014)
11. Zhao, X., Lv, L., Pan, B., Zhang, W., Zhang, S., Zhang, Q.: Polymer-supported nanocomposites for environmental application: a review. Chem. Eng. J. **170**(2), 381–394 (2011)
12. Lin, C.-L., Lee, C.-F., Chiu, W.-Y.: Preparation and properties of poly(acrylic acid) oligomer stabilized superparamagnetic ferrofluid. J. Colloid Interface Sci. **291**(2), 411–420 (2005)
13. Ng, L.Y., Mohammad, A.W., Leo, C.P., Hilal, N.: Polymeric membranes incorporated with metal/metal oxide nanoparticles: a comprehensive review. Desalination **308**, 15–33 (2013)
14. Zeng, X., Yao, Y., Gong, Z., Wang, F., Sun, R., Xu, J., Wong, C. P.: Ice-templated assembly strategy to construct 3D boron nitride nanosheet networks in polymer composites for thermal conductivity improvement. Small **11**(46), 6205–6213 (2015)
15. Bhaumik, M., McCrindle, R.I., Maity, A.: Enhanced adsorptive degradation of Congo red in aqueous solutions using polyaniline/Fe0 composite nanofibers. Chem. Eng. J. **260**, 716–729 (2015)
16. Tang, L., Yang, G.-D., Zeng, G.-M., Cai, Y., Li, S.-S., Zhou, Y.-Y., Pang, Y., Liu, Y.-Y., Zhang, Y., Luna, B.: Synergistic effect of iron doped ordered mesoporous carbon on adsorption-coupled reduction of hexavalent chromium and the relative mechanism study. Chem. Eng. J. **239**, 114–122 (2014)
17. Khezri, K., Mahdavi, H.: Polystyrene-silica aerogel nanocomposites by in situ simultaneous reverse and normal initiation technique for ATRP. Microporous Mesoporous Mater. **228**, 132–140 (2016)
18. Hussain, F., Hojjati, M., Okamoto, M., Gorga, R.E.: Review article: polymer-matrix nanocomposites, processing, manufacturing, and application: an overview. J. Compos. Mater. **40**(17), 1511–1575 (2006)
19. Qu, X., Brame, J., Li, Q., Alvarez, P.J.: Nanotechnology for a safe and sustainable water supply: enabling integrated water treatment and reuse. Acc. Chem. Res. **46**(3), 834–843 (2012)

20. Akharame, M.O., Fatoki, O.S., Opeolu, B.O.: Regeneration and reuse of polymeric nanocomposites in wastewater remediation: the future of economic water management. Polym. Bull. **76**(2), 647–681 (2019)
21. Ummartyotin, S., Pechyen, C.: Role of ZnO on nylon 6 surface and the photocatalytic efficiency of methylene blue for wastewater treatment. Colloid Polym. Sci. **294**(7), 1217–1224 (2016)
22. Yin, J., Zhu, G., Deng, B.: Graphene oxide (GO) enhanced polyamide (PA) thin-film nanocomposite (TFN) membrane for water purification. Desalination **379**, 93–101 (2016)
23. Kim, H.J., Choi, K., Baek, Y., Kim, D.G., Shim, J., Yoon, J., Lee, J.C.: High-performance reverse osmosis CNT/polyamide nanocomposite membrane by controlled interfacial interactions. ACS Appl. Mater. Interfaces **6**(4), 2819–2829 (2014)
24. de Lannoy, C.F., Jassby, D., Gloe, K., Gordon, A.D., Wiesner, M.R.: Aquatic biofouling prevention by electrically charged nanocomposite polymer thin film membranes. Environ. Sci. Technol. **47**(6), 2760–2768 (2013)
25. Pant, H.R., Pandeya, D.R., Nam, K.T., Baek, W.-I., Hong, S.T., Kim, H.Y.: Photocatalytic and antibacterial properties of a TiO_2/nylon-6 electrospun nanocomposite mat containing silver nanoparticles. J. Hazard. Mater. **189**(1), 465–471 (2011)
26. Ravishankar, H., Wang, J., Shu, L., Jegatheesan, V.: Removal of Pb (II) ions using polymer based graphene oxide magnetic nano-sorbent. Process Saf. Environ. Prot. **104**, 472–480 (2016)
27. Reddy, P.M., Chang, C.-J., Chen, J.-K., Wu, M.-T., Wang, C.-F.: Robust polymer grafted Fe_3O_4 nanospheres for benign removal of oil from water. Appl. Surf. Sci. **368**, 27–35 (2016)
28. Yu, L., Hao, G., Gu, J., Zhou, S., Zhang, N., Jiang, W.: Fe_3O_4/PS magnetic nanoparticles: synthesis, characterization and their application as sorbents of oil from waste water. J. Magn. Magn. Mater. **394**, 14–21 (2015)
29. Vaiano, V., Sacco, O., Sannino, D., Ciambelli, P., Longo, S., Venditto, V., Guerra, G.: N-doped TiO_2/s-PS aerogels for photocatalytic degradation of organic dyes in wastewater under visible light irradiation. J. Chem. Technol. Biotechnol. **89**(8), 1175–1181 (2014)
30. Pule, B.O., Degni, S., Torto, N.: Electrospun fibre colorimetric probe based on gold nanoparticles for on-site detection of 17β-estradiol associated with dairy farming effluents. Water SA **41**(1), 27 (2014)
31. Gutierrez-Capitan, M., Baldi, A., Gomez, R., Garcia, V., Jimenez-Jorquera, C., Fernandez-Sanchez, C.: Electrochemical nanocomposite-derived sensor for the analysis of chemical oxygen demand in urban wastewaters. Anal. Chem. **87**(4), 2152–2160 (2015)
32. Sharma, G., Kumar, A., Naushad, M., Pathania, D., Sillanpää, M.: Polyacrylamide@Zr(IV) vanadophosphate nanocomposite: ion exchange properties, antibacterial activity, and photocatalytic behavior. J. Ind. Eng. Chem. **33**, 201–208 (2016)
33. Pathania, D., Gupta, D., Al-Muhtaseb, A.H., Sharma, G., Kumar, A., Naushad, M., Ahamad, T., Alshehri, S.M.: Photocatalytic degradation of highly toxic dyes using chitosan-g-poly(acrylamide)/ZnS in presence of solar irradiation. J. Photochem. Photobiol. A **329**, 61–68 (2016)
34. Kumar, A., Sharma, G., Naushad, M., Singh, P., Kalia, S.: Polyacrylamide/$Ni_{0.02}Zn_{0.98}O$ nanocomposite with high solar light photocatalytic activity and efficient adsorption capacity for toxic dye removal. Ind. Eng. Chem. Res. **53**(40), 15549–15560 (2014)
35. Huang, P., Ye, L.: In situ polymerization of cationic polyacrylamide/montmorillonite composites and its flocculation characteristics. J. Thermoplast. Compos. Mater. **29**(1), 58–73 (2014)
36. Zhao, G., Zhang, H., Fan, Q., Ren, X., Li, J., Chen, Y., Wang, X.: Sorption of copper (II) onto super-adsorbent of bentonite–polyacrylamide composites. J. Hazard. Mater. **173**(1–3), 661–668 (2010)
37. Chen, H., Wang, A.: Adsorption characteristics of Cu (II) from aqueous solution onto poly(acrylamide)/attapulgite composite. J. Hazard. Mater. **165**(1–3), 223–231 (2009)
38. Lee, J., Chae, H.-R., Won, Y.J., Lee, K., Lee, C.-H., Lee, H.H., Kim, I.-C., Lee, J.-M.: Graphene oxide nanoplatelets composite membrane with hydrophilic and antifouling properties for wastewater treatment. J. Membr. Sci. **448**, 223–230 (2013)

39. Abdullah, N., Gohari, R.J., Yusof, N., Ismail, A.F., Juhana, J., Lau, W.J., Matsuura, T.: Polysulfone/hydrous ferric oxide ultrafiltration mixed matrix membrane: Preparation, characterization and its adsorptive removal of lead (II) from aqueous solution. Chem. Eng. J. **289**, 28–37 (2016)

40. Schiffman, J.D., Elimelech, M.: Antibacterial activity of electrospun polymer mats with incorporated narrow diameter single-walled carbon nanotubes. ACS Appl. Mater. Interfaces. **3**(2), 462–468 (2011)

41. He, M.Q., Bao, L.L., Sun, K.Y., Zhao, D.X., Li, W.B., Xia, J.X., Li, H.M.: Synthesis of molecularly imprinted polypyrrole/titanium dioxide nanocomposites and its selective photocatalytic degradation of rhodamine B under visible light irradiation. Express Polym. Lett. **8**(11), 850–861 (2014)

42. Yao, W., Ni, T., Chen, S., Li, H., Lu, Y.: Graphene/Fe_3O_4@polypyrrole nanocomposites as a synergistic adsorbent for Cr(VI) ion removal. Compos. Sci. Technol. **99**, 15–22 (2014)

43. Setshedi, K.Z., Bhaumik, M., Songwane, S., Onyango, M.S., Maity, A.: Exfoliated polypyrrole-organically modified montmorillonite clay nanocomposite as a potential adsorbent for Cr(VI) removal. Chem. Eng. J. **222**, 186–197 (2013)

44. Sahmetlioglu, E., Yilmaz, E., Aktas, E., Soylak, M.: Polypyrrole/multi-walled carbon nanotube composite for the solid phase extraction of lead(II) in water samples. Talanta **119**, 447–451 (2014)

45. Mthombeni, N.H., Mbakop, S., Onyango, M.S.: Magnetic zeolite-polymer composite as an adsorbent for the remediation of wastewaters containing vanadium. Int. J. Environ. Sci. Dev. **6**(8), 602–605 (2015)

46. Chavan, A.A., Li, H., Scarpellini, A., Marras, S., Manna, L., Athanassiou, A., Fragouli, D.: Elastomeric nanocomposite foams for the removal of heavy metal ions from water. ACS Appl. Mater. Interfaces. **7**(27), 14778–14784 (2015)

47. Scott, A., Gupta, R., Kulkarni, G.U.: A simple water-based synthesis of Au nanoparticle/PDMS composites for water purification and targeted drug release. Macromol. Chem. Phys. **211**(15), 1640–1647 (2010)

48. Abdulhussein, A.T., Kannarpady, G.K., Ghosh, A., Barnes, B., Steiner, R.C., Mulon, P.Y., Anderson, D.E., Biris, A.S.: Facile fabrication of a free-standing superhydrophobic and superoleophilic carbon nanofiber-polymer block that effectively absorbs oils and chemical pollutants from water. Vacuum **149**, 39–47 (2018)

49. Jung, G., Kim, H.I.: Synthesis and photocatalytic performance of PVA/TiO_2/graphene-MWCNT nanocomposites for dye removal. J. Appl. Poly. Sci. 131(17) (2014)

50. Mondal, S., Madhuri, R., Sharma, P.K.: Electrochemical sensing of cyanometalic compound using TiO_2/PVA nanocomposite-modified electrode. Journal of Applied Electrochemistry **47**(1), 75–83 (2016)

51. Devi, J.M., Umadevi, M.: Synthesis and characterization of silver–PVA nanocomposite for sensor and antibacterial applications. J. Cluster Sci. **25**(2), 639–650 (2013)

52. Mahdavinia, G.R., Soleymani, M., Sabzi, M., Azimi, H., Atlasi, Z.: Novel magnetic polyvinyl alcohol/laponite RD nanocomposite hydrogels for efficient removal of methylene blue. J. Environ. Chem. Eng. **5**(3), 2617–2630 (2017)

53. Mahdavinia, G.R., Mousanezhad, S., Hosseinzadeh, H., Darvishi, F., Sabzi, M.: Magnetic hydrogel beads based on PVA/sodium alginate/laponite RD and studying their BSA adsorption. Carbohydr. Polym. **147**, 379–391 (2016)

54. Liang, B., Zhang, P., Wang, J., Qu, J., Wang, L., Wang, X., Guan, C., Pan, K.: Membranes with selective laminar nanochannels of modified reduced graphene oxide for water purification. Carbon **103**, 94–100 (2016)

55. Zhang, L., Luo, J., Menkhaus, T.J., Varadaraju, H., Sun, Y., Fong, H.: Antimicrobial nanofibrous membranes developed from electrospun polyacrylonitrile nanofibers. J. Membr. Sci. **369**(1–2), 499–505 (2011)

56. Hir, Z.A.M., Moradihamedani, P., Abdullah, A.H., Mohamed, M.A.: Immobilization of TiO_2 into polyethersulfone matrix as hybrid film photocatalyst for effective degradation of methyl orange dye. Mater. Sci. Semicond. Process. **57**, 157–165 (2017)

57. Rahimi, M., Zinadini, S., Zinatizadeh, A.A., Vatanpour, V., Rajabi, L., Rahimi, Z.: Hydrophilic goethite nanoparticle as a novel antifouling agent in fabrication of nanocomposite polyethersulfone membrane. J. Appl. Polym. Sci. 133(26) (2016)

58. Zinadini, S., Rostami, S., Vatanpour, V., Jalilian, E.: Preparation of antibiofouling polyethersulfone mixed matrix NF membrane using photocatalytic activity of ZnO/MWCNTs nanocomposite. J. Membr. Sci. 529, 133–141 (2017)

59. Bagheripour, E., Moghadassi, A.R., Hosseini, S.M., Nemati, M.: Fabrication and characterization of novel mixed matrix polyethersulfone nanofiltration membrane modified by iron-nickel oxide nanoparticles. J. Membr. Sci. Res. 2(1), 14–19 (2016)

60. Jo, Y.J., Choi, E.Y., Choi, N.W., Kim, C.K.: antibacterial and hydrophilic characteristics of poly(ether sulfone) composite membranes containing zinc oxide nanoparticles grafted with hydrophilic polymers. Ind. Eng. Chem. Res. 55(28), 7801–7809 (2016)

61. Toroghi, M., Raisi, A., Aroujalian, A.: Preparation and characterization of polyethersulfone/silver nanocomposite ultrafiltration membrane for antibacterial applications. Polym. Adv. Technol. 25(7), 711–722 (2014)

62. Basri, H., Ismail, A.F., Aziz, M., Nagai, K., Matsuura, T., Abdullah, M.S., Ng, B.C.: Silverfilled polyethersulfone membranes for antibacterial applications—effect of PVP and TAP addition on silver dispersion. Desalination 261(3), 264–271 (2010)

63. Dzinun, H., Othman, M.H.D., Ismail, A.F., Puteh, M.H., Rahman, M.A., Jaafar, J.: Photocatalytic degradation of nonylphenol by immobilized TiO_2 in dual layer hollow fibre membranes. Chem. Eng. J. 269, 255–261 (2015)

64. Alpatova, A., Meshref, M., McPhedran, K.N., Gamal El-Din, M.: Composite polyvinylidene fluoride (PVDF) membrane impregnated with Fe_2O_3 nanoparticles and multiwalled carbon nanotubes for catalytic degradation of organic contaminants. J. Membr. Sci. 490, 227–235 (2015)

65. Yan, L., Hong, S., Li, M.L., Li, Y.S.: Application of the Al_2O_3—PVDF nanocomposite tubular ultrafiltration (UF) membrane for oily wastewater treatment and its antifouling research. Sep. Purif. Technol. 66(2), 347–352 (2009)

66. Obaid, M., Ghouri, Z.K., Fadali, O.A., Khalil, K.A., Almajid, A.A., Barakat, N.A.: Amorphous SiO_2 NP-incorporated poly(vinylidene fluoride) electrospun nanofiber membrane for high flux forward osmosis desalination. ACS Appl. Mater. Interfaces 8(7), 4561–4574 (2016)

67. Pathania, D., Sharma, G., Kumar, A., Kothiyal, N.C.: Fabrication of nanocomposite polyaniline zirconium(IV) silicophosphate for photocatalytic and antimicrobial activity. J. Alloy. Compd. 588, 668–675 (2014)

68. Haspulat, B., Gulce, A., Gulce, H.: Efficient photocatalytic decolorization of some textile dyes using Fe ions doped polyaniline film on ITO coated glass substrate. J. Hazard. Mater. 260, 518–526 (2013)

69. Gülce, H., Eskizeybek, V., Haspulat, B., Sarı, F., Gülce, A., Avcı, A.: Preparation of a new polyaniline/CdO nanocomposite and investigation of its photocatalytic activity: comparative study under UV light and natural sunlight irradiation. Ind. Eng. Chem. Res. 52(32), 10924–10934 (2013)

70. Tovide, O., Jaheed, N., Mohamed, N., Nxusani, E., Sunday, C.E., Tsegaye, A., Ajayi, R.F., Njomo, N., Makelane, H., Bilibana, M., Baker, P.G., Williams, A., Vilakazi, S., Tshikhudo, R., Iwuoha, E.I.: Graphenated polyaniline-doped tungsten oxide nanocomposite sensor for real time determination of phenanthrene. Electrochim. Acta 128, 138–148 (2014)

71. Khan, A., Khan, A.A.P., Rahman, M.M., Asiri, A.M.: High performance polyaniline/vanadyl phosphate (PANI–$VOPO_4$) nano composite sheets prepared by exfoliation/intercalation method for sensing applications. Eur. Polymer J. 75, 388–398 (2016)

72. Zheng, J., Li, G., Ma, X., Wang, Y., Wu, G., Cheng, Y.: Polyaniline–TiO_2 nano-composite-based trimethylamine QCM sensor and its thermal behavior studies. Sens. Actuat. B: Chem. 133(2), 374–380 (2008)

73. Sharma, G., Pathania, D., Naushad, M., Kothiyal, N.C.: Fabrication, characterization and antimicrobial activity of polyaniline Th(IV) tungstomolybdophosphate nanocomposite material: efficient removal of toxic metal ions from water. Chem. Eng. J. 251, 413–421 (2014)

74. Tian, G., Wang, W., Mu, B., Kang, Y., Wang, A.: Ag (I)-triggered one-pot synthesis of Ag nanoparticles onto natural nanorods as a multifunctional nanocomposite for efficient catalysis and adsorption. J. Colloid Interface Sci. **473**, 84–92 (2016)

75. Mu, B., Tang, J., Zhang, L., Wang, A.: Preparation, characterization and application on dye adsorption of a well-defined two-dimensional superparamagnetic clay/polyaniline/Fe_3O_4 nanocomposite. Appl. Clay Sci. **132**, 7–16 (2016)

76. Muñoz, J., Céspedes, F., Baeza, M.: Modified multiwalled carbon nanotube/epoxy ampero-metric nanocomposite sensors with CuO nanoparticles for electrocatalytic detection of free chlorine. Microchem. J. **122**, 189–196 (2015)

77. Motoc, S., Remes, A., Pop, A., Manea, F., Schoonman, J.: Electrochemical detection and degradation of ibuprofen from water on multi-walled carbon nanotubes-epoxy composite electrode. J. Environ. Sci. **25**(4), 838–847 (2013)

78. Benjwal, P., Kar, K.K.: Simultaneous photocatalysis and adsorption based removal of inorganic and organic impurities from water by titania/activated carbon/carbonized epoxy nanocomposite. J. Environ. Chem. Eng. **3**(3), 2076–2083 (2015)

79. Zhu, Y., Wang, W., Zhang, H., Ye, X., Wu, Z., Wang, A.: Fast and high-capacity adsorption of Rb+ and Cs+ onto recyclable magnetic porous spheres. Chem. Eng. J. **327**, 982–991 (2017)

80. Zhu, Y., Zhang, H., Wang, W., Ye, X., Wu, Z., Wang, A.: Fabrication of a magnetic porous hydrogel sphere for efficient enrichment of Rb+ and Cs+ from aqueous solution. Chem. Eng. Res. Des. **125**, 214–225 (2017)

81. Zhu, Y., Zheng, Y., Wang, F., Wang, A.: Fabrication of magnetic porous microspheres via $(O_1/W)/O_2$ double emulsion for fast removal of Cu^{2+} and Pb^{2+}. J. Taiwan Inst. Chem. Eng. **67**, 505–510 (2016)

82. Zhu, L., Liu, P., Wang, A.: High clay-content attapulgite/poly(acrylic acid) nanocomposite hydrogel via surface-initiated redox radical polymerization with modified attapulgite nanorods as initiator and cross-linker. Ind. Eng. Chem. Res. **53**(5), 2067–2071 (2014)

83. Rafiei, H.R., Shirvani, M., Ogunseitan, O.A.: Removal of lead from aqueous solutions by a poly(acrylic acid)/bentonite nanocomposite. Appl. Water Sci. **6**(4), 331–338 (2014)

84. Liu, P., Jiang, L., Zhu, L., Wang, A.: Novel approach for attapulgite/poly(acrylic acid) (ATP/PAA) nanocomposite microgels as selective adsorbent for Pb(II) Ion. React. Funct. Polym. **74**, 72–80 (2014)

85. Liu, P., Jiang, L., Zhu, L., Wang, A.: Novel covalently cross-linked attapulgite/poly(acrylic acid-co-acrylamide) hybrid hydrogels by inverse suspension polymerization: synthesis opti-mization and evaluation as adsorbents for toxic heavy metals. Ind. Eng. Chem. Res. **53**(11), 4277–4285 (2014)

86. Liu, Y., Zheng, Y., Wang, A.: Enhanced adsorption of Methylene Blue from aqueous solution by chitosan-g-poly(acrylic acid)/vermiculite hydrogel composites. J. Environ. Sci. **22**(4), 486–493 (2010)

87. Zheng, Y., Wang, A.: Enhanced adsorption of ammonium using hydrogel composites based on chitosan and halloysite. J. Macromol. Sci. Part A **47**(1), 33–38 (2009)

88. Zheng, Y., Wang, A.: Evaluation of ammonium removal using a chitosan-g-poly(acrylic acid)/rectorite hydrogel composite. J. Hazard. Mater. **171**(1–3), 671–677 (2009)

89. Liu, Y., Zheng, Y., Wang, A.: Effect of biotite content of hydrogels on enhanced removal of methylene blue from aqueous solution. Ionics **17**(6), 535–543 (2011)

90. Jordan, J., Jacob, K.I., Tannenbaum, R., Sharaf, M.A., Jasiuk, I.: Experimental trends in polymer nanocomposites—a review. Mater. Sci. Eng. A **393**(1–2), 1–11 (2005)

91. Jeon, I.-Y., Baek, J.-B.: Nanocomposites derived from polymers and inorganic nanoparticles. Materials **3**(6), 3654–3674 (2010)

92. Jose, P.J., Malhortra, S.K., Thomas, S., Joseph, K., Goda, K., Sreekala, M.S.: Advances in polymer composites: macro-and microcomposites—state of the art, new challenges, and opportunities. In: Thomas, S., Joseph, K., Malhortra S.K., Goda, K., Sreekala, M.S. (eds.) Polymer Composites. WILEY VCH , Weinheim, Germany (2012)

93. Wang, Y., Li, Q., Bo, L., Wang, X., Zhang, T., Li, S., Ren, P., Wei, G.: Synthesis and oil absorption of biomorphic MgAl layered double oxide/acrylic ester resin by suspension polymerization. Chem. Eng. J. **284**, 989–994 (2016)

94. Sun, X., Chen, F., Wei, J., Zhang, F., Pang, S.: Preparation of magnetic triethylene tetramine-graphene oxide ternary nanocomposite and application for Cr (VI) removal. J. Taiwan Inst. Chem. Eng. **66**, 328–335 (2016)

95. Shenogina, N.B., Tsige, M., Patnaik, S.S., Mukhopadhyay, S.M.: Molecular modeling of elastic properties of thermosetting polymers using a dynamic deformation approach. Polymer **54**(13), 3370–3376 (2013)

96. Garcia, J.M., Jones, G.O., Virwani, K., McCloskey, B.D., Boday, D.J., ter Huurne, G.M., Horn, H.W., Coady, D.J., Bintaleb, A.M., Alabdulrahman, A.M., Alsewailem, F., Almegren, H.A., Hedrick, J.L.: Recyclable, strong thermosets and organogels via paraformaldehyde condensation with diamines. Science **344**(6185), 732–735 (2014)

97. Halley, P.J., Mackay, M.E.: Chemorheology of thermosets—an overview. Polym. Eng. Sci. **36**(5), 593–609 (1996)

98. Cogswell, F.N.: Thermoplastic Aromatic Polymer Composites: A Study of the Structure, Processing, and Properties of Carbon Fibre Reinforced Polyetherketone and Related Materials, p. 6. Butterworth-Heineman, Oxford (1992)

99. Najafi, S.K.: Use of recycled plastics in wood plastic composites—a review. Waste Manag **33**(9), 1898–1905 (2013)

100. Qu, X., Alvarez, P.J.J., Li, Q.: Applications of nanotechnology in water and wastewater treatment. Water Res. **47**(12), 3931–3946 (2013)

101. Chaturvedi, S., Dave, P.N., Shah, N.K.: Applications of nano-catalyst in new era. J. Saudi Chem. Soc. **16**(3), 307–325 (2012)

102. Kango, S., Kalia, S., Celli, A., Njuguna, J., Habibi, Y., Kumar, R.: Surface modification of inorganic nanoparticles for development of organic–inorganic nanocomposites—a review. Prog. Polym. Sci. **38**(8), 1232–1261 (2013)

103. Baranwal, A., Srivastava, A., Kumar, P., Bajpai, V.K., Maurya, P.K., Chandra, P.: Prospects of nanostructure materials and their composites as antimicrobial agents. Front Microbiol. **9**, 422 (2018)

104. Guerra, F.D., Attia, M.F., Whitehead, D.C., Alexis, F.: Nanotechnology for environmental remediation: materials and applications. Molecules 23(7) (2018)

105. Xu, C., De, S., Balu, A.M., Ojeda, M., Luque, R.: Mechanochemical synthesis of advanced nanomaterials for catalytic applications. Chem. Commun. (Camb.) **51**(31), 6698–6713 (2015)

106. Yadav, T.P., Yadav, R.M., Singh, D.P.: Mechanical milling: a top down approach for the synthesis of nanomaterials and nanocomposites. Nanosci. Nanotechnol. **2**(3), 22–48 (2012)

107. Pan, B., Qiu, H., Pan, B., Nie, G., Xiao, L., Lv, L., Zhang, W., Zhang, Q., Zheng, S.: Highly efficient removal of heavy metals by polymer-supported nanosized hydrated Fe (III) oxides: behavior and XPS study. Water Res. **44**(3), 815–824 (2010)

108. Razzaz, A., Ghorban, S., Hosayni, L., Irani, M., Aliabadi, M.: Chitosan nanofibers functionalized by TiO_2 nanoparticles for the removal of heavy metal ions. J. Taiwan Inst. Chem. Eng. **58**, 333–343 (2016)

109. Zhang, Q., Du, Q., Hua, M., Jiao, T., Gao, F., Pan, B.: Sorption enhancement of lead ions from water by surface charged polystyrene-supported nano-zirconium oxide composites. Environ. Sci. Technol. **47**(12), 6536–6544 (2013)

110. Zha, L., Fang, Z.: Polystyrene/$CaCO_3$ composites with different $CaCO_3$ radius and different nano-$CaCO_3$ content—structure and properties. Polym. Compos. **31**(7), 1258–1264 (2010)

111. Yu, Q., Wu, P., Xu, P., Li, L., Liu, T., Zhao, L.: Synthesis of cellulose/titanium dioxide hybrids in supercritical carbon dioxide. Green Chem. **10**(10), 1061–1067 (2008)

112. Musico, Y.L.F., Santos, C.M., Dalida, M.L.P., Rodrigues, D.F.: Improved removal of lead (II) from water using a polymer-based graphene oxide nanocomposite. J. Mater. Chem. A **1**(11), 3789–3796 (2013)

113. Utracki, L., Sepehr, M., Boccaleri, E.: Synthetic, layered nanoparticles for polymeric nanocomposites (PNCs). Polym. Adv. Technol. **18**(1), 1–37 (2007)

114. Cai, L.F., Huang, X.B., Rong, M.Z., Ruan, W.H., Zhang, M.Q.: Fabrication of nanoparticle/polymer composites by in situ bubble-stretching and reactive compatibilization. Macromol. Chem. Phys. **207**(22), 2093–2102 (2006)

115. Li, X., Wang, D., Cheng, G., Luo, Q., An, J., Wang, Y.: Preparation of polyaniline-modified TiO_2 nanoparticles and their photocatalytic activity under visible light illumination. Appl. Catal. B **81**(3–4), 267–273 (2008)
116. Guan, C., Lü, C.L., Liu, Y.F., Yang, B.: Preparation and characterization of high refractive index thin films of TiO_2/epoxy resin nanocomposites. J. Appl. Polym. Sci. **102**(2), 1631–1636 (2006)
117. Tang, E., Cheng, G., Pang, X., Ma, X., Xing, F.: Synthesis of nano-ZnO/poly(methyl methacrylate) composite microsphere through emulsion polymerization and its UV-shielding property. Colloid Polym. Sci. **284**(4), 422–428 (2006)
118. Wang, Z., Lu, Y., Liu, J., Dang, Z., Zhang, L., Wang, W.: Preparation of nano-zinc oxide/EPDM composites with both good thermal conductivity and mechanical properties. J. Appl. Polym. Sci. **119**(2), 1144–1155 (2011)
119. Yang, D., Li, J., Jiang, Z., Lu, L., Chen, X.: Chitosan/TiO_2 nanocomposite pervaporation membranes for ethanol dehydration. Chem. Eng. Sci. **64**(13), 3130–3137 (2009)
120. Luo, Y.-B., Li, W.-D., Wang, X.-L., Xu, D.-Y., Wang, Y.-Z.: Preparation and properties of nanocomposites based on poly(lactic acid) and functionalized TiO_2. Acta Mater. **57**(11), 3182–3191 (2009)
121. Philip, P., Jose, E.T., Chacko, J.K., Philip, K., Thomas, P.: Preparation and characterisation of surface roughened PMMA electrospun nanofibers from PEO-PMMA polymer blend nanofibers. Polym. Testing **74**, 257–265 (2019)
122. Pereao, O., Bode-Aluko, C., Ndayambaje, G., Fatoba, O., Petrik, L.: Electrospinning: polymer nanofibre adsorbent applications for metal ion removal. J. Polym. Environ. **25**(4), 1175–1189 (2017)
123. Baji, A., Mai, Y.-W., Wong, S.-C., Abtahi, M., Chen, P.: Electrospinning of polymer nanofibers: effects on oriented morphology, structures and tensile properties. Compos. Sci. Technol. **70**(5), 703–718 (2010)
124. Wang, C., Fang, C.-Y., Wang, C.-Y.: Electrospun poly(butylene terephthalate) fibers: entanglement density effect on fiber diameter and fiber nucleating ability towards isotactic polypropylene. Polymer **72**, 21–29 (2015)
125. Raghavan, P., Lim, D.-H., Ahn, J.-H., Nah, C., Sherrington, D.C., Ryu, H.-S., Ahn, H.-J.: Electrospun polymer nanofibers: the booming cutting edge technology. React. Funct. Polym. **72**(12), 915–930 (2012)
126. Pereao, O., Bode-Aluko, C., Laatikainen, K., Nechaev, A., Petrik, L.: Morphology, modification and characterisation of electrospun polymer nanofiber adsorbent material used in metal ion removal. J. Polym. Environ. 1–18
127. Putz, K.W., Compton, O.C., Palmeri, M.J., Nguyen, S.T., Brinson, L.C.: High-nanofiller-content graphene oxide—polymer nanocomposites via vacuum-assisted self-assembly. Adv. Func. Mater. **20**(19), 3322–3329 (2010)
128. Wang, D., Kou, R., Choi, D., Yang, Z., Nie, Z., Li, J., Saraf, L.V., Hu, D., Zhang, J., Graff, G.L.: Ternary self-assembly of ordered metal oxide–graphene nanocomposites for electrochemical energy storage. ACS Nano **4**(3), 1587–1595 (2010)
129. Szabó, T., Szeri, A., Dékány, I.: Composite graphitic nanolayers prepared by self-assembly between finely dispersed graphite oxide and a cationic polymer. Carbon **43**(1), 87–94 (2005)
130. Lu, Y., Yang, Y., Sellinger, A., Lu, M., Huang, J., Fan, H., Haddad, R., Lopez, G., Burns, A.R., Sasaki, D.Y.: Self-assembly of mesoscopically ordered chromatic polydiacetylene/silica nanocomposites. Nature **410**(6831), 913 (2001)
131. Ma, P.X., Zhang, R.: Synthetic nano-scale fibrous extracellular matrix. J. Biomed. Mater. Res.: Off. J. Soc. Biomater. Jpn. Soci. Biomate. Aust. Soc. Biomater. **46**(1), 60–72 (1999)
132. Zhou, S., Wu, L.: Phase separation and properties of UV-curable polyurethane/zirconia nanocomposite coatings. Macromol. Chem. Phys. **209**(11), 1170–1181 (2008)
133. Yan, J., Yang, L., Lin, M.F., Ma, J., Lu, X., Lee, P.S.: Polydopamine spheres as active templates for convenient synthesis of various nanostructures. Small **9**(4), 596–603 (2013)
134. Shukla, S., Kim, K.-T., Baev, A., Yoon, Y., Litchinitser, N., Prasad, P.: Fabrication and characterization of gold–polymer nanocomposite plasmonic nanoarrays in a porous alumina template. ACS Nano **4**(4), 2249–2255 (2010)

135. Hulteen, J., Chen, H., Chambliss, C., Martin, C.: Template synthesis of carbon nanotubule and nanofiber arrays. Nanostruct. Mater. **9**(1–8), 133–136 (1997)
136. Wang, Y., Wang, G., Wang, H., Liang, C., Cai, W., Zhang, L.: Chemical-template synthesis of micro/nanoscale magnesium silicate hollow spheres for waste-water treatment. Chem. Eur. J. **16**(11), 3497–3503 (2010b)
137. Anandhan, S., Bandyopadhyay, S.: Polymer nanocomposites: from synthesis to applications. Nanocompos. Polym. Anal. Methods **1**, 1–28 (2011)
138. Nasrollahzadeh, M., Atarod, M., Sajjadi, M., Sajadi, S.M., Issaabadi, Z.: Interface Science and Technology, pp. 199–322, Elsevier (2019)
139. Mittal, V.: Characterization Techniques for Polymer Nanocomposites. Wiley, Weinheim, Germany (2012)
140. Caro, C.A.D.: UV/VIS Spectrophotometry—Fundamentals and Applications. Mettler-Toledo AG, Analytical, Switzerland (2015)
141. Wang, L., Fang, M., Liu, J., He, J., Li, J., Lei, J.: Layer-by-layer fabrication of high-performance polyamide/ZIF-8 nanocomposite membrane for nanofiltration applications. ACS Appl. Mater. Interfaces **7**(43), 24082–24093 (2015)
142. Mourdikoudis, S., Pallares, R.M., Thanh, N.T.: Characterization techniques for nanoparticles: Comparison and complementarity upon studying nanoparticle properties. Nanoscale **10**(27), 12871–12934 (2018)
143. Zhan, F., Wang, R., Yin, J., Han, Z., Zhang, L., Jiao, T., Zhou, J., Zhang, L., Peng, Q.: Facile solvothermal preparation of Fe_3O_4–Ag nanocomposite with excellent catalytic performance. RSC Adv. **9**(2), 878–883 (2019)
144. Wang, W., Tadé, M.O., Shao, Z.: Nitrogen-doped simple and complex oxides for photocatalysis: a review. Prog. Mater. Sci. **92**, 33–63 (2018)
145. Byrne, C., Subramanian, G., Pillai, S.C.: Recent advances in photocatalysis for environmental applications. J. Environ. Chem. Eng. **6**(3), 3531–3555 (2018)
146. Yang, K., Li, X., Yu, C., Zeng, D., Chen, F., Zhang, K., Huang, W., Ji, H.: Review on heterophase/homophase junctions for efficient photocatalysis: the case of phase transition construction. Chin. J. Catal. **40**(6), 796–818 (2019)
147. Rehman, S.U., Siddiq, M., Al-Lohedan, H., Sahiner, N.: Cationic microgels embedding metal nanoparticles in the reduction of dyes and nitro-phenols. Chem. Eng. J. **265**, 201–209 (2015)
148. Khan, S.R., Farooqi, Z.H., Waheeduz, Z., Ali, A., Begum, R., Kanwal, F., Siddiq, M.: Kinetics and mechanism of reduction of nitrobenzene catalyzed by silver-poly(N-isopropylacryl amide-co-allylacetic acid) hybrid microgels. Mater. Chem. Phys. **171**, 318–327 (2016)
149. San Roman, I., Galdames, A., Alonso, M.L., Bartolome, L., Vilas, J.L., Alonso, R.M.: Effect of coating on the environmental applications of zero valent iron nanoparticles: the lindane case. Sci. Total Environ. **565**, 795–803 (2016)
150. Zhu, C.H., Hai, Z.B., Cui, C.H., Li, H.H., Chen, J.F., Yu, S.H.: In situ controlled synthesis of thermosensitive poly(N-isopropylacrylamide)/Au nanocomposite hydrogels by gamma radiation for catalytic application. Small **8**(6), 930–936 (2012)
151. Mogha, N.K., Gosain, S., Masram, D.T.: Gold nanoworms immobilized graphene oxide polymer brush nanohybrid for catalytic degradation studies of organic dyes. Appl. Surf. Sci. **396**, 1427–1434 (2017)
152. Liu, F., Jamal, R., Wang, Y., Wang, M., Yang, L., Abdiryim, T.: Photodegradation of methylene blue by photocatalyst of D-A-D type polymer/functionalized multi-walled carbon nanotubes composite under visible-light irradiation. Chemosphere **168**, 1669–1676 (2017)
153. Rekos, K., Kampouraki, Z.-C., Sarafidis, C., Samanidou, V., Deliyanni, E.: Graphene oxide based magnetic nanocomposites with polymers as effective bisphenol—a nanoadsorbents. Materials **12**(12), 1987 (2019)
154. He, T., Wang, L., Fabregat-Santiago, F., Liu, G., Li, Y., Wang, C., Guan, R.: Electron trapping induced electrostatic adsorption of cations: a general factor leading to photoactivity decay of nanostructured TiO_2. J. Mater. Chem. A **5**(14), 6455–6464 (2017)
155. Wang, H., Wang, Y.-N., Sun, Y., Pan, X., Zhang, D., Tsang, Y.F.: Differences in Sb (V) and As (V) adsorption onto a poorly crystalline phyllomanganate (δ-MnO_2): Adsorption kinetics, isotherms, and mechanisms. Process Saf. Environ. Prot. **113**, 40–47 (2018)

156. Tauqeer, M., Ahmad, M., Siraj, M., Mohammad, A., Ansari, O., Baig, M.: Modern Age Waste Water Problems, pp. 23–46. Springer (2020)
157. Ahmad, J., Deshmukh, K., Hägg, M.B.: Influence of TiO$_2$ on the chemical, mechanical, and gas separation properties of polyvinyl alcohol-titanium dioxide (PVA-TiO$_2$) nanocomposite membranes. Int. J. Polym. Anal. Charact. **18**(4), 287–296 (2013)
158. Šupová, M., Martynková, G.S., Barabaszová, K.: Effect of nanofillers dispersion in polymer matrices: a review. Science of advanced materials **3**(1), 1–25 (2011)
159. Gaaz, T., Sulong, A., Kadhum, A., Nassir, M., Al-Amiery, A.: Surface improvement of halloysite nanotubes. Appl. Sci. **7**(3), 291 (2017)
160. Fagbayigbo, B.O., Opeolu, B.O., Fatoki, O.S., Akenga, T.A., Olatunji, O.S.: Removal of PFOA and PFOS from aqueous solutions using activated carbon produced from Vitis vinifera leaf litter. Environ. Sci. Pollut. Res. **24**, 1–14 (2017)
161. Sadasivuni, K.K., Deshmukh, K., Ahipa, T., Muzaffar, A., Ahamed, M.B., Pasha, S.K., Al-Maadeed, M.A.-A.: Flexible, biodegradable and recyclable solar cells: a review. J. Mater. Sci.: Mater. Electron. **30**(2), 951–974 (2019)
162. Saif, S., Tahir, A., Asim, T., Chen, Y., Adil, S.F.: Polymeric nanocomposites of iron-oxide nanoparticles (IONPs) synthesized using terminalia chebula leaf extract for enhanced adsorption of arsenic (V) from water. Colloids Interfaces **3**(1), 17 (2019)
163. Beyene, H.D., Ambaye, T.G.: Sustainable Polymer Composites and Nanocomposites, pp. 387–412, Springer (2019)
164. Adeleye, A.S., Pokhrel, S., Mädler, L., Keller, A.A.: Influence of nanoparticle doping on the colloidal stability and toxicity of copper oxide nanoparticles in synthetic and natural waters. Water Res. **132**, 12–22 (2018)
165. Ghaemi, N.: A new approach to copper ion removal from water by polymeric nanocomposite membrane embedded with γ-alumina nanoparticles. Appl. Surf. Sci. **364**, 221–228 (2016)
166. Chan, W.-F., Marand, E., Martin, S.M.: Novel zwitterion functionalized carbon nanotube nanocomposite membranes for improved RO performance and surface anti-biofouling resistance. J. Membr. Sci. **509**, 125–137 (2016)
167. Aditya Kiran, S., Lukka Thuyavan, Y., Arthanareeswaran, G., Matsuura, T., Ismail, A.F.: Impact of graphene oxide embedded polyethersulfone membranes for the effective treatment of distillery effluent. Chem. Eng. J. **286**, 528–537 (2016)
168. Xie, P., de Lannoy, C.F., Ma, J., Wang, Z., Wang, S., Li, J., Wiesner, M.R.: Improved chlorine tolerance of a polyvinyl pyrrolidone-polysulfone membrane enabled by carboxylated carbon nanotubes. Water Res. **104**, 497–506 (2016)
169. Lee, J., Ye, Y., Ward, A.J., Zhou, C., Chen, V., Minett, A.I., Lee, S., Liu, Z., Chae, S.-R., Shi, J.: High flux and high selectivity carbon nanotube composite membranes for natural organic matter removal. Sep. Purif. Technol. **163**, 109–119 (2016)
170. Goei, R., Dong, Z., Lim, T.-T.: High-permeability pluronic-based TiO$_2$ hybrid photocatalytic membrane with hierarchical porosity: fabrication, characterizations and performances. Chem. Eng. J. **228**, 1030–1039 (2013)
171. Palit, S.: Nanomaterials for industrial wastewater treatment and water purification. In: Handbook of Ecomaterials, pp. 195–235 (2019)
172. Harsanyi, G.: Polymeric sensing films: new horizons in sensorics? Sens. Actuat. A **46–47**, 85–88 (1995)
173. Raoufi, N., Surre, F., Rajarajan, M., Sun, T., Grattan, K.T.V.: Optical sensor for pH monitoring using a layer-by-layer deposition technique emphasizing enhanced stability and re-usability. Sens. Actua. B: Chem. **195**, 692–701 (2014)
174. Hsu, L., Selvaganapathy, P.R., Brash, J., Fang, Q., Xu, C.-Q., Deen, M.J., Chen, H.: Development of a low-cost hemin-based dissolved oxygen sensor with anti-biofouling coating for water monitoring. IEEE Sens. J. **14**(10), 3400–3407 (2014)
175. Pelaez, M., Nolan, N.T., Pillai, S.C., Seery, M.K., Falaras, P., Kontos, A.G., Dunlop, P.S., Hamilton, J.W., Byrne, J.A., O'shea, K.: A review on the visible light active titanium dioxide photocatalysts for environmental applications. Appl. Catal. B **125**, 331–349 (2012)

176. Andersen, N.I., Serov, A., Atanassov, P.: Metal oxides/CNT nano-composite catalysts for oxygen reduction/oxygen evolution in alkaline media. Appl. Catal. B **163**, 623–627 (2015)

177. Yuan, Y., Liu, F., Xue, L., Wang, H., Pan, J., Cui, Y., Chen, H., Yuan, L.: Recyclable *Escherichia coli*—specific-killing AuNP-polymer (ESKAP) nanocomposites. ACS Appl. Mater. Interfaces. **8**(18), 11309–11317 (2016)

178. Alonso, A., Munoz-Berbel, X., Vigues, N., Macanas, J., Munoz, M., Mas, J., Muraviev, D.N.: Characterization of fibrous polymer silver/cobalt nanocomposite with enhanced bactericide activity. Langmuir **28**(1), 783–790 (2012)

179. Keshavarz, A., Zilouei, H., Abdolmaleki, A., Asadinezhad, A.: Enhancing oil removal from water by immobilizing multi-wall carbon nanotubes on the surface of polyurethane foam. J. Environ. Manage **157**(Supplement C), 279–286 (2015)

180. Tran, M.T., Nguyen, T.H.T., Vu, Q.T., Nguyen, M.V.: Properties of poly(1-naphthylamine)/Fe_3O_4 composites and arsenic adsorption capacity in wastewater. Frontiers of Materials Science **10**(1), 56–65 (2016)

181. Fischer, H.C., Chan, W.C.: Nanotoxicity: the growing need for in vivo study. Curr. Opin. Biotechnol. **18**(6), 565–571 (2007)